Technology

Technology

Critical History of a Concept

ERIC SCHATZBERG

The University of Chicago Press   Chicago and London

The University of Chicago Press, Chicago 60637
The University of Chicago Press, Ltd., London
© 2018 by The University of Chicago
All rights reserved. No part of this book may be used or reproduced
in any manner whatsoever without written permission, except
in the case of brief quotations in critical articles and reviews.
For more information, contact the University of Chicago Press,
1427 E. 60th St., Chicago, IL 60637.
Published 2018
Printed in the United States of America

27 26 25 24 23 22 21 20 19 18   1 2 3 4 5

ISBN-13: 978-0-226-58383-9 (cloth)
ISBN-13: 978-0-226-58397-6 (paper)
ISBN-13: 978-0-226-58402-7 (e-book)
DOI: https://doi.org/10.7208/chicago/9780226584027.001.0001

Library of Congress Cataloging-in-Publication Data

Names: Schatzberg, Eric, 1956– author.
Title: Technology : critical history of a concept / Eric Schatzberg.
Description: Chicago ; London : The University of Chicago
    Press, 2018. | Includes bibliographical references and index.
Identifiers: LCCN 2018011131 | ISBN 9780226583839 (cloth :
    alk. paper) | ISBN 9780226583976 (pbk. : alk. paper) |
    ISBN 9780226584027 (e-book)
Subjects: LCSH: Technology—History. | Technology—Philosophy.
Classification: LCC T15 .S343 2018 | DDC 601—dc23
LC record available at https://lccn.loc.gov/2018011131

♾ This paper meets the requirements of ANSI/NISO Z39.48–1992
(Permanence of Paper).

To my sister Sharon Hormby,

who was so excited to see this book.

May her memory be a blessing.

Contents

ONE

Introduction:
An Odd Concept

Technology is everywhere. The word permeates discourse high and low, from television advertisements to postmodern theories. In terms of word frequency, *technology* ranks on a par with *science*, right in the middle of other key concepts of modernity (see figure 1). In many ways, technology has displaced science as the main concept for making sense of modern material culture, as seen in phrases such as "information, bio-, and nanotechnology."

But the definition of *technology* is a mess. Rather than helping us make sense of modernity, the term sows confusion. Its multiple meanings are contradictory. In popular discourse, *technology* is little more than shorthand for the latest innovation in digital devices.[1] Leading public intellectuals, such as Thomas Friedman, produce an endless stream of shallow prose on this theme in their best-selling books.[2]

Academics do only a bit better. Some scholars define *technology* as "all the many ways things are in fact done and made."[3] Such definitions are so broad as to be almost useless, covering everything from steelmaking to singing. Other academics define *technology* narrowly as the application of science, often pointing to technologies like the atomic bomb and the transistor, both of which depended heavily on prior scientific discoveries. Yet historians of technology have spent decades criticizing this definition, arguing that science is at most one factor in technology.[4] Cultural critics and philosophers, in contrast, often view

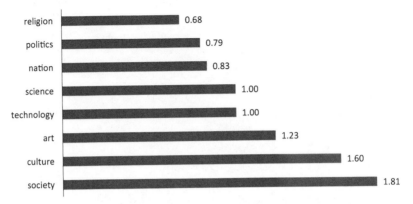

FIGURE 1 Frequency of major keywords relative to *technology* in the year 2000. Calculated from an n-gram search via Google Books Ngram Viewer, February 12, 2017.

technology as an oppressive system of total control that turns means into ends, seeking only its own perpetuation, what Lewis Mumford called the "megamachine."[5] Poststructuralists use *technology* in a similar but more positive way to refer to methods and skills in general, as when Michel Foucault writes about "technologies of the self" and "technologies of power."[6] *Technology* can also refer to material artifacts, from prehistoric stone tools to nuclear power stations.[7] And finally, other scholars, including me, define *technology* as the set of practices humans use to transform the material world, practices involved in creating and using material things.[8]

Given this welter of contradictory meanings, some scholars have suggested jettisoning the term completely.[9] Yet I believe that a concept like technology is necessary to make sense of human history. Since the beginning of the species, humans have consciously shaped the material world to sustain life and express culture. Without stone tools, woven baskets, and other artifacts, early human civilization would have been impossible. And without the material forms that express our values, dreams, and desires, human culture would barely exist.[10] These diverse material practices lie at the heart of human history, and technology provides the best concept to describe them.

Technology in the sense of material practices is particularly central to modernity, the era that gave birth to the industrial age and all its consequences. "Technology made modernity possible," proclaims the philosopher Philip Brey. Yet as Brey notes, theorists of modernity have done remarkably little to incorporate technology into their theories. One searches in vain for a clearly articulated concept of technology among leading theorists of modernity such as Jürgen Habermas, Michel Foucault, Anthony Giddens, Jean-François Lyotard, and others.[11] Even scholars who focus explicitly on technology often treat the concept as unproblematic.[12]

The marginal status of technology as a concept isn't the result of historical accident. Instead, this marginality is rooted in a fundamental problem. In the division of labor that accompanied the rise of human civilization, people who specialized in the use of words, namely scholars, grew distant from people who specialized in the transformation of the material world, that is, technicians. Our present-day concept of technology is the product of such tensions between technicians and scholars. Since the time of the ancient Greeks, technicians have fit uneasily into social hierarchies, especially aristocratic hierarchies based on birth.[13]

This tension between scholars and technicians has produced two sharply divergent traditions about the nature of technology and antecedent concepts. On one side, defenders of technicians view technology as a creative expression of human culture. In this view, technology is imbued with human values and strivings in all their contradictory complexity. I term this position the cultural approach to technology. The cultural approach is epitomized by the American public intellectual Lewis Mumford. In the 1930s, Mumford argued that technology (*technics*, in his terminology) "exists as an element in human culture and it promises well or ill as the social groups that exploit it promise well or ill." German engineers around the turn of the twentieth century made similar claims, insisting that technology (*Technik*) was an essential component of culture and a product of the human spirit.[14]

Technologies thus express the spirit of an age, just like works of art. The invention of the mechanical clock in western Europe at the end of the thirteenth century, for example, did not itself create the sense of time. Instead, the mechanical clock reflected a prior consciousness of time, rooted in monasteries and medieval towns, that motivated people to invent, improve, and embrace this new instrument. Chinese craftsmen had created far more sophisticated astronomical clocks in the eleventh century, but these devices failed to spread because they did not reflect a widespread desire for timekeeping.[15]

In contrast to the cultural approach, other scholars take what I term the instrumental approach to technology. Supporters of this approach, often humanist intellectuals, insist that technology is a mere instrument that serves ends defined by others. This vision portrays technology as narrow technical rationality, uncreative and devoid of values.

The American sociologist Talcott Parsons, for example, described technology as "the simplest means-end relationship," the choice of the best methods for achieving a specific goal. In this narrow view, even cost becomes irrelevant. For example, an engineer seeking corrosion resistance might choose gold rather than iron for an oil pan.[16] Two millennia earlier, Aristotle had made a similar argument, dividing the practical arts (*techne*) from both ethical and philosophical knowledge. According to Aristotle, ethics and philosophy were inherently virtuous, while the practical arts acquired virtue only by serving ends external to themselves.[17]

Both these conceptions of technology express fundamental truths. As a set of concrete material practices, technologies are always both cultural and instrumental, similar to what we find in works of fine art. Artistic expression requires both aesthetic sensibility and technique, that is, cultural creativity along with the instrumental means to express this creativity. Although we can distinguish between aesthetic creativity and technique, fine art can never be reduced to either, even though the art world often denigrates technical skill over aesthetic expression.[18]

Discourse about modern technology favors the instrumental over the cultural viewpoint. An entire tradition of philosophical critique is based on precisely this reduction of technology to technique, that is, instrumental rationality.[19] But enthusiasts also embrace the instrumental understanding of technology. For enthusiasts, our modern technological civilization represents the embodiment of reason in the world, with new technologies as the vanguard of progress.[20] In contrast, the cultural understanding of technology is definitely a minority view. It is found, for example, among historians of technology who connect technological choices to specific aspects of culture and society, and among a few thoughtful engineers who seek to defend the dignity and autonomy of their profession.[21]

More than intellectual clarity is at stake here. The tension between the instrumental and the cultural understanding of technology has concrete implications for the role of technology in late modernity. Most fundamentally, the instrumental concept of technology effaces the role of human agency. It focuses on innovation rather than use, treating actual technologies as natural objects, stripped of creativity and craft, subordinate to scientific knowledge, mere means to ends. When the instrumental view grants a role to human agency, it restricts this agency to a narrow technical elite or the rare inventive genius. In contrast, the cultural concept of technology is human centered, stressing use rather than novelty. It views technology as a creative, value-laden human practice, a practice that relies irreducibly on craft skills as well as formal knowledge. In this view, all humans are the rightful heirs to technology, not just technical elites.[22]

Unfortunately, historical amnesia has obscured the tensions between the cultural and instrumental views of technology, even among historians of technology. Because these tensions remain largely unrecognized, they have produced a concept of technology that is, ironically, ill suited for understanding late modernity. Most of the present-day pathologies in the concept of technology are, I argue, rooted in this clash between the cultural and the instrumental approaches. If we ever hope to make sense of modernity, scholars need to acknowledge their unique and at times pernicious role in the history of this concept.

The Marginality of Technology in Scholarly Discourse

This lack of attention to technology is all the more surprising when compared with the massive literature on other central concepts of modernity, such as art, science, culture, and politics.[23] There are two main reasons for this neglect. The first lies in the human propensity to take most technologies for granted. The philosopher Langdon Winner terms this attitude of not-seeing "technological somnambulism," the tendency to sleepwalk through the material processes that constitute much of what it means to be human.[24] Similarly, Paul Edwards describes how everyday technologies disappear in a fog of familiarity, particularly "mature technological systems" like sewers or electric power, which "reside in a naturalized background, as ordinary and unremarkable to us as trees, daylight and dirt."[25] Only when such technologies fail do people become aware of their powerful presence.

As a result of this somnambulism, the technologies that garner the most attention are new devices and processes. Yet as David Edgerton has shown, daily life relies far more on old technologies like cotton fabrics and asphalt roads than on the latest smartphone. The focus on novelty has become embedded in the very concept of technology itself, thus rendering invisible most of what makes technology significant both today and in the past.[26]

But invisibility of the quotidian only partly explains the relative neglect of technology in scholarly discourse. Language is surely less visible than the infrastructures and artifacts of modern technology. Yet since the ancient Greeks, scholars have worked tirelessly to make language visible as an object of inquiry.[27] Why has technology, both in substance and as a concept, not received similar attention in Western intellectual history?

Most scholarship is produced by intellectuals who experience technology primarily from the outside. Since ancient times, much of this discourse has displayed an appalling ignorance if not outright hostility toward the practical arts. In the 1920s, John Dewey criticized this attitude of "profound distrust of the arts" and "disparagement attending the idea of the material," which was expressed philosophically in "the sharp division between theory and practice."[28] (Note that Dewey was using the term *arts* in its broader meaning, which includes all forms of making, not just the fine arts.) In the Western philosophical tradition, knowing has been consistently ranked higher than making, if making was considered at all. This prejudice continues in our late-modern era, when humanist intellectuals often see technology as a threat, and most natural scientists elevate pure knowledge over practical applications, a bias that also dominates philosophy of science.[29]

Despite the general distrust of the practical arts and the denigration of practice, alternative scholarly traditions have dealt sympathetically with topics that would now be termed technology. One tradition has affirmed the dignity of the mechanical arts and technology, from the twelfth-century theologian Hugh of St. Victor through Francis Bacon and Karl Marx. Another approach, more critical of technology's pervasive role in the culture of modernity, has also been deeply engaged with technological practices and ideas, from Mary Shelley to Lewis Mumford. Some of these thinkers, most important Mumford, played key roles in the history of the concept of technology.

More recently, scholars in science and technology studies (STS) have tried to remedy the scholarly neglect of technology, producing significant theoretical insights in addition to their empirical works.[30] However, STS scholars still tend to collapse technology into science, as demonstrated by the widely used concept of technoscience.[31] Historians of technology also have made major contributions to the discourse of technology, particularly through the Society for the History of Technology and similar academic organizations. And philosophers of technology have generated an impressive body of literature that grapples with fundamental questions of theory.[32] But the history and philosophy of technology have for the most part remained intellectual ghettoes, even within their respective scholarly disciplines. Few papers in these fields are presented at national meetings of historians and philosophers.[33] When scholars outside these subfields discuss technology, they often ignore the contributions of historians and philosophers of technology.[34]

Some historians and philosophers of technology have internalized the prejudices that make them feel marginal. Historians of technology in particular have developed a Rodney Dangerfield complex, plaintively lamenting their lack of respect in the broader scholarly community.[35] Such thinking may explain why this field has failed to clarify the concept of technology itself, a failure directly linked to the lack of historical consciousness about *technology* the term. If academics who specialize in the study of technology can't figure out what it means, how can we expect others to do so?

This lack of historical consciousness is especially clear when philosophers try to get a handle on the concept of technology.[36] Philosophers tend to prefer prescriptive to descriptive definitions; that is, they make claims about how terms ought to be used.[37] Such claims are usually justified on logical rather than historical grounds. For example, a number of philosophers writing in English have sought to differentiate *technology* from *technique* (or *technics*). One such philosopher is Larry Hickman, a creative thinker who applies insights from John Dewey's pragmatism to the philosophy of technology. Hickman notes that etymologically, *technology* should refer to the study of technical things, but that instead the term usually refers to the technical things themselves, or "technique." Hickman defines *technique* rather oddly as "habitualized skills together with tools and artifacts."[38]

Hickman calls for separating these two terms, *technique* and *technology*. He ascribes to *technology* all the higher cognitive qualities involved in using tools and artifacts to solve problems. When a problem is solved, what remains is *technique*, the stable, largely noncognitive solution. In this schema, *technology* is "active, reflective and creative," while *technique* is "for the most part passive, non-reflective and automatic."[39]

Hickman's approach falls short, however, in its lack of any historical basis for his definitions of *technique* and *technology*. He seems unaware that almost all continental European languages maintain a distinction between *technique* and *technology*, but one that does not deny creativity to *technique*. Since the 1890s, all attempts to strip away the creative, cognitive components of the Continental concept of technique have been fiercely resisted by technical practitioners and their intellectual allies.[40] In a sense, Hickman has reproduced, in his artificial division between technique and technology, the same historical separation of mind and hand, theory and practice, that Dewey himself sought to overcome.

The History of the Concept of Technology: An Overview

The dominant definitions of *technology* are fundamentally at odds with its etymology. The *-ology* suffix suggests that *technology* should refer to an academic field or a system of formal knowledge, a meaning derived from the ancient Greek term *logos*, or "reasoned discourse." However, in present-day usage, *technology* refers more to things than ideas, to material practices rather than a scholarly discipline. Similar terms for other fields of knowledge can also refer to the object of study, such as the use of *ecology* to refer to biological communities themselves. Yet *technology* is different. Its original meaning as a field of study has almost completely disappeared, at least in English. This original meaning survives here and there, primarily in the phrase "Institute of Technology" in names of schools of higher technical education.

Perhaps this divergence from etymology is what Heidegger meant when he said, "The essence of technology is by no means anything technological." But Heidegger never said that. He said: "So ist denn auch das Wesen der Technik ganz und gar nichts Technisches." The word that Heidegger used in his essay, which is translated into English as "technology," is actually *Technik*.[41] In fact, almost all Continental languages have a cognate of *technique* that can be translated into English as "technology." For example, "history of technology" in French is *l'histoire des techniques*, in German *Technikgeschichte*, in Dutch *techniekgeschiedenis*, in Italian *storia della tecnica*, and in Polish *historia techniki*.[42]

What is even odder, however, is that cognates of *technology* also exist in all Continental languages, for example *die Technologie* in German and *la technologie* in French. Through most of the twentieth century, these *technology*-cognates were less common than *technique*-cognates, but they were still present. However, all these terms, the cognates of *technique* and *technology*, are translated as "technology" in English: Continental languages use two words where English uses one. Continental languages have maintained, at least until recently, a distinction between *technique* and *technology* that is lost in English translation.[43]

What, then, is this distinction between *technique* and *technology* in Continental languages? Quite simply, it is exactly what we would expect from etymology. *Technology* is the science of *technique*; in this regard, *technique* refers to the principles and processes of the useful arts, a meaning rooted in the Greek word *techne*. In French and German, this distinction between *technology* and *technique* remained fairly clear until the 1980s, when the English concept of technology began to influence

Continental usage.[44] And just to confuse the matter further, the Continental cognates of *technique*, such as *die Technik* and *la technique*, also have another primary meaning as skills and methods for achieving any goal.[45] This meaning is best translated into English as "technique."

The English word *technology* was not always so at odds with its etymology. Before World War I, English-language dictionaries invariably defined *technology* as either the science of or the discourse about the useful arts. The 1911 edition of the *Century Dictionary*, an impressive twelve-volume work, provides a typical definition of *technology*: "that branch of knowledge which deals with the various industrial arts; the science or systematic knowledge of the industrial arts and crafts." This definition of *technology* is very similar to contemporary definitions of the term's French and German cognates.[46]

Yet just at this historical moment, the turn of the twentieth century, the meanings of the English-language term *technology* began to diverge from its Continental cognates. This moment also marks the emergence of the present-day meanings of the English term. As I noted above, by the late twentieth century *technology* had become about as common as *science*. Yet in 1925, *technology* occurred about eighty times less frequently than *science*. As figure 2 shows, only in the mid-1950s did the use of *technology* begin its rapid rise, reaching parity with *science* in the mid-1980s.

Before the 1960s, the term *technology* was largely confined to scholarly discourse. It first appeared in sixteenth-century academic Latin as *technologia*; by the seventeenth century, the term was listed in at least

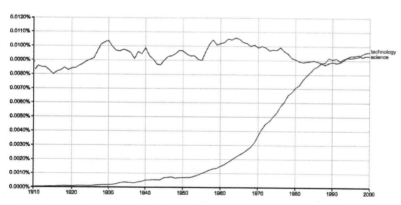

FIGURE 2 Word frequency of *technology* and *science*, 1910–2000. Google Books Ngram Viewer, accessed 1/17/2017, https://books.google.com/ngrams/graph?content=science %2Ctechnology&year_start=1910&year_end=2000&corpus=15&smoothing=1.

one dictionary in its English form. For the next few centuries, *technology* occupied recondite corners of scholarship, occasionally bubbling to the surface in some narrow context only to disappear again, until it found a place in the consciousness of Anglo-American scholars after World War I. Even then, *technology* remained uncommon among non-specialists, except in reference to technical education. Not until the 1960s did *technology* rise to the status of keyword in popular discourse.

Before World War II, people made sense of material culture primarily with other terms, such as *invention, industry, manufactures, machinery, science,* and especially *art.* None of these terms encompassed all of what we would now categorize as technology, except perhaps for *art.* Through most of its history, *art* primarily referred not to aesthetics, but rather to all forms of making. The concept of art remained central to discourse about material culture into the early twentieth century, though often modified by *mechanical, useful,* or *industrial* to distinguish it from the concept of fine art.

The concept of art has a long history as a keyword of Western philosophy, providing a direct link through the Latin term *ars* to the Greek concept of techne. In ancient Greek, *techne* referred not just to crafts but also to medicine and even rhetoric. But the Greeks also engaged in struggles over the meaning of *techne,* struggles closely connected to the social status of technicians and their knowledge. These struggles focused on two key questions that have survived into the present. First is the issue of boundaries. What forms of action and knowledge deserve to be classified as techne (or technology)? The second question concerns the moral status of techne and the social status of its practitioners. Are *technai* (technologies) neutral means, or do they possess inherent moral value? Are technicians virtuous or base?

These questions about moral status are key to the tensions between the instrumental and the cultural views of technology. Aristotle set the tone in ancient Greece when he developed an instrumental approach to techne. His approach made the virtue in techne subordinate to ends defined by nontechnical elites. Aristotle also articulated a hierarchy of knowledge with techne at the bottom, below moral and theoretical knowledge. This hierarchy remains relevant even today.

These tensions continued through the medieval and early modern eras. Most medieval authors maintained some version of Aristotle's sharp distinction between philosophical and productive knowledge, *episteme* and techne, sometimes framed as the division between liberal and mechanical arts. But this situation began to change in the fifteenth century in a way that partly bridged the gap between episteme and techne, between science and the arts. This shift was rooted in what Pamela Long calls the alliance between techne and praxis as new forms of technical knowledge became increasingly crucial to political power.[47] This alliance encouraged a surge in authorship about the mechanical arts, with works written by both humanist scholars and artisan-practitioners. Francis Bacon drew from this tradition when he began arguing, two centuries later, for a closer connection between natural philosophy and practical applications. Yet as Peter Dear and Steven Shapin have shown, respect for the mechanical arts did not imply respect for the artisan. Instead, natural philosophers maintained the conceptual hierarchy of mind over hand that mirrored the social hierarchy of the philosopher over the artisan.

The concept of technology was completely absent from this early modern discourse, which focused instead on the relationship between science and art. How, then, did the modern meanings of *technology* arise? Between the late nineteenth and early twentieth centuries, a fundamental change occurred in the meaning of *technology*. By this time, *art* had become increasingly restricted to the fine arts, eliminating it as a term for "material culture" in general. Thus, at the end of the nineteenth century, no English-language term seemed suitable to describe the momentous changes associated with the Second Industrial Revolution, with its new industries of electricity, synthetic chemicals, and steel. In the early twentieth century, *the machine* became a popular term for this new industrial complex. But *the machine* never became common in scholarly discourse.[48] Instead, there arose what Leo Marx has called a "semantic void," the absence of adequate concepts to describe the new industrial era.[49]

This void was eventually filled by new meanings of *technology* that emerged between 1900 and the mid-1930s. During this time, a variety of scholars, mainly in the social sciences, transformed *technology* into a suitable replacement for *industrial arts*. But Marx's "semantic void" did not call forth these meanings, vacuuming them up from dusty corners in the house of language. What motivated these changes, at least at first, were intellectual dynamics within the social sciences.

The shift began when American social scientists appropriated the German discourse of *Technik*. As used by German engineers and some social scientists, *Technik* referred to the methods, tools, and instruments used to create and maintain material culture. In their search for social status, elite German engineers developed a theoretical discourse about Technik that was explicitly cultural.

But scholars faced a dilemma when translating *Technik* into English.

Some used the English *technique*, but this was a mistranslation, at least when *Technik* was used in the sense of the industrial arts. Instead, beginning around 1900 with the work of Thorstein Veblen, American social scientists began translating the German *Technik* as "technology." By the 1930s, "technology" had become the most common translation for the industrial meaning of *Technik*.

About the same time that Veblen redefined *technology* as "the state of the industrial arts," other scholars began equating *technology* with the nineteenth-century concept of applied science. Like Veblen's redefinition, this move represented a profound shift in meaning. No nineteenth-century dictionary ever defined *technology* as applied science. Technology was a science, the systematic study, teaching, or discourse about the industrial arts. But as recent scholars have argued, a fundamental ambiguity existed (and remains) in the concept of applied science. It could refer to a kind of science, science suited to application. Or it could refer to the application itself, that is, the results of applying science. Both these meanings were common in the nineteenth century. However, the use of *technology* as a synonym for *applied science* became widespread only after World War II, in part because of the prominent role played by scientists in creating the atomic bomb.

The meanings of *technology* were also influenced by the other main definition of *Technik*: skills and methods for achieving a specific goal. German dictionaries clearly distinguish this broadly instrumental definition of *Technik* from its definition as industrial arts. *Klaviertechnik*, for example, means "piano technique," that is, skill in playing the piano. In contrast, *Bautechnik* means "building technology" or "structural engineering."

Yet the boundary between these two meanings remained porous. It was easy to move from viewing modern Technik as the material expression of an era's culture (that is, industrial arts) to seeing Technik as technique, instrumental rationality. In the early twentieth century, Continental critics of modernity often invoked *Technik* in this sense to explain the dehumanizing tendencies of the industrial era.[50] So in the 1930s, when *technology* became a standard translation for the industrial meanings of *Technik*, some scholars also began translating the instrumental meaning of *Technik* as "technology." In essence, these scholars redefined *technology* as technique. Although this translation practice remained fairly uncommon until the 1960s, it further confused the concept of technology, encouraging the same combination of industrial and instrumental meanings that exists in Technik. However, in contrast to the German *Technik*, these two meanings of *technology* in English were never clearly distinguished.

Thus by World War II, three main meanings of *technology* had become established in scholarly discourse: *technology* as industrial arts, as applied science, and as technique.[51] After the war, however, the tremendous cultural authority of science tended to push *technology* into the background. It was only in this era that an understanding of *technology* as the application of science became dominant. But public awareness of *technology* in the 1950s was also encouraged by critics of modern industrial civilization. In philosophy and religious studies, American scholars in that decade began to take notice of the philosophical discourse of Technik, which was translated into English as a discourse about technology.

From the 1960s until today, the discourse of technology has become pervasive, particularly as English has become the lingua franca of global communication. Yet confusion between these three principal meanings of *technology* remains common in both scholarly and popular discourse. This confusion pervades the most recent entry on *technology* in the *Oxford English Dictionary*. In a convoluted, multipart definition, the *OED* defines *technology* as a "branch of knowledge," as "the application of such knowledge," as "the product of such application," as "a technological process, method, or technique," and as "a branch of the mechanical arts or applied sciences."[52] In a sense, the *OED* captures all the principal meanings of the word *technology*. But without awareness of the history of technology as a concept, the authors of this entry remain unable to disentangle these meanings.

Method in the Muddle

Like many historians, I start with an empirical question and then decide on the most relevant theories and methods. In that spirit, I use several approaches in this book. First and foremost, my analysis of the meanings of *technology* is an exercise in the history of concepts, a subfield of intellectual history most thoroughly developed by German scholars under the name *Begriffsgeschichte*. This approach is associated with the historian Reinhart Koselleck. Koselleck focused on political concepts, but had little to say about those in science or technology.[53]

Second, my approach also bears affinity to what Foucault, drawing from Nietzsche, has termed "genealogy." Foucault used this term to suggest that the history of concepts should be understood as a process of descent rather than origin. Despite occasional sweeping generalizations, Foucault did capture the essence of an argument against essences. Not only did he denounce the notion of a historical moment that gives birth to the essence of a concept; he also rejected the very notion that concepts have essences. Historians of ideas, in his view, should not seek "to go back in time to restore an unbroken continuity," nor should they attempt to reveal the "predetermined form" of the past that still lives in the present. Rather than building foundations, the genealogist undermines them, "fragment[ing] what was thought unified," revealing not a point of original truth but messy details, ironic reversals, and chance combinations.[54]

Foucault's genealogy does not imply that history bears no relevance to understanding present-day concepts. To the contrary, genealogy makes the past even more essential for questioning the present, for uncovering the biases and assumptions embodied in concepts that we take for granted.[55] However, the denial of essences in concepts does create a problem for the historian: the problem of continuity. Unlike an earlier generation of scholars, we cannot assume the existence of fairly stable concepts over time. So how can someone write a history of concepts at all? Etymology is part of the answer. But even when words have a clear past, their meanings can change fundamentally from their roots. Stark changes have occurred, for example, in the translation of the Greek *techne* into the Latin *ars* and then into the present-day English *art*.[56] Thus, historians of concepts cannot find continuity solely in etymology and translation practices. Instead, continuity arises in part from the historical actors themselves, who create their own predecessors, repeatedly drawing support from real or imagined traditions.[57]

The concept of technology is well suited to such an analysis. Rarely the focus of explicit theorizing, technology migrated among scholarly fields and across linguistic borders, shifting its meanings and, through translation, the words that denoted the concept. Technology is not a stable entity; it has no essence that scholars can uncover through a correct definition. Neither does the concept have an origin, a single point from which it emerged, like early humans, to spread throughout the scholarly world. Technology is a bastard child of uncertain parentage, the result of a twisted genealogy cutting across multiple discourses. No scholarly discipline owns this term.

But words aren't everything. As a historian of technology, I am wedded to an ontological distinction between words and things, a distinction that certain scholars within science and technology studies seem to disdain.[58] Ideas, expressed in language, can indeed transform the material world, but not on their own. This transformation requires human mediation, the action of humans who in their own materiality provide the link between words and things. Therefore, how we think about technology on an abstract level, that is, how we understand the concept, can make a difference in our material interactions with our fellow humans and the world we inhabit.

My history of this concept is not merely disinterested scholarship, if such a thing were possible. Instead, I've written a work of historical critique, following the political scientist Terence Ball's call for "critical conceptual histories." Ball has argued that the history of political concepts cannot be politically neutral. I believe that technology has also become a political concept, and thus my analysis is likewise not neutral. I explicitly reject any reduction of technology to instrumental action. While efficacy must be an element of all human action, I insist on understanding technology as an expression of human culture, as *art* in the older sense of the term, as a union of thought and action, mind and hand. And, following Veblen, I see technology not as the product of a narrow technical elite, but rather as the handiwork and rightful heritage of all humans.

Although academics can't easily change the world, a critical history of concepts has that potential. According to the moral philosopher Alasdair MacIntyre,

Philosophy leaves everything as it is—except concepts. And since to possess a concept involves behaving or being able to behave in certain ways . . . , to alter concepts, whether by modifying existing concepts or by making new concepts available or by destroying old ones, is to alter behavior. . . . A history which takes this point seriously, which is concerned with the role of philosophy in relation to actual conduct, cannot be philosophically neutral.[59]

A transformed concept of technology will not solve the technological problems of late modernity. But a shift from an instrumental to a cultural understanding of technology would, I believe, help humans exert more conscious control over their technological futures.

TWO

"The Trouble with *Techne*": Ancient Conceptions of Technical Knowledge

The ancient Greek concept of techne, so difficult to translate, lies at the core of our preeminently modern term *technology*. From ancient Greece to the nineteenth century, a continuous line of scholarly discourse links techne, the Latin term *ars*, and the English term *art*. Techne first emerged as a key philosophical concept with Plato, who bequeathed its heritage to the West and beyond. The progeny of techne reached a broader audience through Latin, which translated *techne* as *ars*. From *ars* came *art* and related terms, such as *artful* and *artificial*, along with the concepts of liberal arts and mechanical arts. Not until the nineteenth century did writers reduce art, in most usage, to the much narrower concept of fine art. In effect, this shift in meaning, the narrowing of art to fine art, ended a millennia-old tradition of philosophical discourse about productive knowledge and action.

Thus there is no direct path from ancient ideas about techne to our modern concept of technology. Concepts have no origins for historians to uncover, nor do they contain essences to be revealed. The meanings of *techne* changed substantially between the Homeric age and the classical Athens of Plato, while the *ars* of medieval Europe was not the same as *art* in Enlightenment encyclopedias. And even within a given time and place, meanings were often contested; for example, practitioners often disagreed with literary scholars about the nature of techne and ars.

In this messy world of contested concepts, scholars often drew from past philosophical traditions to justify how they used key terms. Many scholars continued to read classical texts in their original languages. Latin remained essential for European scholarship from the Middle Ages into the nineteenth century, and many nineteenth-century universities required that students learn ancient Greek.[1] Scholars drew from the well of Greek and Latin sources to refresh the philosophical arguments of their day, appropriating and retranslating classical terms. Such appropriation continued into the twentieth century. Heidegger, for example, explored the classical Greek meanings of *techne* in his essay, "The Question concerning Technology" (1954).[2] Thus, the past of *techne* remains part of the present.

But drawing from the past is only part of the story. Scholars are usually only dimly aware of the historical lineage of concepts like technology. Historical continuity comes less from core concepts than from core questions, questions about the place and role of technical knowledge and action in complex human societies. Discussions of techne, ars, and art were united by two related questions that emerged, at least in incipient form, in ancient Greece. First is the question of scope, of boundaries. What forms of action and knowledge deserve to be classified as art or technology? The question of boundaries is connected to a second question about the moral status of the arts and the social status of their practitioners. Are arts morally neutral means to ends given by others, or do arts contain inherent moral values or virtues? Are technicians, practitioners of the arts, virtuous, or are they base? These questions arose, somewhat surprisingly, in ancient Greece, a premodern society that did not especially value technological practitioners. As the historian Sarafina Cuomo has argued, even though artisans were socially marginal in classical Greek society, the question of techne was not.[3]

The Boundaries of *Techne* in the Ancient World

What counts as a techne, and what doesn't? What characteristics do the various *technai* share? These questions about the term's scope go back to its earliest recorded uses. Etymologically, the word derives from the Indo-European root *tek*, a term that probably referred to the building of wooden houses by wattling, that is, weaving sticks together. This root is related to the Sanskrit *takṣ*, "to fashion," and the Latin *texere*, "to weave." In early Greek, *tekton* denoted a woodworker, and *techne* the skill of working with wood. By the time the Homeric epics were recorded, probably in the eighth century BCE, the meaning of *techne* and its derivative forms had broadened to include the skills of the smith as well as the woodworker. But in Homer the term could also refer to skills of doing as well as making, for example piloting a ship. This Homeric extension to doing represents a crucial broadening of the concept. Perhaps more significantly, *techne* began to take on a figurative meaning as well, implying ruse, wile, or deceit, much like the English *craftiness*. Still, the dominant meaning of *techne* in the Homeric epics focused on productive crafts, that is, making things.[4]

By the era of classical Athens, from roughly the fifth to the fourth centuries BCE, the boundaries of *techne* broadened further, embracing medicine, divination, and even rhetoric. This change partly reflected a shift away from a simpler society defined largely by kinship and mutual reciprocity to one with a complex social structure featuring many specialized occupations. *Techne* became, in this context, a socially valued category that attested to specialized expertise in particular occupations, from carpentry to medicine. As a cultural good, its boundaries were sometimes contested.[5]

Greek writers in this era agreed on key aspects of techne. One was its artificiality; techne produced a result that would not have existed without the intervention of a technician, a practitioner of techne. Techne was fundamentally about how to do things, "knowing-how rather than simply knowing-that." Furthermore, techne was not innate but teachable.[6]

Nevertheless, Greek authors, especially philosophers like Plato, rarely addressed the boundaries of *techne* explicitly. Instead, *techne* was usually invoked as a self-evident term useful for clarifying other concepts of more interest to philosophers. In *Gorgias*, for example, Socrates repeatedly used craft analogies in his critique of rhetoric, to the point that one of his interlocutors, Callicles, exploded with irritation. "By the Gods," complained Callicles, "you're literally always talking nonstop about leather workers and drycleaners and cooks and doctors, as if our speech had anything to do with *them*."[7] Indeed, struggles over the meaning of *techne* usually arose with regard to specific trades and professions. These debates are difficult to reconstruct, because little evidence from practitioners survived. Two exceptions are the fields of medicine and rhetoric.

The principal source for the status of medicine as a techne is the

Hippocratic Corpus, a set of essays from the physician's perspective written from about 450 BCE to 340 BCE. Several Hippocratic works explicitly examined whether medicine deserved to be termed a techne. In the essay *On Techne*, the Hippocratic author argued, against the objections of a hypothetical critic, that medicine is a techne. The critic claimed otherwise, insisting that a techne required the ability to produce desired results reliably. Medicine lacked the reliability of other arts; patients sometimes died despite the ministrations of the physician, and they sometimes got better on their own. Reliability implied human mastery over the random forces of nature, *tyche* (luck), but, according to the critic, many cures were often no more than luck.[8]

The author of *On Techne* accepted the critic's characterization of medical practice, but questioned his definition of a techne. One could not expect various technai to be equally precise. Medicine dealt with the complex and variable human body, whose workings were hidden from direct observation. What distinguished the physician from a lay practitioner was the physician's knowledge of and ability to articulate the causes of health and disease. Rather than overcoming *tyche*, the physician worked with it, finding the right moment to treat a patient.[9]

The physicians' embrace of techne in the Hippocratic Corpus shows that the concept covered more than just the production of material artifacts. In the same period, Greek rhetoricians also began arguing that rhetoric was a techne, thus asserting the status of rhetoric as a profession. Rhetoricians often made this claim in treatises about techne intended to codify the rules of their art. But rhetoric was particularly slippery as a techne, given the abstractness of its object and its indeterminate product. Leading philosophers disagreed about its status. Plato, for example, emphatically denied that rhetoric could be a techne, while for Aristotle, rhetoric's status as a techne was unproblematic.[10] This debate over the status of rhetoric as techne (or art) continued into the late twentieth century.[11]

Was techne a form of knowledge, and if so, what kind of knowledge? Greek writers repeatedly acknowledged that techne involved logos, that is, language and reason, as well as productive acts. In *Gorgias*, written about 380 BCE, Plato made this point explicitly. When Gorgias claimed that the subject of rhetoric is "speech" (*logos*), Socrates replied, "Surely Gorgias, the rest of the arts are the same way; each of them is about just those speeches that have to do with the subject matter" of the art. Although Socrates admitted that some technai, like painting, require little or no speech, others, like arithmetic, accomplish their goals entirely through logos.[12]

Socrates then denied that rhetoric is a techne, because it relied on unsystematized experience rather than logos. Rhetoric, he argued, is "not art but a matter of experience, because it has no speech to give about the nature of the things it makes use of or what it uses them on, . . . and therefore it can't state the cause of any of them. I don't call any proceeding that's irrational an art."[13] In effect, Plato was making a claim about the relationship between theory and practice, arguing that a true techne required rational thought about cause and effect, a form of knowledge that we might classify as science. Plato's comments in *Gorgias* cannot be translated as a statement about the relationship between science and technology; such modern categories would have made no sense in ancient Greece. Nevertheless, Plato's comments in *Gorgias* did question the nature of the knowledge involved in techne.

Plato never focused on techne as a concept, and he did little to distinguish techne from other forms of knowledge. Particularly in his early dialogues, he often used *techne* and *episteme* (scientific knowledge) interchangeably, using both in the sense of expertise.[14] For Plato's student Aristotle, in contrast, techne and episteme become fundamentally distinct categories.

Aristotle made two crucial distinctions that profoundly shaped Western thought. These distinctions appeared most clearly in his *Nicomachean Ethics*, written about 350 BCE. In this work, Aristotle sharply distinguished *techne* from moral knowledge, *phronesis*, and from scientific knowledge, *episteme*.[15] These demarcations elevated *episteme* and *phronesis* at the expense of *techne*.

In chapter 6 of the *Nicomachean Ethics*, Aristotle defined *episteme*, *phronesis*, and *techne* as three categories of reason. These Greek terms are traditionally translated as "science," "prudence," and "art" respectively, but their meaning in the Greek differs substantially from their English counterparts. Episteme, argued Aristotle, deals with the realm of necessity, with knowledge of the eternal, what cannot be otherwise. In contrast, both techne and phronesis deal with "those things that do admit of being otherwise," that is, the realm of contingency.[16]

Aristotle described techne as "artfully contriving," an activity that originates from the person doing the making, not the thing made. Episteme was devoid of techne, because it dealt with things that arise from necessity, out of their own nature. In Aristotle's worldview, then, episteme and techne are completely separate, the one dealing with the natural and the other with the artificial. Aristotle was not arguing that techne lacked reason (logos), but rather that techne concerned reasoning about a completely different subject matter from episteme. Like

Plato, Aristotle explicitly linked *logos* and *techne*, defining *techne* as "a characteristic bound up with making [*poiesis*] that is accompanied by true reason [*logos*]."[17] Thus, techne is not just rote activity or manual dexterity, like riding a bicycle. The true technician, such as a physician or a carpenter, can give reasons for his actions.

Unlike episteme (science), both techne (art) and phronesis (prudence) deal with contingent human choices rather than what exists of necessity. But Aristotle also argued for a sharp boundary between the two: techne is about poiesis (making); phronesis is defined by praxis (action). Poiesis and praxis are fundamentally distinct, Aristotle insisted. For making, "the end is something other than the making itself," while for action, "acting well [*eupraxia*] itself is an end."[18] Therefore, "since making and action are different, an art [*techne*] is necessarily concerned with making but not with action."[19] Thus, techne serves ends that are external to itself, while phronesis embodies its own end, eupraxia.[20]

Aristotle did little to justify this distinction between techne and phronesis. But the implications are profound. By distinguishing these two forms of reason, Aristotle in effect excluded all ethical content from the process of making itself.

The Moral Status of *Techne* and *Ars*

Aristotle's definition of *techne* has direct bearing on its moral status, the second key question in the discourse of techne. This moral question links techne to the social order. As Cuomo has argued, technologies are both essential to social stability and potential disruptors of the social order. People with technical skills can threaten status hierarchies founded on birth or property. Because skills are teachable, even the hoi polloi can master them. Debates over the moral status of both technology and techne are thus also about the nature of the social order and the role of technicians within it.[21]

Aristotle explicitly denied techne moral virtue while simultaneously asserting a hierarchy of knowledge suited to aristocratic values. At the beginning of the *Nichomachean Ethics*, Aristotle argued that all human inquiry and activities are directed toward ends that involve some form of the good. But the good involved in making, poiesis, lies in its products, not in the activity itself. For poiesis, therefore, "the works are naturally better than the actions."[22] In other words, when a carpenter builds a ship, the excellence (*arete*) of the activity lies only in the final product, the ship, rather than in the process of its construction. In this way, poiesis is inferior to praxis (action), which directly serves the good as an end in itself. Aristotle extended this hierarchy to the forms of knowledge associated with poiesis and praxis, namely techne and phronesis. Because techne serves ends outside itself, it lacks virtue (arete). As Martin Ostwald explains in his notes on the *Nicomachean Ethics*, "Practical wisdom [*phronesis*] is itself a complete virtue or excellence [*arete*], while the excellence of art [*techne*] depends on the goodness or badness of its product."[23] *Techne*, defined as serving ends outside itself, becomes morally neutral, to be judged by standards of virtue external to it.

Aristotle thus established a hierarchy of knowledge strongly linked to social class. At the top was episteme, knowledge of the eternal and necessary, followed by phronesis, knowledge of how to act well in specific circumstances, and at the bottom techne, knowledge that serves ends outside itself.[24] This hierarchy was well suited to the aristocratic values that became dominant after the decline of Periclean Athens in the late fifth century BCE. Episteme was theoretical knowledge, understood in the Greek sense of contemplation, an activity suited to free men who had time for leisure. Aristotle was explicit about this point. Theoretical knowledge became available to men only after "practically all the necessities of life were already supplied." People sought such knowledge "not for any practical utility, . . . [but] with a view to recreation and pastime." Such knowledge provided "no extrinsic advantage; . . . for just as we call a man independent who exists for himself and not for another, so we call this the only independent science, since it alone exists for itself." Pursuit of theoretical knowledge thus served as a marker of aristocratic status, of the man whose material needs were provided by the labor of others.[25]

Phronesis, the knowledge of moral action, also served aristocratic values, even though it was inferior to episteme because it involved action rather than contemplation. Moral virtue implied a freedom of action that presupposed a degree of wealth and social standing, as well as citizenship, membership in the polis. As Alasdair MacIntyre noted, slaves and barbarians (that is, non-Greeks) had no place in the polis, and therefore no freedom to exercise moral virtue.[26]

Aristotle articulated a hierarchy of theory over practice that continues to this day. This hierarchy endures particularly among academics, who almost invariably value general principles over practical applications. But Aristotle's sharp division between episteme and techne went beyond hierarchy: it implied a complete separation between the two types of knowledge. For Aristotle, technical and theoretical knowledge can have nothing to do with each other as a matter of definition. Techne and episteme concern fundamentally distinct forms of being. Techne deals with the contingent and artificial, while episteme is about the necessary and natural. This division does not mean that techne lacks general principles, but rather that the principles involved in techne are fundamentally different from the principles of episteme. Because episteme has no practical utility, there can be no "applied science" in the sense of the application of scientific knowledge to technical problems. As Nicholas Lobkowicz remarked in his historical survey of theory and practice, "What today we would call 'applied science' was more or less unknown to the Greeks."[27] This separation of contemplative and productive knowledge remained remarkably durable. In fact, the concept of applied science did not arise until the middle of the nineteenth century.[28] The modern concept of technology represents, in one sense, the overcoming of Aristotle's isolation of episteme from techne.

Something fundamental gets lost, however, in more recent discussions of the science-technology relationship, namely Aristotle's tripartite distinction that also divides techne from phronesis, technical knowledge from moral knowledge. Even Heidegger, who grounded his critique of modern technology in Aristotle, ignored the role of phronesis.[29] Aristotle's distinction between techne and phronesis is just as categorical as his separation of techne and episteme, and just as consequential. This separation reduces technical knowledge to means devoid of any inherent moral value. As I argue in later chapters, the cultural concept of technology overcomes, at least implicitly, this Aristotelian separation.

Aristotle's denial of inherent moral value in techne was not universal in classical Athens, particularly in the early fifth century BCE. Pamela Long notes that the technical arts prospered in fifth-century Athens, both materially and in social status. Athenian democracy, despite its many flaws, did broaden public participation in the political process. In this context, argues Long, "technical and craft skills came to be admired as significant human achievements." These attitudes brought about what Long terms an alliance between techne and praxis, between craft knowledge and political action. The great civic monuments of Periclean Athens, such as the Parthenon, are tangible products of this alliance.[30]

Techne was at times celebrated in Greek literature of this era. A choral ode from Sophocles' *Antigone* (ca. 441 BCE) proclaims the wonders of man, who tames the wild forces of nature with his ships, plows, yokes, nets, cities, and even language. Only death is inescapable, but even here man "has devised escapes from hopeless diseases." Yet Sophocles concluded the ode by revealing a dark side to techne. Even though man's techne is "clever beyond hope," it also "brings him now to ill, now to good." The good arises from adhering to human laws and to the justice of the gods, while negative consequences follow from ignoble actions, which render man "citiless."[31]

By both celebrating techne and warning against it, the ode is ambiguous. Some commentators see it as evidence that techne was already viewed as morally neutral in fifth-century Athens.[32] But such an interpretation conflates moral neutrality with moral ambiguity. A morally ambiguous techne can still teem with moral values, from good to evil, such as the arts of healing and the arts of torture.[33] But to portray techne as morally neutral, devoid of inherent moral value, is far more radical. This is the step that Aristotle took in the *Nicomachean Ethics*, a century after Sophocles, when he divorced means from ends in techne.

The separation of means from ends, and the consequent extirpation of moral value from techne, devalued knowledge associated with productive activities. This devaluation implied a loss of status for technicians. According to MacIntyre, by separating poiesis from praxis and techne from phronesis, Aristotle made excellence in "craft skill and manual labor . . . invisible from the standpoint of Aristotle's catalog of the virtues."[34] In other words, productive activity could not provide a path to moral virtue for the technician, whatever the technical excellence of the product.[35]

Regardless of their moral value, technai were clearly essential to civilization, despite the threat that technicians posed to the social order. This resulting tension, in which elites viewed techne as both a necessity and a threat, led to discussions throughout the classical world about the social and moral status of technicians and their technical knowledge. Cuomo refers to this tension as "the trouble with *techne*."[36] Although literary elites dominated this discourse, their distrust of techne was at times challenged by technicians themselves.

Contempt for manual trades was widespread in the Greek and Roman worlds, at least among the elites who defined cultural norms. It was expressed in the concept of the "banausic arts," those base or vulgar occupations shunned by aristocrats. The Greeks labeled as *banausos* those arts that corrupted the ideal of the male body, an ideal highly valued in Greek culture. Occupations that required sitting for long periods of time, for example, could fall into this category. Baseness was also the result of work performed for wages. In his *Politics*, Aristotle summarized this view: "A deed, *techne* or learning must be considered base [*banausos*] if it makes the body or soul or mind of free men useless for the uses and actions of virtue. Therefore we call base those *technai* that render the body worse and the activities done for pay."[37] It is rather ironic that Greek and Roman elites, who valued the finely wrought products of skilled craft workers, had contempt for the workers themselves.[38]

Although some scholars may have exaggerated Greek and Roman contempt for manual labor and craft workers, such views were in fact widespread among the literary elites from the time of classical Athens through the Roman Empire.[39] During the most open period of Athenian democracy, for example, quite a few citizen-craftsmen or their sons rose to positions of political authority, particularly in the second half of the fifth century. But they achieved their success despite, not because of, their craft connections. Much of what we know about their craft origins comes from insults hurled at them by their political opponents.[40] Plato clearly knew a great deal about the crafts, and as discussed above, he repeatedly used craft metaphors in his dialogues, often in a way that suggested respect for craft methods.[41] But in his mature dialogues, he codified the low social status of the craftsman in order to preserve social stability. In the *Republic*, for example, Plato proposed his "noble lie," a myth that tied each person to a social role according to his inborn metallic constitution: gold for the guardians, silver for the soldiers, and bronze for the craftsmen. Nothing would be worse, thought Plato, than a craftsman trying to do philosophy.[42]

Among the Greeks, not all arts were banausic. What made a techne base was inherent not necessarily in the techne itself but in how it was practiced. According to the historian Elspeth Whitney, what defined an art as vulgar was not "physicality alone" but rather its use "*merely* to satisfy physical needs or pleasures." And as Cuomo suggests, contempt for manual trades did not necessarily imply distrust of technical expertise, that is, techne in general.[43]

This subtlety among the Greeks was largely lost among the Romans, who replaced the category of banausos with a division between liberal and vulgar arts. According to Roman writers, almost all the arts associated with skilled manual labor, especially arts that produced material objects, were classified as vulgar. Writing in 44 BCE, Cicero dismissed a whole range of occupations as unsuited to free men, in particular those involving artisanal production. In a well-known passage, he declared that "the trades practised by all artisans are also vulgar, for there can be nothing in a workshop which befits a gentleman." In contrast, he considered professions like medicine and architecture "honorable," at least "for those of the appropriate social class." He also praised agriculture, declaring that "of all profit-making activities none is . . . more worthy of a free man than agriculture."[44] As Long notes, Cicero was referring not to agricultural work itself but rather to the management of estates by large landowners.[45]

Not surprisingly, technicians generally rejected depictions of their skills as lacking moral virtue, at least in those rare cases where their voices have been preserved. The most spirited defense of the moral value of a techne is found in the Hippocratic Corpus. "The Hippocratic treatises," states Cuomo, "abound in statements about the virtues of the doctor." Proper conduct was essential to the techne of medicine. Through their training, physicians learned which potions harmed as well as helped their patients. However, according to the Hippocratic writers, using medical knowledge to do harm fell outside the scope of techne. In this conception, kindness to patients, humility in bearing, and restraint in the pursuit of wealth were all integral to the techne of medicine, not external to it. Someone could not, in fact, be a good physician without these moral virtues.[46]

Lower-status artisans also contested their social status. Little textual evidence survives, but other forms of evidence do exist. For example, Cuomo argues that artisans in ancient Rome used funerary symbols to express their sense of self and assert their position in Roman society. These symbols frequently portrayed instruments of the artisan's trade, such as the carpenter's square. Only the most successful artisans could afford custom-made funerary monuments. By displaying the tools of their craft, these prosperous craftsmen chose to celebrate rather than suppress their technical origins, commemorating the technical skills that had given them wealth and social status. Cuomo argues that such funerary symbols not only expressed pride in technical practices but also conveyed a message of equality, a claim to social status based on skill rather than birth.[47]

Shifting Boundaries in the Roman Era

Under the Roman Empire, scholars writing in both Greek and Latin continued to develop the concepts of techne and ars. The complex Aristotelian distinctions between techne, phronesis, and episteme were largely lost, as techne and then ars grew to an expansive category that embraced all types of learning, from the most practical to the most theoretical. With this broadening, a sharper division emerged between liberal and banausic arts, although writers differed on the boundary. The Greek physician Galen, writing around the second century CE, characterized a wide range of fields as techne, dividing them into the honorable (σεμναί) and the contemptible (εὐκαταφρόνητοι). The contemptible arts were those involving physical labor, including the handicrafts. Among the honorable arts he included "medicine, rhetoric, music, geometry, arithmetic, philosophy, astronomy, literature, and jurisprudence," as well as "sculpture and painting," which "do not demand great strength" even though "they are associated with manual labor."[48] Galen's somewhat grudging inclusion of sculpture and painting is actually less significant than his embrace of purely intellectual fields among the arts, which provided a foundation for what became, during the Middle Ages, the liberal arts (*artes liberales*).

Some version of Aristotle's tripartite division of knowledge survived among Roman authors, but they did not maintain his sharp epistemological boundaries between technical knowledge, moral knowledge, and theoretical knowledge. As with Galen, these writers broadened the concept of ars, the Latin translation of *techne*, to encompass nonproductive forms of knowledge. The Roman rhetorician Quintilian provides an early example. Writing a century before Galen, Quintilian still embraced a form of the threefold Aristotelian division of knowledge. When he invoked these terms, he wrote them in Greek rather than Latin, making their origins in Greek philosophy clear. Yet even while maintaining Aristotle's tripartite division, Quintilian characterized all these fields, including the theoretical and practical disciplines, using the Latin term *ars*. The exemplar of a theoretical art is astronomy, "which demands no action, but is content to understand the subject of its study." Rhetoric and dance, in contrast, are practical arts, which produce no finished product but achieve their end in action. Productive arts include activities like painting.[49] Quintilian's classification differed significantly from that of Aristotle, who considered rhetoric and performing arts to be technai, not praxis disciplines aiming at moral virtue.

Whitney has argued that tripartite divisions like Quintilian's continued in the Roman world through the fifth century CE, but disappeared in the Latin West thereafter, possibly because Aristotle's works became largely inaccessible. She also notes the lack of any standard Latin translation for the Greek term ποιητικαί (*poietikai*), "productive," an absence that helped obscure Aristotle's distinction between productive and practical action. The tripartite division survived primarily in the

27

Greek-speaking Eastern Roman Empire, and then in the Arabic intellectual tradition, only to return to western Europe from the Arabs in the twelfth century.[50]

Martianus Capella codified the seven liberal arts in the early fifth century CE as the Roman Empire was collapsing under attacks from Germanic tribes. By this time, Aristotle's definition of *techne* had largely collapsed as well, along with his division between techne and episteme.[51] Despite continued hierarchical ranking of the arts, Roman classifications included such varied fields as geometry and sculpture under the single conceptual rubric of ars. Such classifications ensured that in late antiquity, "the difference between sciences and arts was . . . always vague and indefinite."[52]

Questions about technical knowledge and its practitioners persisted in the ancient world even as terminology shifted from one era to the next. A range of writers continued to discuss the relationship between technical and theoretical knowledge, as well as that between technical knowledge and moral action. These questions persisted in part because technical specialization poses a problem for all complex societies. For this reason, questions about techne and technicians remain as germane today as they were in classical Athens.

Nevertheless, a vast gap separates the classical debates about techne and ars from the twentieth-century concept of technology. Just as the Greeks and Romans had no word for "art" in the present-day sense of *fine art*, they also had no comparable term for "technology."[53] Neither the Greek term *techne* nor the Latin *ars* is equivalent to the concept of technology that emerged in the early twentieth century. Techne was obviously much broader than the present-day concept of technology, encompassing, for example, rhetoric and medicine. It was also in many ways narrower, confined to a specific kind of knowledge embodied in a wide variety of skillful practices.

This gap between the past and present persisted in the medieval world. Nevertheless, significant conceptual changes occurred in that world, even as "the trouble with *techne*" continued.

THREE

The Discourse of *Ars* in the Latin Middle Ages

By the end of the Roman Empire, the Latin concept of ars had already become an expansive category encompassing various forms of knowledge. Late Roman and early medieval writers did bracket off one set of the arts, the liberal arts, which received special attention in terms of pedagogy. But medieval scholars also created a new category, the mechanical arts, which encompassed crafts and other types of productive knowledge. Historians disagree about the cultural significance of this new category. Some argue that it represented a new appreciation for the moral status of crafts and craftsmen, while others insist that medieval scholars, with their otherworldly concerns, had no more appreciation for the crafts than their ancient predecessors had. But even if the mechanical arts were the lowest form of art, they nevertheless remained art, and thus shared with the liberal arts the general attributes of reason and virtue.

Unfortunately, we still lack a comprehensive history of the concept of ars and its counterparts in the medieval and early modern periods. There is a substantial scholarly literature on medieval ideas about the arts, but most scholars interpret the concept of ars in relation to later categories like fine arts and technology. Many discussions of the arts in medieval thought focus on the relationship between art and aesthetics, even when scholars acknowledge that the Latin ars implied nothing comparable to the later concept of fine arts.[1] Other historians focus on either liberal arts or mechanical arts. But scholars rarely address the concept of art as a whole, failing to explain how this concept united such diverse fields of knowledge.[2]

Claims about ancient and medieval attitudes toward technology also reflect the spread of the concept of technology after World War II and its spectacular rise in cultural prestige. In the postwar era, many historians tried to rectify what they saw, often with justice, as a dismissive attitude toward the technological contributions of the periods they studied. These historians typically denied that their eras were hostile to technological innovation. Much scholarship on medieval technology exemplifies this tendency, in particular the pioneering work of Lynn White.[3]

These historians tried to raise the cultural status of the people they studied by arguing that these people did indeed value technology. This pro-technology bias has produced immensely valuable research on the history of premodern technology. But this research can be misleading. There was no concept of technology before the nineteenth century, despite the existence of things, institutions, and forms of knowledge that were clearly technological under present-day definitions. There were, however, tensions over the meanings of technical knowledge and practice, especially among the elites most likely to leave records. Two key topics emerge from these tensions during the medieval era. First is the rise of the mechanical arts as a distinct category encompassing productive crafts that we would now classify as technological. Second are continued debates about the moral status of the arts and their practitioners.

The Medieval Concept of Mechanical Arts

The mechanical arts emerged as a distinct category in medieval Europe during the ninth century. Greek and Roman writers had no distinct category for the manual arts, even though they viewed some arts as noble or liberal and others as vulgar, servile, or banausic. Yet the classical distinction between noble and vulgar arts was not about categories of knowledge. Cicero, for example, praised the management of agricultural estates as ideally suited to the free man. Nevertheless, agricultural supervision was clearly a productive activity even if the estate owner did not get his hands dirty.[4]

With the final fall of the western half of the Roman Empire at the end of the fifth century, population, economic activity, and political

organization all declined markedly. Although there were wide variations in local conditions, this decline flattened social hierarchies and
reduced social distinctions. These changes undoubtedly brought literate elites, now primarily clerical, in closer contact with craft work
and craft workers. Such contacts undoubtedly encouraged some writers to reflect more deeply on the productive arts.[5] From these reflections there emerged a new category within the arts, the mechanical
arts (*artes mechanicae*).

The mechanical arts were rooted in Christian ideas about manual labor and productive arts. Lewis Mumford, Ernst Benz, and Lynn White
all argued that Latin Christianity encouraged a positive view of labor
and craft that partly explains the economic and technological "rise of
the West."[6] Positive evaluations of manual labor were certainly widespread among Christian writers in late antiquity and the early Middle Ages. George Ovitt has suggested that the marginal status of early
Christians "encouraged a toleration of labor and laborers not found
in the Greco-Roman world."[7] The image of God as artisan (*artifex*) was
common among the church fathers and also in commentaries on the
biblical Creation story (the "hexameral" literature).[8] In late antiquity,
Christian monastic texts urged monks to devote themselves to manual
as well as spiritual labor, both to provide for the material needs of the
community and to avoid the temptations that arise from idleness.[9]

Nevertheless, as Ovitt and other scholars have concluded, Christian
attitudes toward labor and crafts remained profoundly ambivalent.[10]
Although manual labor was a necessary adjunct to the work of salvation, medieval theologians insisted that manual labor remain subordinate to spiritual work. Early monastic writers counseled monks against
letting labor distract them from spiritual tasks. Most Christian authors
saw labor as a consequence of original sin and the expulsion of Adam
and Eve from the Garden of Eden.[11] Likewise, the metaphor of the
craftsman God did not imply that human crafts were similarly divine.
Hexameral commentaries rarely examined the creative processes of
craft production in any detail, often stressing instead how God's om-
nipotence made divine creation fundamentally different from human
crafts. In contrast to human labor, God created through the power of
his word alone. Ovitt has noted that one ninth-century commentator,
John the Scot (Johannes Scotus Eriugena), depicted the God of creation
as more rhetorician than craftsman.[12]

Scholars disagree on how to assess this conflicting evidence about
medieval attitudes toward craft skills and manual labor. Some claim
an intellectual rupture with the classical world, while others insist on

continuity.[13] Benz and White clearly overstate the case for a more positive view of labor and the mechanical arts, at least among the literate elites. There was widespread ambivalence toward technical knowledge in medieval as well as ancient thought. But there were significant changes. To be sure, work and technical knowledge remained subordinate to spiritual goals, but craft work also gained a role in the path to salvation, especially in the monasteries. Early medieval acceptance of manual labor and craft work was stronger than anything preceding it in ancient Greece and Rome.[14]

By the twelfth century, artes mechanicae became a major division of knowledge and remained so for eight centuries, until the concept was finally shunted aside by technology in the twentieth century. The category became firmly centered on specific knowledge and practices, those used to transform and control the material world. Nevertheless, ascriptions of social status remained deeply embedded within the concept of mechanical arts.

According to Elspeth Whitney, the term *artes mechanicae* first appeared in the writings of John the Scot, a ninth-century philosopher who referred to seven mechanical arts as adjuncts to the seven liberal arts. John never named these mechanical arts, but in another passage he referred to "architecture and certain other arts." In this passage he contrasted the liberal arts, which were "understood naturally by the soul," with these other arts, which arise from "some imitation or human devising."[15] Whitney has argued that John's analysis made the mechanical arts parallel to the liberal arts. Although John clearly established a hierarchy, linking the mechanical arts to the human and the liberal arts to the divine, Whitney insists that he did so without "the pejorative tone associated with the banausic, or illiberal, arts."[16]

Although Whitney has perhaps concluded a bit much from John the Scot's brief comments, the term *mechanical arts* did rise to prominence several centuries later in the *Didascalicon*, a work written around 1120 by the theologian and educator Hugh of St. Victor. Hugh's influential manuscript helped define the mechanical arts as a distinct category within medieval classifications of knowledge. Lynn White described the work of Hugh and his successors as "giving an unprecedented psychic dignity and speculative interest to the mechanic arts."[17] Similarly, Whitney has credited Hugh with producing a "vision of the mechanical arts as part of man's religious and philosophical quest."[18] Both White and Whitney exaggerate Hugh's positive attitude toward the mechanical arts, given that Hugh explicitly and repeatedly stressed the subordinate role of mechanical arts in his philosophy.[19] Nevertheless, his inclusion of the mechanical arts as a fundamental category of knowledge does represent a major event in the conceptual prehistory of technology.

Hugh's *Didascalicon* (from the Greek *didaskalia*, meaning "teaching" or "instruction") set out a division of philosophy into categories of knowledge to study through reading.[20] Such classification schemes go back to ancient times. What differed in Hugh's classification was his explicit inclusion of the mechanical arts as an integral part of philosophy, parallel to the liberal arts.[21] Hugh's book had an "immediate and penetrating" influence, and it was widely reproduced, surviving in almost one hundred manuscript copies.[22] After the *Didascalicon*, the concept of mechanical arts became widespread among scholars.

Hugh devised several lines of argument to support making the mechanical arts a fundamental division of knowledge. Hugh started by quoting Boethius, who defined philosophy as contemplative knowledge, thus excluding craft knowledge. On the contrary, Hugh insisted, philosophy is "concerned with the theoretical consideration of *all* human acts and pursuits," even if the "actual performance" of an art was excluded. Thus, "the theory of agriculture belongs to the philosopher, but the execution of it to the farmer." Hugh also claimed that because the crafts imitate nature, craft products partake of nature's forms, and thus are also appropriate material for philosophy. Therefore, he argued, all human action aims toward a shared goal "to restore in us the likeness of the divine image." To accomplish this restoration, man has to "take thought for the necessity of this life," that is, rely on the mechanical arts.[23]

Hugh situated the mechanical arts within a fourfold scheme of knowledge comprising the theoretical, the practical, the mechanical, and the logical. He divided theoretical knowledge into theology, mathematics, and physics. The practical dealt with ethics of the individual, ethics of the family (economics), and ethics of the state (politics). The seven liberal arts were split across two divisions. Logic consisted of the three arts of the trivium: grammar, rhetoric, and dialectics, while mathematics, a subdivision of the theoretical, contained the four arts of the quadrivium: arithmetic, music, geometry, and astronomy. Although these categories are a bit idiosyncratic, they represent a synthesis of medieval knowledge up to the early twelfth century, before the rediscovery and translation of Aristotle's works in Roman Catholic Europe.[24]

The principal innovation in Hugh's classification was his explicit inclusion of mechanical arts as a fundamental division of knowledge.

"Mechanical" knowledge referred to knowledge pertaining to "the manufacture of all articles." Hugh patterned this category on the liberal arts, positing seven subdivisions of mechanical arts parallel to the seven liberal arts, including a mechanical trivium and quadrivium. Hugh's list of mechanical arts consisted of "fabric making, armament, commerce, agriculture, hunting, medicine, and theatrics."[25] This rather odd list resulted from Hugh's attempt to fit all the mechanical arts within seven categories.

In seven short chapters, Hugh described each of these mechanical arts.[26] Fabric making referred to all the arts related to coverings in general, including blankets, saddles, carpets, and nets in addition to clothing. Armament dealt with fortifications as well as weapons of all sorts. Commerce concerned trade in all goods, foreign and domestic. Agriculture encompassed all knowledge related to growing plants. Hunting was a broad category that included the preparation of food in general. Medicine covered the means for preserving health (including food), as well as the interventions of the physician. Finally, there is the odd category of theatrics, which Hugh defined as "the science of entertainments." This category included competitive sports, dance, drama, singing, and even dice.[27]

Hugh's descriptions of the mechanical arts show that the category was equivalent to neither the ancient concept of techne nor the modern concept of technology. Its roots in techne are clear from his inclusion of theatrics and medicine, even though these fields did not create material objects. Hugh also excluded all forms of poetry (that is, fictional writing) from the arts, such as "tragedies, comedies, satires, heroic verse and lyric," and more.[28] In excluding poetry from the arts, Hugh resembles Plato, who denied that poetry was a techne because it was rooted in divine inspiration more akin to madness. In contrast, Aristotle had treated poetry as "a complete *techne*, a rational productive activity whose methods can be both defined and justified."[29] Yet Hugh's category of the mechanical differed most significantly from techne by excluding the arts of language, in particular rhetoric. Hugh elevated the arts of language by placing them at the top of his hierarchy in the category of logic, which was the basis for all the other divisions of knowledge.[30] Similarly, his mechanical arts cannot be translated into the present-day concept of technology. On the one hand, by defining *mechanical* as concerned with the manufacture of physical objects, Hugh does come close to at least one core meaning of *technology*. On the other hand, his inclusion of theatrics and medicine points to key differences with the present-day meanings of *technology*.

Although Hugh's definition of the mechanical arts as a fundamental division of knowledge represented a significant conceptual shift, his *Didascalicon* did not signify a major change in the social status of these arts or in elite attitudes toward artisans. In many ways, the *Didascalicon* continued the ancient ambivalence toward technical knowledge, "the trouble with *techne*."

Because Hugh described mechanical arts as necessary yet subordinate to the goal of restoring man's divine nature, their place in his system of knowledge is complex. His discussion is not entirely consistent, suggesting that he was struggling with the problem of how to classify technical knowledge. Most of the time, the mechanical arts seem like a bastard child accepted only grudgingly by its legitimate siblings. For example, in one brief chapter Hugh divided actions into two types: those that aim at restoration of man's divine nature, and those "which minister to the necessity of this life." Divine actions consist of "the contemplation of truth and the practice of virtue"; such actions he termed "understanding" (*intelligentia*). Actions providing the means of life, that is, purely human actions, Hugh labeled "knowledge" (*scientia*). Understanding consists of a theoretical and a practical (that is, ethical) part, while "knowledge, since it pursues merely human works, is fitly called 'mechanical,' that is to say adulterate."[31]

In the *Didascalicon*, Hugh repeatedly labeled the mechanical arts as adulterate because they imitate nature and use human labor. This characterization arose in part from a false etymology that derived *mechanica* from the Greek word for "adulterer," *moichos*. In context, Hugh's usage of the term *adulterate* is not explicitly pejorative, and in one chapter he lavished praise on the mechanical arts immediately after calling them adulterate.[32] However, Hugh's characterization of the mechanical arts as adulterate was no mere "holdover" from ancient prejudices, as Whitney has claimed. Instead, it reflected what Birgit van den Hoven has called the "enormous gap" that separated mechanica from the higher divisions of philosophy.[33] As Hugh explained, the other divisions of philosophy put man directly in touch with the divine, either through contemplation or through ethical practice. Mechanical arts, in contrast, helped alleviate the human suffering brought about by the Fall. Because the corporeal body is part of the temporal world, it belongs to the "lowest category of things." Thus, actions that "minister to the necessity of this life" are merely human, not divine.[34]

In his evaluation of practical crafts, Hugh was certainly more positive than the late Romans with their disdain for the banausic arts. By including the mechanical arts as an integral though subordinate part of philosophy, he did elevate their intellectual standing. But Hugh did almost nothing to elevate the status of the practitioners of the mechanical arts. He never connected the mechanical arts with moral virtue, that is, ethical action. Admittedly, he did in one passage praise the "infinite" ways that the mechanical arts imitate nature, which led him to "look with wonder not at nature alone but at the artificer as well."[35] But this is faint praise; the artisan becomes like nature, an object of contemplation.

Hugh in fact expressed little curiosity about how artisans actually worked, and his descriptions of the seven mechanical arts are little more than lists. Instead, his primary focus in the *Didascalicon* was to provide a guide to reading and interpreting texts, especially biblical texts. To that end, knowledge of the mechanical arts was primarily an aid to understanding and making craft metaphors. Guy Allard has suggested that this literary purpose was the main driver of scholarly interest in the crafts during the medieval period. Hugh himself provided a clear example of such metaphorical use, in the only detailed discussion of artisanal work in the *Didascalicon*. In an elaborate and rather confusing metaphor, Hugh likened the Bible to a building, describing in some detail the mason's process of laying courses of stone. Hugh drew from this metaphor to ascribe eight layers of meaning to the divine text. His language reveals that he had spent time observing stonemasons at work, probably in his supervisory role at the Abbey of St. Victor near Paris. But such observation does not contradict van den Hoven's conclusion that such literary appreciation of the artisan's work had "no consequence for the *status* of the skilled craftsman."[36]

Hugh's influence was significant, but it should not be exaggerated.[37] As Allard has insisted, Hugh did little to create a true "pensée mécanologique," that is, serious scholarly reflection on the nature of craft knowledge and practice.[38] In the work of restoring human perfection, according to Hugh, the mechanical arts play only a supporting role. These arts have no direct connection to the divine, but rather serve mainly as means to an end, supplying the needs of the corporeal body and thus freeing the mind for the real work of salvation. In his understanding of mechanical arts as a means to an end, Hugh was recreating the instrumental vision of technical knowledge that was so clearly present in Aristotle.[39]

Hugh's work established the *artes mechanicae* as a standard term in descriptions of the arts and sciences in the Latin West.[40] In the late twelfth century, one of Hugh's successors at the monastery of St. Victor, Godfrey, argued explicitly for the moral value of mechanical arts.

Godfrey rejected the notion of the mechanical arts as an adulterine imitation of nature, affirming the inherent moral value of crafts and attributing any resulting evils to improper application.[41] Hugh's ordering of the mechanical arts remained widespread in the thirteenth century, despite the influence of alternative classifications based on Arabic interpretations of Aristotle. For example, Bonaventure (1221–1274) drew from Hugh's classification of the mechanical arts to argue for their connection to divine wisdom. In what later became a common trope in the discourse of the mechanical arts, Bonaventure analogized the works of man to the works of God, comparing human and divine acts of creation.[42]

Thomas Aquinas and the Moral Character of the Arts

Hugh of St. Victor's concept of the mechanical arts owed little to the recovery of Aristotle that was beginning in the Latin West in the twelfth century. When Thomas Aquinas began contributing to medieval thought a century after Hugh, all of Aristotle's key surviving works had become available.[43] Whitney has suggested that the influence of Aristotle in the thirteenth century represents a step backward in the status of the mechanical arts, given Aristotle's general neglect of techne.[44] Indeed, in contrast to Hugh, Aquinas did not view the mechanical arts as a distinct epistemological category. Yet he was more explicit than Hugh in articulating the moral character of the arts. More significantly, Aquinas helped legitimate the mechanical arts by explicitly including them in the category of arts in general.

Even before Aquinas, medieval Latin scholars had moved away from Hugh's notion of the mechanical arts as a principal division of knowledge, in part under the influence of Arabic scholarship. Arabic scholars were the true heirs to classical Greek thought, and it was through their translations that Aristotle was reintroduced in the medieval West.[45] In both scholarship and craftsmanship, the leading centers of the Arab world were far more sophisticated than their medieval European counterparts, even after the sack of Baghdad by the Mongols in 1258.[46] There were extensive technological exchanges between medieval Christians, Muslims, and Jews, especially in the Iberian Peninsula, where practitioners of these three faiths lived side by side.[47] The exchange of ideas really took off around the twelfth century, partly due to the acceleration of Christian conquests in Spain, which were for the most part accomplished without widespread destruction.

Like medieval Latin writers, Islamic scholars did little to develop a theory of productive knowledge grounded in the technologies of their day. Following Aristotle, Islamic classifications were hierarchical, with the crafts at the bottom. But these classification schemes differed from their Aristotelian heritage in a key way. Instead of Aristotle's sharp division between productive and theoretical knowledge, between techne and episteme, Islamic scholars often viewed the mechanical arts as expressions of theoretical sciences. Although this idea was far from a theory of applied science, it did support a new classification system that located specific arts within a related science. For example, the tenth-century Baghdadi scholar al-Farabi, whose classification of knowledge was highly influential in the Islamic world, placed carpentry under practical geometry and medicine under physics (in the Aristotelian sense of the term).[48]

Al-Farabi's key works were translated into Latin in the later twelfth century by the philosopher and translator Domingo Gundisalvo in Toledo. Gundisalvo then produced his own influential classification of knowledge that drew heavily from Arabic sources. In sharp contrast to Hugh, Gundisalvo did not group the mechanical arts together but rather scattered specific crafts throughout his schema, always placing them at the bottom of the hierarchy. Mechanical arts appeared under three subdivisions: physics, mathematics, and economics (in its original meaning of household management). Following al-Farabi, Gundisalvo included practical applications with some theoretical sciences, such as optics. He acknowledged that certain arts, medicine for example, had a theoretical as well as a practical side. Although Gundisalvo discussed the mechanical arts only in passing, his approach to classifying them, along with his recognition that the arts involved both theory and practice, persisted well into the eighteenth century.[49]

Aristotle's influence deepened in the mid-thirteenth century, when Thomas Aquinas made his mark on medieval philosophy, becoming the most celebrated theologian of the era. More than any other medieval scholar, Thomas brought the recently recovered works of Aristotle into a Christian theological context.[50] But this appropriation also transformed Aristotle by reconciling Greek ideas with medieval Latin philosophy. This reconciliation is quite clear in the translation of Aristotle's ideas about *techne* into the Latin *ars*. In accord with the Latin usage of his time, Thomas accepted ars as a broad category encompassing intellectual as well as mechanical capacities. He followed Aristotle in placing art at the bottom of the hierarchy of knowledge, but he also insisted that art in all its forms is virtuous, and that the virtue of art is of the same type as the virtue of scientific knowledge (*scientia*). With regard to the mechanical arts specifically, Thomas had little to say. In contrast to Hugh of St. Victor, he did not grant the mechanical arts any epistemological distinctiveness.

Some of Thomas's key ideas on art are found in the section on the virtues in his *Summa Theologiæ*.[51] He grounded this analysis in Aristotle's *Nicomachean Ethics*. According to Thomas, virtue is a habit that is oriented toward acting well. Following Aristotle, he distinguished intellectual from moral virtues. The intellectual virtues include scientific knowledge (*scientia*), wisdom (*sapientia*), understanding (*intellectus*), prudence (*prudentia*), and art (*ars*). Intellectual virtues create only the capacity for good acts; they do not ensure that the person who possesses them will actually act well. Moral virtues are virtues in actuality, guiding humans directly toward good acts by governing the will. Knowledge of an art like grammar or masonry makes a person good only in a particular sense, as a good grammarian or a good mason. In contrast, someone who possesses a moral virtue like justice or temperance is good in an absolute sense, a good person.[52]

Thomas expounded these ideas further in the *Summa* when he compared art to speculative (that is, theoretical) forms of knowledge. He subscribed to the traditional hierarchy of knowledge inherited from the Greeks. He rated liberal above mechanical arts, and theoretical above practical forms of knowledge.[53] Despite this prejudice, Thomas insisted that art was an intellectual virtue as much as speculative fields like scientific knowledge. He noted that the good of an art exists not in the state of mind of the craftsman but only in the goodness of the product. "What is relevant to a craftsman's praiseworthiness insofar as he is a craftsman is not the sort of act of willing by which he makes his work . . . , but the quality of the work which he makes." Similarly, the truth of a geometric proof does not depend on the will of the geometer but on the result. From this logic, Thomas concluded that liberal and mechanical arts are equally virtuous: "It is not necessary that if the liberal arts are more noble, the concept art should apply to them to a higher degree."[54]

In treating art as comparable to speculative intellectual virtues, Thomas blurred an important difference implicit in his analysis. Speculative knowledge is about contemplation of what is true; such contemplation, when directed by the will, gives order to the intellect and thus shapes the self.[55] Practical knowledge, in contrast, shapes the world outside the self. For both practical and speculative knowledge, ultimate virtue lies not in the capacity to produce a good result but in the use to which the result is put, because good use is the product of moral virtues that guide the will.[56] However, the good use of speculative knowledge—that is, contemplation of the true—results from the combination of intellectual and moral virtues in a single person. In contrast, because the product of art is external to the craftsman, its good use arises from the moral virtue of the user rather than the maker. In other words, the ultimate virtue of an art lies outside the practitioner of the art.[57] This distinction echoes a key point also implicit in Aristotle's *Nicomachean Ethics*, that art serves ends external to itself.

In the end, Thomas expressed, at a general level, a continued ambivalence toward the mechanical arts. This ambivalence included both instrumental and moral visions of the mechanical arts. On the one hand, Thomas elevated these arts by explicitly treating them as virtuous in the same way as the liberal arts and scientific knowledge. The mechanical arts embody the same virtue as all arts, a virtue that rests on the excellence of their product. And because art belongs to the mind, not the body, art thus belongs with the intellectual virtues as much as scientific knowledge.[58]

On the other hand, Thomas separated the arts, viewed as means to an end, from the moral virtues.[59] He argued that moral virtue does not depend on some key intellectual virtues, such as scientific knowledge and art. Nevertheless, a connection between intellectual and moral virtue is necessary, because moral action requires both desire to achieve a good end and knowledge of the means to that end. This connection is achieved by two specific intellectual virtues, prudence and understanding.[60] Of these two, prudence is central.[61] Moral virtues direct humans toward good ends; prudence provides knowledge of suitable means to achieve those ends. In fact, Thomas insisted, prudence cannot exist apart from moral virtue; that is, one cannot be prudent about evil ends. In this way, prudence differs from art, because the products of art are judged by reason alone, not morals. That is, the fact that an artifact is well made does not preclude its immoral use. In Thomas's own words, "unlike prudence, an art or craft does not require a virtue that perfects the appetite."[62] As a virtue, prudence perfects the person, while art perfects the product.

In terms of their relevance to the history of the concept of technology, there is more continuity than discontinuity between Hugh of St. Victor and Thomas Aquinas. Yes, Hugh did make the mechanical arts one of the key divisions of knowledge, a move that Thomas rejected in his return to a division of knowledge grounded in Aristotle. Yet this rejection made little difference to their understanding of the mechanical arts. Both scholars were heir to the sharp divisions of antiquity between productive and theoretical knowledge, and also between the instrumental ends of making and the moral aims of action. Neither scholar developed anything approximating a theory of productive knowledge, a science of art, a logos of techne, that is, technology. The first steps on that path would be taken by philosophers in the early modern era, in their analysis of the relationship between art and science.

FOUR

Natural Philosophy and the Mechanical Arts in the Early Modern Era

The concept of applied science was foreign to the medieval mind. Medieval scholars understood the arts as a form of knowing. Even though both art and science included knowledge of the material world, most scholars believed that artisans and philosophers knew the material world in fundamentally different ways. Even mathematical methods, when applied to practical purposes, were excluded from philosophy, being labeled "mixed mathematics," as Peter Dear and others have argued.[1] There are some exceptions, of course. Some individual scholars are now seen as precursors of a more modern view of the relationship between scientific theory and technical practice, such as Archimedes or Roger Bacon. But this is a post hoc construction. Neither Archimedes nor Roger Bacon did much to create a viable alternative to the divorce between techne and episteme, a divorce codified by Aristotle and perpetuated by Thomas Aquinas.

In the fifteenth century, economic and technical changes began to reshape the relationship between episteme and techne, between science and the arts. Historians have long argued about whether there was a rupture in the development of technology between the medieval and the modern eras, and if there was, then exactly when and where this occurred. No serious scholar of the era continues to believe that the Middle Ages were technologically stagnant. Some historians have gone so far as to argue for a medieval "industrial revolution."[2] While such claims overstate the case, medieval Europeans developed and adopted an impressive array of technologies, especially as commerce and population grew rapidly after 1000 CE. These technologies included machines well known to the Romans, such as water-powered grain mills, which were nevertheless far more widespread in the High Middle Ages than they had been in the Roman Empire. Many new technologies came from China, such as paper, gunpowder, and the compass, while others, such as the mechanical clock, were indigenous European innovations.[3] The skills of traditional artisans also increased with urbanization and the rise of craft guilds. These changes helped make artisanal skills essential to the military and economic power of princes and merchants, especially in the increasingly powerful city-states of northern Italy.[4]

However, before the fifteenth century, these technological developments had almost no influence on scholarly concepts. Ancient and medieval rulers, along with the scholars they supported, recognized the potential threat that dependence on artisanal skills posed to established social hierarchies. Even when medieval scholars treated the mechanical arts sympathetically, they demonstrated no more than a remote awareness of the economic and technological changes that were beginning to reshape their world. This lack of interest was mutual. Although many elite artisans were literate, few works written by them survive. On the rare occasions when artisans wrote their own manuscripts, these did little to connect scholars and craftsmen.[5]

The Early Modern Alliance of Techne and Praxis

The fifteenth century marked the beginning of one of the most profound intellectual changes in European history, fundamentally altering the place of technical knowledge in Western thought. This transformation had the potential to free the mechanical arts from their subordinate intellectual status. This potential was never fulfilled, however, in part because early modern scholars remained trapped by instrumentalist conceptions of the mechanical arts.

A new alliance between techne and praxis, between artisanal skills and political power, started this transformation. These changes coincided with the resumption of European commercial expansion, which had been derailed by the famines, plagues, and wars of the fourteenth century. After 1400, new forms of technical knowledge became increasingly crucial to political power, especially in the city-states of northern Italy and the principalities of central Europe. Most of these states relied on trade for their economic power, and they protected this trade with innovations in military organization and technology, such as gunpowder weaponry.[6] Military and economic power also depended on supplies of metals for both specie and weapons, thus making expertise in mining and metallurgy more valuable. Finally, political legitimacy drew heavily from public displays of wealth and power, which often took the form of spectacular architecture and visual arts. These displays could also require new types of expertise, as epitomized by Brunelleschi's construction of his magnificent dome for the great cathedral of Florence.[7]

The alliance between techne and praxis helped transform the relationship between natural philosophy and the mechanical arts. This transformation was based on a surge in authorship about the mechanical arts starting in the fifteenth century. Unlike earlier medieval works, however, fifteenth-century writings on the mechanical arts were framed less by theology than by politics, that is, by relations of patronage linked to secular power. Both artisans and humanists sought patronage from princes by writing treatises about technical subjects relevant to state power. As Pamela Long notes, such texts helped rationalize these arts, "turning them into more discursive, learned subjects."[8] Humanist scholars also attacked the Scholastic divide between theory and practice, arguing that natural philosophy should be useful, not just contemplative. As Francis Bacon put it somewhat later, natural philosophy should become "a rich storehouse, for the glory of the Creator and the relief of man's estate."[9] In other words, Bacon claimed that natural philosophy could serve both theological and utilitarian ends. Such sentiments expressed a fundamental change in the understanding of the relationship between science and the arts.

In *Openness, Secrecy, Authorship*, Long provides a masterly analysis of this early modern literature on the mechanical arts, discussing writings by ambitious artisans as well as humanist scholars.[10] The authors of these works had to bridge two distinct worlds, the world of Latin scholarship and the artisanal workplace. In the fifteenth century, Latin was still the universal language of European intellectuals, but few artisans knew Latin. In fact, formal texts about technical arts in any language were generally foreign to craft workers, who conveyed knowledge mostly through oral instruction and physical demonstration, usually in the context of apprenticeships. University-trained scholars, in contrast, acquired knowledge in a fundamentally different way, by producing Latin commentaries and conducting disputations about canonical texts.[11]

The barriers between scholars and craftsmen thus remained formidable in the early modern world. Nevertheless, numerous writers on both sides of the divide managed to produce texts that bridged the gap. These works circulated as manuscripts before the spread of the printing press in the second half of the fifteenth century. By the sixteenth century, technical writings proliferated through the medium of print. Among scholars writing in Latin, physicians penned many technical works. These men included Conrad Keyser and Giovanni Fontana, who both wrote on military arts in the early fifteenth century, and more famously Agricola (Georg Bauer), whose book on mining, *De re metallica*, is probably the best-known technical work of the sixteenth century. Physicians had the advantage of combining a university education, conducted in Latin, with at least some degree of contact with the world of practice—even if, as Vesalius complained, early modern physicians rarely got their hands dirty.[12] Many of these scholarly authors were insecure in their social standing; Agricola, for example, came from a family of artisans.[13] Their writings, by serving the needs of the state, helped enhance their social status.

On the other side of the scholar-craftsman divide were artisan-practitioners who wrote in the vernacular, drawing directly from their professional experience. Some were ambitious men with military backgrounds who used their technical writings to seek patronage, for example Vannoccio Biringuccio, who wrote the military manual *De la pirotechnia*,[14] or Michael of Rhodes, whose extensive manuscript from the 1430s constitutes the earliest surviving treatise on shipbuilding.[15] Other artisan authors emerged from elite crafts like goldsmithing and other visual arts; these elite practitioners used authorship to stake claims for painting and architecture as liberal arts.[16]

These fifteenth-century writings demonstrate new links between men of letters and artisans, between theory and practice, connections that did much to raise the status of the mechanical arts. By the middle of the sixteenth century, as shown by Paolo Rossi, a wide range of writers across Europe hailed these developments as the coming of a new order. These writers included many scholars and noblemen who defended the dignity of the mechanical arts and argued for their role in the progress of knowledge. Elite authors also condemned the ancient prejudices that favored contemplative over practical or mechanical knowledge.[17] For example, the Valencian humanist Juan Luis Vives called on scholars to abandon their contempt for the mechanical arts, to "enter into the workshops and into the factories, asking questions of the artisans" in order to understand their work.[18] Agricola explicitly defended the dignity of mining and metallurgy against critics who considered these arts base because they had once been performed by slaves.[19] The Elizabethan adventurer Sir Humphrey Gilbert included practical subjects in his proposal for an academy to educate young noblemen in the queen's service. Gilbert's curriculum went beyond the traditional skills appropriate to a gentleman, such as horsemanship and music, to include navigation, gunnery, shipbuilding, fortifications, and other practical fields useful to the state.[20]

A more full-throated defense of the mechanical arts, or at least specific arts, emerged in writings produced by artisans themselves. Like the technicians of ancient Greece, these writers defended not just the dignity of their art but also the moral worth of the artisan. One example is an anonymous manuscript on gunpowder weapons written about 1420, the *Feuerwerkbuch*. This work circulated widely in manuscript before being printed in 1529. The manuscript was clearly the work of an artisan-practitioner, although it was addressed not to fellow artisans but to potential patrons, the princes and city rulers who needed expertise in the new gunpowder weaponry. The work is filled with details about the manufacture of gunpowder and techniques for firing cannons. The author also included natural-philosophical discussions about causal principles, such as the nature of the chemical reactions between the components of gunpowder. The author went beyond technical details, however, when he stressed the moral character required of a master gunner, who should be godly, steadfast, literate, honest, and friendly. This was a clear assertion of moral dignity and social status, a claim further implied by the absence of "obedience" from the list of desirable traits. Although the *Feuerwerkbuch* was anonymous, later fifteenth-century manuscripts by artisan gunners featured portraits of the author on their title pages, another claim to social status.[21]

Similar writings emerged from fifteenth-century Italian artisan-practitioners, elite artisans we would now classify as painters, sculptors, and architects. Many of these artisans were just as likely to engage in military engineering as they were to paint; Leonardo da Vinci was no exception among these "Renaissance men."[22] Most of their activities produced public works that served the needs of the state and often symbolized the power and wealth of the ruler. Their writings emerged in a context where elite artisans rubbed shoulders with humanist scholars. Artisan authors, such as the goldsmith and sculptor Lorenzo Ghiberti, insisted on the need for both classical learning and artisanal practice while also stressing, like the *Feuerwerkbuch*, the need for the artisan, in this case the sculptor, to be of upright moral stature. Similarly, the Florentine architect Antonio Averlino, writing under the name Filarete, presented a lofty vision of the architect in his utopian treatise describing the ideal city. Filarete portrayed the architect as nearly equal to his patron, worthy of conversing and dining with the nobility. Like Humphrey Gilbert, Filarete proposed a school where the liberal and the mechanical arts were taught jointly, and he insisted that all the crafts were "necessary and noble."[23]

Early Modern Science and "The Trouble with *Techne*"

Examples like these, which could be multiplied many times, demonstrate a new appreciation of the mechanical arts among European elites. It is tempting to see this new appreciation, this alliance of techne and praxis, as the cusp of our modern world, the germ of the idea that science serves not merely to understand the material world but also to transform it. We must take care, however, not to exaggerate the significance of the early modern connections between artisans and scholars. As with Hugh of St. Victor, an appreciation of the mechanical arts did not necessarily translate into higher status for the craftsman. Whatever the aspirations of artisans, aristocratic values continued to dominate Europe. It was hard enough for a wealthy merchant to achieve acceptance among aristocratic elites, let alone a craft worker.[24]

In other words, the alliance of techne and praxis, along with the new belief in the utility of natural philosophy, did not eliminate the trouble with techne. As Peter Dear notes, natural philosophers placed themselves in an awkward position when they sought to raise the status of the mechanical arts, which were practiced by men of a social status far beneath the natural philosopher. Scholarly elites did not embrace an alliance of equals in which artisanal and philosophical knowledge would combine toward a joint project to improve the human condition. Instead, according to Dear, advocates of the mechanical philosophy "set themselves up as prophets of a kind of value-added practical knowledge wherein the untutored artisan would be disciplined by the literate gentleman overseer."[25] Thus was solved the dilemma of accommodating the mechanical philosophy to a social order that despised mechanicals.

This "solution" was not dictated by any necessary relationship between theoretical and practical knowledge. Instead, it was a social choice. However, the nature of this choice is often obscured. Too often, the history of modern science and technology shows science bearing theoretical fruits that are then harvested by lowly technicians, migrant farmworkers who toil in the orchards of science.[26] Yet an alternative path was possible, one that artisans (and later engineers) frequently proposed. This path was to use natural philosophy to elevate the artisan, introducing the benefits of systematic thought and experimental methods to craft workers themselves. In effect, this approach would have turned every craftsman into a natural philosopher, though without destroying their roots in practice. Such a change was conceivable because, as some scholars of the time recognized, craftsmen already used experimental methods and theoretical principles in their own work.[27]

But an egalitarian relationship between philosopher and craftsman was simply incompatible with the existing social order, given the entrenched hierarchies of the early modern era. Instead, proponents of the new mechanical philosophy, starting with Francis Bacon, took another path to reshaping the relationship between science and the arts. These elite scholars did reject the categorical separation of science and material practice, but they did so without rejecting the existing hierarchy of head over hand. Rather than an equal and fruitful exchange between philosophers and artists, proponents of the new science argued for a subordination of the mechanical arts to philosophy, and of practice to theory.

This subordination is quite clear in the work of Francis Bacon (1561–1626). In some ways Bacon was the father of modern science, but in other ways he missed the boat entirely. His reputation has gyrated wildly over the centuries. He was a respected figure in English letters during his lifetime. In contrast, one of his predecessors, Giordano Bruno, had the misfortune to be burned for heresy.[28] In the mid-seventeenth century, Bacon's writings became a rallying point for proponents of the new mechanical philosophy. Advocates of the new science repeatedly invoked his name when they founded the Royal Society in 1660. For Enlightenment scholars in the eighteenth century, Bacon had an almost God-like reputation, especially among the French *philosophes*. What he missed, however, was the increasing importance of quantitative measurement and mathematical models, which had become central to the physical sciences by the end of the seventeenth century. For this reason, leading scientific spokesmen in the nineteenth century downplayed Bacon's significance to modern science. His scientific reputation declined further after World War II; his insistence that science be useful carried a Marxist taint when viewed through the Cold War ideology of pure science. More recently, Bacon's standing in the history of science has risen as scholars acknowledge his originality.[29]

Whatever Bacon's role in the rise of modern science, his writings are highly relevant to the status of the mechanical arts. He drew from a literature about the mechanical arts that had flourished for over a century. He also relied on authors who stressed the direct observation of and intervention in nature. As Eric Ash argues, Bacon's philosophy was closely tied to the "humanist emphasis on practical knowledge that so pervaded English education and learned culture" in the late sixteenth century. Bacon differed from his predecessors, however, in building an entire project for the reform of philosophy on the basis of this new understanding of natural knowledge, and in representing this approach as a forceful rejection of Aristotelian natural philosophy.[30]

Bacon rejected the Aristotelian divide between episteme and techne, that is, between science as contemplative and art as productive. He believed that the most esoteric principles of natural philosophy should yield practical applications ("works" or "fruits") to improve the human condition. Practical applications testified to the truth of the theoretical principles behind them. "For fruits and works are as it were sponsors and sureties for the truth of philosophies."[31] Bacon praised the mechanical arts for their contributions to both knowledge and human needs, in contrast to the barrenness of Aristotelian natural philosophy. "The sciences stand still in their own footsteps," while "we see the opposite in the mechanical arts, which are . . . always progressing."[32] He contrasted the "savage and barbarous regions of New India" with the "civilised provinces of Europe," a difference due "not to soil, climate or bodily qualities, but to Arts." Like many of his contemporaries, he invoked the three great inventions of printing, gunpowder, and the magnetic compass, inventions "unknown to antiquity," which together "have changed the face and condition of things all over the globe."[33] Bacon insisted that the knowledge produced by the mechanical arts was of value to natural philosophy. He wanted the history of the mechanical arts to be an integral part of natural history. Despite those who consider it a "dishonour unto learning" to study these vulgar arts, he insisted, "the use of history mechanical is, of all others, the most radical and fundamental towards natural philosophy."[34]

Some historians, most important Paolo Rossi, have presented Bacon's praise of the mechanical arts as a turning point in modernity. Rossi argued that Bacon took key aspects of technical knowledge, "collaboration, progressiveness, perfectibility, and invention," and used them "to define the whole field of human knowledge." Bacon, "by taking the mechanical arts as a model for culture," described a society amenable to progress, in contrast to the static ancient world.[35] In a sense, Rossi found in Bacon the same desire to rehabilitate the mechanical arts that Elspeth Whitney has attributed to Hugh of St. Victor over four centuries earlier.

But Bacon's call for a useful natural philosophy and his praise of the mechanical arts did not eliminate tensions between literary elites and craft practitioners. Practitioners tended to embrace a cultural view of technology, defending the social status of artisans and the moral value of their practices. In contrast, literary elites typically praised the knowledge embodied in the arts, but not the artisans themselves.[36]

Bacon's praise of the mechanical arts was indeed a major shift in Western thought. But as Romano Nanni argues, a careful survey of the *New Organon*, Bacon's best-developed philosophical work, reveals a profound ambivalence toward the mechanical arts, and also a deep ignorance of their practitioners' methods. Overall, Bacon was not particularly impressed with the mechanical arts, describing them as superficially empirical. They "merely glide over the variety of things on the surface" without benefit from the "true observations" of natural philosophy. Furthermore, he argued, the apparent fecundity of the mechanical arts is as misleading as the "immense variety of books" in libraries. For just as books endlessly repeat the same arguments, works of the mechanical arts are mere variations on a few "axioms of nature," which are "neither many nor profound." Beyond this handful of axioms, most known to the ancients, the development of the mechanical arts "is due simply to patience and the subtle, ordered movement of hand and tool." Inventions like the mechanical clock were discovered "easily . . . by ready opportunities and casual observations." Rather than marveling at the products of the arts, the careful observer would "rather pity the human condition" for its "dearth of objects and discoveries."[37]

In these passages, Bacon was engaged in a variant of what Thomas Gieryn has called "double boundary-work." Bacon praised the mechanical arts to differentiate his new science from Aristotelian natural philosophy, yet at the same time insisted on the primacy of philosophy. This allowed him to claim the best of both worlds, taking from the mechanical arts their progressive, empirical, and collaborative character, while retaining the cultural authority of philosophy. Such "double boundary-work" has in fact remained part of modern science into the present. In nineteenth-century Britain, as Gieryn has shown, advocates sought to distinguish science from both religion and technology. They emphasized practical utility when defending science from theology, while they stressed cultural value and intellectual purity when demarcating science from engineering.[38] Similar boundary work continued into the twentieth century, with apologists for science focusing on practical utility when seeking funding, while insisting on the purity of science when defending their professional autonomy.[39]

What are we to make of Bacon's ambivalence, his fervent praise for the mechanical arts in one section of *The New Organon* counterpoised with his disparagement in nearby passages? His praise of the mechanical arts helped align his new philosophy with Elizabethan culture, allowing him to tap into the era's esteem for practical knowledge, especially when practiced by expert mediators in the service of the state.[40] But he also needed to set a boundary between the mechanical arts and his reformed system of natural philosophy, a boundary that maintained the existing hierarchy of knowledge. Bacon did so by arguing that the progress in the mechanical arts could continue only through his own method of discovering fundamental axioms, that is, laws of nature. Hence Bacon's claim in the *New Organon* that "the discovery of useful works has ceased" because of centuries of sterility in natural philosophy. This argument directly contradicted his claims a few pages earlier about the progressive character of the mechanical arts.[41] In a sense, Bacon wanted to expropriate the technical arts for natural philosophy, establishing a hierarchy that firmly subordinated the mechanical arts to scientific principles. This hierarchy is explicit in Bacon's *New Atlantis*, where philosophers ruled over artisans.[42]

Bacon thus revealed his ignorance of the nature of technical knowledge and the process of change in the mechanical arts. By undermining Aristotle's sharp distinction between episteme and techne, Bacon denied that autonomous progress in the mechanical arts was possible, as Nanni points out. Bacon insisted that progress in the mechanical arts required progress in natural philosophy, failing to grasp that mechanical invention requires more than just correct axioms. An entire generation of science and technology studies (STS) scholars has stressed the essential role of skills, tacit knowledge, and interpretive practices in science as well as technology.[43] These insights from STS would hardly have been foreign to authors in the generations before Bacon, from Agricola to Ghiberti, authors with direct experience of the artisans' world. These writers understood the vast gulf separating theory and practice. For Bacon, in contrast, progress in the mechanical arts was very much a top-down affair, a sharp break with the past based on the discovery of new principles. Bacon insisted that only his method would permit the speedy discovery of the "many exceedingly useful things . . . still hidden in the bosom of nature" once their principles were uncovered.[44] There was little room, in Bacon's philosophy, for the steady accumulation of technical knowledge.

Some historians have argued that artisanal knowledge and practices played a key role in early modern science. As early as 1906, Thorstein Veblen claimed that during the early modern era "the concepts of the scientists came to be drawn in the image of the workman," meaning that scientists borrowed their ideas about cause and effect from the world of the artisan.[45] Between the world wars, a few scholars, including some heterodox Marxists, embraced the thesis that modern science had artisanal roots. The best-developed version of this argument was presented in 1942 by the refugee scholar Edgar Zilsel in his classic paper, "The Sociological Roots of Science." Zilsel traced the origins of modern science to the rise of capitalism and development of technology in the early modern era. But more important, he argued that the key process in the rise of modern science was the breakdown around 1550 of the barriers between scholars and "superior craftsmen." Before this barrier came down, "science in the modern meaning of the word was impossible."[46] But with the rise of the Cold War in the late 1940s, the Marxist-tainted "Zilsel thesis" became taboo among historians of science. With the end of the Cold War and the decline of visceral anti-Marxism, claims about the role of artisans in modern science are no longer met with ritual denunciation.[47]

Long's argument about the early modern alliance of techne and praxis provides an important addition to the Zilsel thesis, helping explain why scholars and craftsmen became more willing to collaborate. This alliance between technicians and the state began in the fifteenth century and later became a key element of absolutism.[48] This political alliance helped reduce barriers between scholars and craftsmen. Mechanical arts were always included in the programs of early scientific societies, as illustrated by the seemingly bizarre range of topics discussed in the early decades of the Royal Society.[49] Philosophers and artisans engaged in lively exchanges and even collaboration, especially on technical topics of interest to the state, such as navigation or gunnery.[50] Pamela Smith and Deb Harkness describe the vibrant early modern world of elite artisans who embraced both science and art, a world of painters, engravers, instrument makers, practicing physicians, and small merchants.[51] This evidence all supports the idea that artisans and their technologies played a key role in the rise of modern science.

Yet the structure of early modern science conflicted with the social structure of the time. No matter how collaborative their intellectual exchanges might be, philosophers and artisans would never interact as equals, and their social differences were reflected in the concepts that framed the interaction. When Bacon broke down the barrier between episteme and techne, he made possible a fundamental rethinking of the relationship between theory and practice, mind and hand, philosophy and the arts. Yet early modern natural philosophers were not, in fact, free to rethink these categories in a radical way. Despite their praise of direct observation and experimentation, men of science did not seek to transform existing social structures and the sharp distinction between mental and manual activity. As Lissa Roberts and Simon Schaffer note, "In a society of orders, the steeply graded hierarchy of head and hand was vital to defining persons and their social places."[52]

Perversely, the increasing importance of technical expertise in early modern Europe may well have encouraged supporters of the new science to defend their aristocratic status with greater fervor. Early modern science remained largely a gentleman's pastime, as Steven Shapin has shown, in which notions of trust were essential. For men of science, the veracity of their empirical claims was guaranteed by their status as gentlemen. Artisans did not have access to this culture of trust, because they did not have the status of "free actors." Only free actors had enough autonomy to be trusted to convey truth, as they did not act out of pecuniary interest or at the behest of others.[53]

Even though craft workers were largely excluded from this system of trust, their labor remained essential to seventeenth-century experimental science. The result was what Shapin has termed the "invisible technician," the erasure of craft labor from accounts of experiments. In the second half of the seventeenth century, Robert Boyle transformed Bacon's philosophy into a systematic program of experimental science. Boyle stressed his willingness to get his hands dirty while engaging directly in experimental work. But in reality Boyle rarely had to dirty his hands, because his staff of paid assistants performed most of the work, including dangerous tasks like experimenting with explosive materials. These skilled assistants not only performed the experiments but also recorded the results and at times even drafted the analyses, complete with inferences about cause. Yet as Shapin has noted, these technicians remained largely invisible or at least anonymous, even in Boyle's extremely detailed accounts of his experiments. When a scientific gentleman claimed to have performed experiments himself, it meant that he had initiated and supervised the work, and that he vouched for its authenticity. The technicians, however, remained voiceless, or more precisely they spoke only through the philosopher, who had the epistemic authority as a gentleman to make scientific claims.[54]

This erasure of craft labor extended beyond the technicians who assisted in experiments. Accounts of experiments also erased the contribution of the highly skilled instrument makers, whose devices were essential for experimental science. These artisans were among the most elite in the early modern era. Yet their work too was largely absent from accounts of experiment, revealing another way in which skilled labor was erased from natural philosophy.[55]

This invisibility of craft knowledge was reflected in the way men of science understood the relationship between natural philosophy and the mechanical arts. To put it crudely, scientific gentlemen understood the relationship as one of master to servant. They believed that craft knowledge needed to be appropriated by the disinterested gentleman and transformed into philosophical principles. Only then, as Shapin has argued, did such "systematized knowledge acquire the capacity to *improve* the arts." By appropriating, purifying, and systematizing craft knowledge, gentlemen of science could claim a degree of epistemological authority over the increasingly important work of elite artisans.[56] In this way, early modern science solved the problem posed by Peter Dear: how gentlemen could claim utility for natural philosophy while preserving its elite social status, a status essential for its truth claims.[57]

Natural philosophers also defended the elite status of their knowledge by arguing that the new science served theological as well as utilitarian ends. The elite scholars who formulated the new sciences of the seventeenth century believed that natural philosophy could help satisfy earthly needs. But at its core, natural philosophy remained a spiritual enterprise, demonstrating God's handiwork in nature. As a form of natural theology, early modern science was an end in itself, imbued with moral virtue, much like episteme for Aristotle.[58] In contrast, for natural philosophers the mechanical arts remained purely instrumental, serving ends defined by others. In many ways, the pure-science ideal of the twentieth century drew from the spirit of natural theology, though stripped of God as maker.

The basic tension in early modern science between social structures and knowledge practices would continue into the eighteenth century and beyond. These tensions played out in terms of an Enlightenment discourse about the relationship between art and science, a discourse that would break down in the nineteenth century.

FIVE

From *Art* to *Applied Science*: Creating a "Semantic Void"

As the scientific revolution of the seventeenth century gave way to the Enlightenment of the eighteenth, collaboration between artisans and men of letters strengthened, despite persistent aristocratic prejudice against the mechanical arts. Joel Mokyr terms this movement the "industrial enlightenment." As many critics have noted, there are serious problems with Mokyr's understanding of the relationship between formal knowledge and technical practice.[1] Ursula Klein offers a better way to frame this collaboration, emphasizing the emergence of artisanal-scientific experts, social hybrids who were at home among both men of letters and skilled workers. Especially on the Continent, mercantile states encouraged these hybrid experts to apply their skills to key industries.[2] In France, for example, military engineers often played the role of hybrid experts, staking out what Ken Alder has termed a "middle epistemology" that united "the universal knowledge of the savant with the particularistic knowledge of the skilled craftworker."[3]

This harnessing of technical expertise to the state represents another chapter in Long's alliance of techne and praxis, an alliance reflected in the world of ideas. Encouraged by Francis Bacon's widespread influence in Britain and France, Enlightenment authors explored the relationship between natural knowledge and productive activity as expressed in terms of a discourse of art and science. Many Enlightenment authors rejected the hierarchy that placed science above art, and instead viewed the relationship in terms of mutual dependence. Of course, we must remember that the terms *art* and *science* had far broader meanings in the eighteenth century than now. *Art* encompassed the mechanical, liberal, and fine arts, while *science* referred to not just natural science but systematic knowledge in general. Furthermore, there was significant overlap between the two categories; for example, liberal arts were often classified under science.[4]

Yet these eighteenth-century intellectual trends did not create a utopian merger of mental and manual labor. Although the spread of the market economy weakened the traditional society of orders,[5] it also encouraged two key changes that aborted a deeper reconceptualization of the art-science relationship. The fine arts arose as a clearly defined category, which in effect stripped aesthetic creativity from the mechanical arts. And with the rise of industrial capitalism in the late eighteenth century, industrialists emerged as new allies for men of science.

These two factors, the rise of fine arts and the alliance between science and industry, helped keep mechanical arts subordinate to philosophy. Existing concepts like art and science acquired new meanings, and the term *applied science* became an important new concept in the nineteenth century.[6] Although the term *technology* remained irrelevant to these debates, this background would prove central in shaping the meanings of *technology* in the early twentieth century.

The Discourse of Science and Art in Enlightenment Encyclopedias

The discourse of art and science dates to Aristotle's distinction between techne and episteme. Writers in the eighteenth century took increasing interest in art, science, and the relationship between them. This interest arose in part from the sense that the period was an "age of improvement," with widespread innovation in art and industry. It also arose from the era's mania for comprehensive encyclopedias, which typically included all the "arts and sciences" within their purview.[7]

The Enlightenment discourse of art and science found its fullest expression in two of the best-known eighteenth-century encyclopedias, Ephraim Chambers's *Cyclopaedia* of 1728 and the more famous *Encyclopédie* of Diderot and d'Alembert, published from 1751 to 1772, with its many detailed and well-illustrated articles on the mechanical arts. Encyclopedias of the time typically included in-depth discussions about the classification of knowledge, a topic closely related to the structure of the encyclopedia itself.

Chambers's *Cyclopaedia; or, An Universal Dictionary of Arts and Sciences* was the most successful English-language encyclopedia of its era.[8] In his preface, Chambers enshrined art and science as the two main branches in his system of knowledge. *Science* he defined as the deductive use of reason, with Euclid's *Elements* as its paragon. *Art*, in contrast, encompassed knowledge that reason alone could not attain, requiring the addition of sense perception. Science was therefore universal; art particular to the *artist* (that is, a practitioner of an art). But the difference between art and science had a more fundamental basis, namely the distinction between works of God and those of man. In science, the mind was passive, and the contents of science flowed from God, unshaped by human agency. Art, in contrast, started with the stuff of science, which was then "directed and applied by us, to particular Purposes and Occasions of our own." In other words, art was knowledge shaped by human purposes, which therefore gave it a moral dimension.[9]

This analysis was more or less consistent with Aristotle's discussion of episteme and techne in the *Nicomachean Ethics*. But when Chambers examined the relationship between art and science, he shifted from Aristotle to Bacon. There was, according to Chambers, no sharp boundary between art and science. Science and art existed on a continuum defined by the purity of reason, with substantial overlap between the two extremes. Art, in particular, had its principles, which constituted a "doctrinal Part [that] is of the nature of Science." Such principles arose in part from reflection about the arts, reflection that then took the form of science. Such science, when applied to particular purposes, reassumed the character of art, so that development of an art always involved a reciprocal movement between artistic and scientific aspects.[10] By viewing science and art as interacting categories, Chambers put art on the same epistemological level as science, and by extension also raised the cultural status of art.

Chambers's discussion of art places him more on the cultural than the instrumental side, which is clear from his analysis of human agency and creativity in art. Even though Chambers showed little interest in artisans, human agency was central to his concept of art, though secondary to his concept of science. There was no art without an artist. Chambers rejected the reduction of art to instrumental action; he insisted instead that creative inspiration was the essence of all the arts. He posited poetry as the archetypical art, but he did not separate creative arts from the arts in general. Instead, he endorsed a broad understanding of creativity, linking poetry with the Greek concept of ΠΟΙΕΣΙΣ (*poiesis*). Chambers defined *poiesis* as "making, feigning, [or] inventing," implying that the activities involved in poetry were applicable to the arts in general. He did, however, embrace hierarchy within the arts, viewing manual arts like "sculpture, architecture, [and] agriculture" as inferior to poetry. Yet this hierarchy was not about the degree of science in an art. Instead, for Chambers both art and science partook in aspects of the divine, and therefore neither could be subordinated to the other.[11] He conceptualized the relationship between science and art as both intimate and nonhierarchical. This framing clearly reflected the rapprochement between elite artisans and scholars during the previous two centuries.

Chambers's *Cyclopaedia* inspired a French imitation just two decades later that provided a much more explicit defense of artisans and their knowledge. The *Encyclopédie* of Diderot and d'Alembert endorsed an egalitarian relationship between art and science, but also reflected key conceptual changes that undermined its egalitarianism.

The *Encyclopédie*'s most detailed discussions of art are found in two of its best-known documents, both published in 1751: the *Preliminary Discourse*, written mainly by d'Alembert, and the article "Art," written by Diderot and published separately from the first volume.[12] These two publications have many similarities with Chambers's *Cyclopaedia*, which is not surprising, since the *Encyclopédie* began as a French translation of the *Cyclopaedia*. In the *Encyclopédie*, art and science are divisions within knowledge in general, distinguished by the nature of their object: art is knowledge directed toward action; science, toward contemplation. Thus, wrote Diderot, "Metaphysics is a science and Ethics an art; it is the same with Theology and Pyrotechnics" (the last pair presumably an anticlerical joke). Like Chambers, Diderot ascribed a speculative and a practical side to every art. "The speculative side consists of nothing more than knowing the rules of an art without using them, while the practical side is but the habitual and unreflective employment of these same rules." Echoing a Baconian theme, he argued that one cannot advance the practical side of an art without speculation, or grasp fully the speculative side without practice. Diderot and d'Alembert accepted the division of the arts into liberal and mechanical, but they defended the virtue of the mechanical arts. Diderot gave prominence to the mechanical arts in detailed articles on particular crafts, which he researched by directly visiting artisan workshops, where he both observed and conversed with the workers. Both men stressed that knowledge of the arts was embodied in artisans (*artistes*). As Diderot confidently proclaimed, "Let us finally restore to artists the justice that is their due."[13]

But another change in the concept of art weakened the *Encyclopédie's* defense of the mechanical arts: the new category of fine arts. As the art historian Larry Shiner argues, this new category split aesthetic creativity from "mere" craft skill, so that "all the 'poetic' attributes—such as inspiration, imagination, freedom, and genius—were ascribed to the artist and all the 'mechanical' attributes—such as skill, rules, imitation, and service—went to the artisan." Drawing from the work of Paul Kristeller, Shiner describes in detail how the category of fine art arose in the mid-eighteenth century, reflecting social changes connected to the growing cultural consumption of the middle classes. Although advocates of the new category disagreed on its boundaries, its core embraced poetry, music, painting, sculpture, and architecture. Such a grouping was far from self-evident. Chambers, for example, had no category to unite such disparate arts as music and sculpture. Instead, he classified poetry with rhetoric and grammar, painting under optics, music under phonics, and sculpture with trades and manufactures. As bizarre as this organization might seem today, it was thoroughly conventional at the time, with roots in the twelfth-century writings of Gundisalvo (see chapter 3).[14]

Although there were antecedents for the category of fine arts stretching back to the Italian Renaissance, the idea of segregating one set of the arts according to aesthetic criteria became accepted only in the second half of the eighteenth century. The first clear articulation of this new category was Charles Batteux's book, *Les beaux arts réduits à un meme principe* (1746). According to Shiner, Batteux connected "the term 'beaux-arts' with a restricted set of arts," including specific arts "on the basis of an explicit principle, the imitation of beautiful nature." The book's influence extended well beyond France, as it was quickly translated into both German and English.[15]

The new category of beaux arts received a big boost from the *Encyclopédie*. Batteux's book appeared just five years before the *Preliminary Discourse* and Diderot's article on art. While Diderot's article did not mention fine art, d'Alembert's *Preliminary Discourse* embraced it. D'Alembert placed the fine arts among the liberal arts, noting that fine arts were commonly distinguished from other liberal arts by having "pleasure for their principal object." But, he wrote, the fine arts differed in another respect from traditional liberal arts. "The practice of the Fine Arts consists principally in an invention which takes its laws almost exclusively from genius."[16]

Following this logic, Diderot and d'Alembert created a new classification of knowledge displayed in the famous chart "Système figuré des connoissances humaines," which appeared in a foldout page in the first volume of the *Encyclopédie*. In contrast to Chambers, who divided knowledge into two supercategories, the artificial and the natural, Diderot and d'Alembert invoked a threefold schema based on Bacon's division of learning into history, philosophy, and poesy. For Bacon, each type of learning corresponded to a faculty of the mind: history to memory, philosophy to reason, and poesy to imagination. Poesy was essentially a category for works of fiction, which Bacon described rather dismissively as "feigned history." He cautioned his readers that "it is not good to stay too long in the theatre."[17]

In contrast to Bacon, Diderot and d'Alembert embraced poetry (*poésie*) as a worthy category. Bacon had restricted poesy primarily to literary works, but Diderot and d'Alembert expanded *poésie* in the image of the beaux arts, including under this heading "architecture, music, painting, sculpture, engraving," and other unnamed arts, in addition to literature. Following Batteux, they defended their classification by insisting that all these arts "imitate and counterfeit nature," differing only in their materials, be they words, paints, bronze, or the human voice.[18]

By embracing the new category of fine arts, Diderot and d'Alembert significantly weakened their project to elevate the mechanical arts. By separating fine art from mechanical art, they removed the creative element from the mechanical arts, in effect reducing them to mere technique. This shift in meaning was reflected in their ambivalent rhetoric about artisans, which contradicted their proclamations about the value of the mechanical arts.

Diderot and d'Alembert adopted a decidedly paternalist tone toward artisans. D'Alembert, for example, complained about the inarticulate artisans who "work only by instinct" and could not explain their tools and methods. He could have made similar disparaging remarks about practitioners of the fine arts, who are famously inarticulate about their creative processes. But fine artists produced masterworks through ineffable acts of creative genius, and thus were not obliged to explain their methods to philosophers.[19] In the end, the *Encyclopédie* did less to build bridges between scholars and artisans than to encourage scholars to appropriate artisanal knowledge for scholarly purposes.

The discourse of art and science continued throughout the nineteenth century and even into the twentieth, despite the emergence of a parallel discourse of pure and applied science.[20] Philosophers, most notably John Stuart Mill, were still writing about art in relation to science in the broader sense of both terms well into the second half of the nineteenth century.[21] Nevertheless, despite the continued use of *art* in its broad sense through the nineteenth century, the adoption of the concept of fine art involved a parallel deprecation of what remained. New dichotomies emerged and hardened, such as the sharp division between fine art and craft. The mechanical arts were shorn of their creativity, becoming mere craft, bound by rule. The terms *artist* and *artisan*, used interchangeably in English through most of the eighteenth century, gradually grew distinct in the nineteenth century.[22] This distinction allowed the artist to make claims to middle-class status, but not the artisan.[23] In some ways, the Arts and Crafts movement, especially as articulated by William Morris, was a futile struggle against these new boundaries.[24] Ultimately, these distinctions eliminated art as a useful concept for understanding industrial technology and its relationship to new forms of natural knowledge.

The Genius of Invention and the Devaluation of the Mechanical Arts

The late eighteenth century marked the beginning of the Industrial Revolution in Great Britain, a set of related technological, economic, and social changes that include the rise of mechanized industry, widespread use of mechanical power, rapid urbanization, and sustained economic growth. This revolution created opportunities for fresh thinking about the mechanical arts.

The success of the Industrial Revolution did not depend on whether scholars understood its nature or causes. In some ways, its middle-class beneficiaries had an interest in denying its origins so that they could legitimate it as a process rooted in nature.[25] The concept of fine art played an indirect role in this process of concealment. By the early nineteenth century, fine art had become the model for creativity in all the arts. In the mechanical arts, this model encouraged a cult of the inventor, who was portrayed as a lone genius (invariably male) giving unconstrained expression to his creativity.[26]

According to an older historiography of industrialization, the rise of modern industry was premised on the destruction of craft skills. In this view, the factory arose from a highly developed internal division of labor, mechanization of handwork, and use of central power sources.[27] Through the work of Berg, Zeitlin, Scranton, and other recent scholars of industrialization, historians no longer see the rise of modern industry as premised on the destruction of craft skills. Major new technologies almost always call forth new skills and create new categories of workers, even as old skills are lost. In European industrialization, the accumulation and transmission of artisanal knowledge remained a key factor.[28]

Thus, the Industrial Revolution depended deeply on skilled craftsmen to create, improve, and maintain the new machinery of industry. In early nineteenth-century Britain, however, middle-class theorists of industrialization rarely acknowledged and often actively disparaged the role of artisanal knowledge. Instead, they praised the division of labor and skill-destroying mechanization. For these theorists, progress in the useful arts did not depend on the artisan. Rather, artisans were seen as obstacles to progress whose limitations could be remedied by science. Science, from this perspective, was a form of middle-class knowledge that could be applied to manufacturing. This devaluation of artisanal knowledge was not new; Bacon was similarly ambivalent. Nevertheless, this nineteenth-century approach decisively reframed the connection between science and art in terms of the application of science to art. The reframing was a major shift from the Enlightenment view of science and art as interacting along a continuum.

Perhaps the most striking example of this reconceptualization is Andrew Ure's *The Philosophy of Manufactures* (1835). Ure's *Philosophy* was both a technical survey and an apologia for the new mechanized factories of the Industrial Revolution, especially textile mills. He stridently defended existing practices in the textile mills, especially the use of child labor, while also articulating a theory of the ideal automatic factory.[29] Like other writers of the era, Ure did not have access to a twentieth-century concept of technology to explain the rise of the factory. Instead, he invoked the phrase "arts and manufactures," in effect reducing the arts to manufacturing. This reduction is explicit in his *Dictionary of Arts, Manufactures, and Mines* (1844), where he defined "an art or manufacture" as "that species of industry which effects a certain change in a substance, to suit it for the general market, by combining its parts in a new order and form, through mechanical or chemical means."[30] Human agency had no role in Ure's definition of *art*.

In *The Philosophy of Manufactures*, Ure made it clear that skilled artisans deserved no place in this system of arts and manufactures. "The more skilful the workman, the more self-willed and intractable he is apt to become, and, of course, the less fit a component of a mechanical system." Therefore, "it is . . . the constant aim and tendency of every

improvement in machinery to supersede human labour altogether, or
to diminish its cost, by substituting the industry of women and children for that of men; or that of ordinary labourers, for trained artisans."[31] Although Ure's concept of skill may have been "more than
a little paranoid" for his time, his general approach was endorsed in
more sober terms by other middle-class theorists of industry, such as
the Cambridge mathematician Charles Babbage, one of the most prominent British scientists of his time.[32]

Ure was quite explicit that in the modern factory, craft skills were
to be replaced by science. "The principle of the factory system then is,
to substitute mechanical science for hand skill." Through the "union
of capital and science," work would be reduced to the "exercise of vigi-
lance and dexterity." He also praised the ability of self-acting machinery to suppress strikes. One such machine, Robert's spinning mule,
demonstrated "that when capital enlists science in her service, the refractory hand of labour will always be taught docility."[33] Karl Marx,
who despised Ure yet drew heavily from his analysis of modern industry, famously quoted this passage in the machinery chapter of *Capital*.[34]

Ure framed his broader project in terms of the application of science. He surveyed British factories after spending years teaching "practical men . . . the application of mechanical and chemical science to
the arts." Yet, he claimed, these sciences were not the natural philosophy of the universities. Instead, he praised the "science of the factory"
as understood by "enlightened manufacturers" over the "theoretical
formulae" of the "recluse academician." Ure's science was eminently
practical. He referred, for example, to "mill architecture" as a "science
of recent origin." He also described the "modern cotton mill" as an
exemplar of "exact mechanical science."[35] Ure was clearly using *science*
to describe the same type of autonomous practical knowledge that Babbage, writing just three years earlier, defined as applied science.[36] Unlike Babbage, Ure felt no need to differentiate this practical knowledge
from science generally. This is not surprising, because his career as an
industrial consultant placed him far from Babbage's elite world of university science.[37]

In Ure's thinking, this broad concept of science transferred agency
from artisans to factory owners, inventors, and engineers. Drawing
from the fine arts, he granted these middle-class agents of industrialism a degree of creativity denied to artisans, likening the cotton mill
to "individual master-pieces" that are studied in the "philosophy of
the fine arts." Ure mentioned a number of heroic inventors, in particular Richard Arkwright, whom he praised more for his willingness "to

subdue the refractory tempers of work-people" than for the brilliance of his technical contributions.[38] Similarly, Babbage also described mechanical invention in terms borrowed from the fine arts. He saw the invention of machinery as rooted less in craft skills than in the creative process. Although new machines were common, he opined, "the more beautiful combinations are exceeding rare." Such machines "are found only amongst the happiest productions of genius."[39] As Christine MacLeod shows in her award-winning study of invention in Britain, this view of mechanical invention as an act of genius became widespread in the early nineteenth century.[40]

The Discourse of Pure and Applied Science

In the mid-nineteenth century, a new discourse emerged around the concept of applied science. This discourse displaced older language about the relationship of science and art, and elevated science as a key agent in the rise of modern industry. Such a conceptual shift, however, involved fundamental changes in the meanings of both *art* and *science*.

Robert Bud describes in detail the rise of the concept of applied science in Britain. He shows that supporters of applied science routinely presented it as an alternative to craft knowledge, which was usually referred to disparagingly as "rule of thumb."[41] Furthermore, as Bud and other historians have shown, *applied science* had multiple meanings, though these were rarely contested. Most important was a fundamental ambiguity: *applied science* could refer on the one hand to a type of science, that is, science suited to application, or on the other hand to the process of applying science to practical problems. This semantic flexibility allowed the term to do boundary work for both engineers and scientists, especially as their professions solidified in the second half of the nineteenth century. American engineers in particular based their claim to social status on the concept of applied science as an autonomous body of knowledge. Scientists, in contrast, often defined *applied science* as the application of pure science in order to claim credit for modern wonders of the industrial age.[42]

However, some nineteenth-century scientists objected to claims about applied science, which they saw as undermining the ideal of pure science. These objections were strongest among professionalizing scientists who sought to define the scope of their field in order to gain professional autonomy and financial support. By defending pure science, advocates for the profession helped restrict the meaning of *science* while also enhancing its social status in a way that paralleled changes in the concept of art. Somewhat ironically, these changes produced concepts of art and science that were less useful for understanding the vast changes in material culture occurring at the end of the nineteenth century, changes that very much involved both science and (mechanical) art.

Just as with *art*, present-day meanings of *science* cannot be projected back into the nineteenth century. This principle applies especially to the present-day assumption that *science* refers to natural science unless explicitly modified, as in the term *social science*. Most humanities scholars know that the present-day English term *science* is far narrower than their French and German equivalents. As the historian of science James Turner argues in a remarkable article, the nineteenth-century American term *science* was in fact similarly broad, and did not refer exclusively to natural science. According to Turner, historians of science, including him, have erred in equating the discourse of science in nineteenth-century America with discourse about natural science. He insists that before 1900, *science* did not refer to "laboratory methods or the quest for general laws," but rather to something "more general, a kind of 'systematic knowledge' or 'rigorous thought'" that could include most social sciences and sometimes even theology. By the later nineteenth century, the increasing professionalization of academic disciplines and the growing prestige of technology helped natural scientists claim a monopoly on science. When historians view the nineteenth century in terms of this monopoly, accepting the primacy of natural science over other forms of knowledge, "we are ourselves victims of a linguistic coup d'état carried out by specialists in the natural sciences at the end of the 19th century."[43]

Turner's argument is important for understanding the discourse of science and art in the nineteenth century, and the eventual emergence of ideas about applied science and, later, technology. Historians continue to argue over the role that scientific knowledge played in European industrialization.[44] Yet statements from that period about the relationship between science and art cannot simply be interpreted as equivalent to present-day discussions of science and technology.[45] When nineteenth-century writers discussed the application of science to the arts, they had a far broader conception of science than we have today, even when they were explicitly focused on natural science. In everyday usage, in fact, *art* and *science* were often used interchangeably, forcing writers to take pains to distinguish them.[46] Both terms frequently encompassed broad areas of useful knowledge.

There are numerous examples of the blurred boundary between science and art in the nineteenth century, often found under the heading "Science and Art," which was common in popular periodicals of that time. This heading encompassed an eclectic selection of items, from improvements in agriculture and mechanical inventions to discoveries of new planets and events in the fine arts. One such column, from a single issue of the *Christian Advocate* in 1882, discussed a talking canary, a new chemical compound produced by electrolysis, a nebula in Orion, and a prize for inventors.[47] Such a jumble of topics, so incongruous to our modern minds, had been common in early modern scientific journals like the *Philosophical Transactions of the Royal Society*, although by the nineteenth century curiosities and monsters had largely disappeared from such journals. Popular periodicals like *Scientific American*, however, continued to include a diverse range of topics, drawing no firm distinction between art and science. Popular handbooks and encyclopedias also embraced a variety of topics under the terms *art* and *science*, continuing the tradition of Enlightenment encyclopedias.[48] Furthermore, when nineteenth-century writers sang the praises of technical progress, they typically framed it as progress in both art and science. For example, Edward Everett—Massachusetts congressman, orator, and later president of Harvard—pronounced that "art and science are, in themselves, progressive and infinite. Nothing can arrest them which does not plunge the entire order of society into barbarism."[49]

The breadth of the nineteenth-century concept of science, and the lack of a clear line between science and art, puts claims about the utility of scientific knowledge in a new light. In general, these claims did not represent the intellectual imperialism of natural scientists seeking credit for progress in technology, what Paul Forman tendentiously misidentifies as the "preposterous primacy of science relative to technology."[50] Praise of science may have implied the elevation of theory over practice, and of intellectual over manual work, but not necessarily the primacy of natural science and its institutions. Instead, most of these expansive claims about science drew from the broad meanings of the term, even into the twentieth century.

One example is an eleven-volume *History of Science* published from 1904 to 1910 by the American physician and popular science writer Henry Smith Williams. This work has little scholarly merit, yet it nicely captures the broad conception of science still operating in the early twentieth century. Williams cited Herbert Spencer's definition of science as "organized knowledge," and used this broad definition to discuss not only the natural sciences but also all manner of folk knowledge. He claimed that even primitive peoples possessed science. "A barbarian who could fashion an axe or a knife of bronze had certainly gone far in his knowledge of scientific principles and their practical application." The encyclopedia's first five volumes focused more or less on "pure" sciences, but the next four examined topics that would now be classified as technology, such as the steam engine, the dynamo, printing, papermaking, the telegraph, and aviation. Williams was perfectly happy to portray these topics as manifestations of science, having no apparent concern about distinctions between science and art, or any need for the concept of technology. In order to justify his inclusion of both abstract and practical knowledge in the work, he argued that the supposedly "radical distinction between theoretical and practical aspects of science" was little more than the "differences between two sides of a shield."[51] Williams's views were hardly anomalous at the time. Marcel LaFollette has found that Thomas Edison was the most frequently discussed "scientist" in American popular magazines in the first half of the twentieth century, and that other inventors and engineers were also featured prominently as "scientists."[52]

Nevertheless, as Turner argues, the concept of science did undergo a profound transformation in the nineteenth century, especially in scholarly usage. When "men of science" began to professionalize in Britain and the United States, they made a concerted effort to define the boundaries of science. They not only restricted science to the natural sciences but also defined the practical arts as subordinate to the sciences. Natural scientists, as the sociologist Thomas Gieryn has shown in his analysis of "boundary work," worked self-consciously to narrow the scope of the sciences by drawing boundaries with religion on the one side and the practical arts on the other. In the process, they embraced a new concept of pure science that undermined the relationship between science and art.[53]

Gieryn focuses on John Tyndall, a prominent British physicist and science popularizer. According to Gieryn, Tyndall's goal was to construct a space for science that would give scientists professional autonomy, "unfettered by the competing authorities of religion and mechanics." In drawing boundaries between science and religion, Tyndall frequently highlighted the empirical basis and practical utility of science, as demonstrated by new technologies. He contrasted these aspects of science with religion, which he described as speculative and not useful. Yet this argument created a dilemma for scientists. If too much stress were placed on practical utility, it might imply that science was subordinate to industrial progress, a culturally powerful symbol in Victorian England. To defend the epistemological space of science, Tyndall rejected the idea of science and art as a continuum. Instead, he posited a radical separation between the empiricism of the "mechanician" and the disinterested theories and experiments of the scientist. This sphere of abstract, disinterested knowledge, he argued, provided the necessary principles for progress in the practical arts.[54]

Tyndall applied this argument to the history of electric telegraphy, which involved a complex interaction of theory and practice, as recent historians have shown.[55] Tyndall, in contrast, insisted that scientific principles preceded practical application. "Before your practical men appeared on the scene, the force had been discovered, its laws investigated and made sure, the most complete mastery of its phenomena had been attained—nay, its applicability to telegraphic purposes demonstrated—by men whose sole reward for their labors was the noble excitement of research, and the joy attendant on the discovery of natural truth."[56] In Tyndall's unrealistic summary of the history of the telegraph, we can see the roots of the twentieth-century definition of *technology* as the application of science.

As Gieryn has pointed out, Tyndall's concept of science was internally inconsistent in multiple ways. Like future defenders of the boundaries of science, Tyndall "exploited the ambivalences inherent in the meaning(s) of science" depending on the adversary being addressed. He emphasized the status of science as theory to distinguish it from industrial application, while separating science from religion on the basis of its empirical character. Like fellow British scientist Thomas Huxley, Tyndall rejected the category of applied science, leading him to assert that the utility of science was rooted in its purity and disinterestedness.[57] This claim was conceivable only because of the broader meaning of *science* as organized knowledge, a meaning that did not distinguish between types of knowledge. In effect, Tyndall switched back and forth between two notions of science in order to credit all progress in the industrial arts to a narrow concept of professionalized science. His argument may have made sense as a strategy for gaining professional autonomy. As a theory about the relationship between science and the practical arts, however, it was far less adequate than Diderot's homilies from a century earlier about the need for both theory and practice.

In the United States, different rhetoric emerged around the term *applied science*, but it led to similar results. With the simultaneous professionalization of both science and engineering in post–Civil War America, *applied science* took on contested, ideologically charged meanings. Scientists deployed *pure science* to distinguish their work from the *applied science* of engineers and inventors, while still arguing that applied science remained dependent on and subordinate to pure science. Many engineers, in contrast, adopted definitions of *applied science* that lent their profession the prestige of science while asserting their autonomy from scientists.[58]

The prominent physicist Henry Rowland popularized the idea of pure science in a remarkably arrogant and bitter lecture to the American Association for the Advancement of Science in 1883. In this lecture, Rowland sought to draw a sharp boundary between the "noble pursuit" of pure science and the "vulgarity" required for its application, denouncing the common tendency "to call telegraphs, electric lights, and such conveniences, by the name of science." Yet he never questioned the utility of pure science. Like Tyndall, Rowland credited all industrial progress to the application of prior discoveries in pure science. Rowland held up as an exemplar of pure science the selfless work of the English scientist Michael Faraday, who "died a poor man" despite the vast wealth that lesser minds gained from his discoveries.[59] (The Faraday trope was repeatedly invoked well into the twentieth century by scientists arguing for the utility of pure science.)[60]

Rowland insisted that science should be disinterested, uncorrupted by commerce or concern for application. According to Michael Dennis, he drew from the rhetoric of American republicanism to distinguish virtuous science from the corrupt commerce of its application. But Rowland was not really a defender of republican virtue. His talk was explicitly antidemocratic, unabashedly arguing for an elite system of science based in well-funded universities. The poor state of American higher education, he claimed, "could only exist in a democratic country, where pride is taken in reducing every thing to a [common] level." Like Tyndall, Rowland simultaneously employed multiple visions of science, portraying science both as a form of high culture to be pursued as its own end, and as a source of wealth and progress. "Like the rain of heaven, this pure science has fallen upon our country, and made it great and rich and strong."[61]

American natural scientists rarely echoed Rowland's explicit hostility toward inventors and other appliers of science, nor did they claim all progress in the industrial arts as the fruits of science. To be fair to Rowland, he was hostile not to inventors and appliers as such, but to the fact that practical work was accorded high cultural status and at times given the name of science. He just wanted the public to acknowledge that practical work was subordinate to pure science and not a worthy pursuit for university professors. But by insisting on the virtue of unfettered science along with the utility of its application, Rowland managed to eat his cake and have it too, making a case both for professional autonomy and for financial support of science. With but minor variations, his arguments have been repeated by scientists up to the present day.[62]

American engineers also embraced the pure science model espoused by professionalizing scientists. Beginning in the late nineteenth century, engineers accepted a division of labor that, at least in principle, accorded them a subordinate cultural status as practitioners of applied science. Forman claims that when American engineers explicitly embraced this subordinate role, they were enacting the universal ideological supremacy of pure science over technology.[63]

Forman simplistically equates technology and engineering, which is highly problematic, as engineers' rhetoric actually represented a complex dance of professional boundaries. Ronald E. Kline has shown that American engineers had a sophisticated understanding of applied science specifically and of their relationship to science more generally. As he has argued, leading engineering thinkers, such as the Cornell mechanical engineer Robert H. Thurston, defined *applied science* in a way that gave more autonomy to engineers, thus reflecting the realities of engineering practice. Thurston sometimes used *applied science* to refer to the use of scientific methods in engineering, such as systematic experimentation. At other times, *applied science* for Thurston meant an autonomous body of specialized knowledge, which engineers produced through research and innovation and transmitted in schools of engineering. A similar idea was developed by the British academic engineer William Rankine, who argued for an "intermediate" form of knowledge that linked scientific principles to practical applications.[64]

Nevertheless, as Kline has shown, leading spokesmen for American engineering repeatedly embraced a simplistic model of applied science closer to that of Rowland than Thurston, particularly as the cultural authority of science increased during the Progressive Era. This model was also a form of boundary work, in this case blurring the distinctions between science and engineering to help engineers "raise their occupational status above that of artisans" by lending objective authority to their professional expertise.[65] This strategy was effective, because American engineers had little to fear from the far smaller number of scientists in their country, at least before World War II. In this context, the epistemological subordination of engineering to science did not imply professional subordination.

The rhetoric of applied science had no discernible effect on actual technological change. Inventors and engineers did not bow down to newly professionalized natural scientists. Instead, these lowly appliers of science happily drew from the knowledge of scientists when it was relevant, and generated their own general principles when scientists had little to offer.[66]

Nevertheless, all this boundary work had real consequences for the concept of science into the early twentieth century. These consequences included a radical separation of science from art, along with an elitist subordination of practical knowledge to pure science. There were occasional dissenters such as Rankine who endorsed, at least in principle, the idea that "men of science" and "men of practice" could each benefit from the other's knowledge. The British social scientist Herbert Spencer also defended the older, broader concept of science, which allowed for the interpenetration and mutual dependence of art and science.[67]

In effect, this new discourse of science represented a reversal of Francis Bacon's praise of the arts over philosophy, and a rejection of Enlightenment discourse that argued for a balance between art and science. Taken together, these changes in the concepts of art and science made it difficult to analyze their relationship, which had been so central to Enlightenment thought, but which disappeared from technical dictionaries and encyclopedias by the end of the nineteenth century.[68] Instead, with the rise of applied science after about 1850, theorists of material progress could use the new category in place of the older concept of mechanical arts. The frequency of the term *mechanical arts* declined steadily after 1850, while *applied science* increased in almost equal measure, surpassing the older term in the first decade of the twentieth century.[69]

Furthermore, the elevation of science over practical knowledge had real implications for American engineers, as exemplified by the work of Frederick W. Taylor, the founder of "scientific management." Although historians give little credit to Taylor's claim to have reduced metalworking to a science, this claim drew from the same logic used by professionalizing scientists. Taylor's approach was certainly scientific in Herbert Spencer's broad definition, since scientific management did indeed organize and systematize craft knowledge about metalworking. In this usage, Taylor was very much following in the footsteps of

Andrew Ure, who also used science as a cudgel against skilled workers. But in contrast to Ure, he was able to borrow from almost a century of boundary work by natural scientists intent on raising the prestige of science. This boundary work helped Taylor define *science* as a sphere of elite, professional knowledge that belonged only to engineers, not to the skilled workers who had in fact generated most of this knowledge, or the foremen and plant managers who traditionally set the technical conditions of production in negotiation with skilled workers.[70]

The "Semantic Void" in the Second Industrial Revolution

By the end of the nineteenth century, the terms *art* and *science* had lost most of their utility for explaining changes in what we now call technology. This loss occurred in part because scientists and fine artists had redefined these concepts as a way of asserting their professional autonomy. Art was aestheticized, while mechanical and industrial arts were debased. Science was made the property of a small elite, excluding actual producers. The vast reservoirs of skill and knowledge lodged in the shop floors, foremen's offices, inventors' workshops, and even trade journals were rendered invisible, as was the complex process of translating arcane formal knowledge (science) into effective material practices.

This lack of appropriate concepts made little practical difference during the Second Industrial Revolution, the period from about 1870 to 1930, perhaps the most profound technological transformation in human history. But the loss of meaning in *art* and *science* did have intellectual consequences, contributing to what Leo Marx has termed a "semantic void." For Marx, this void refers to the lack of adequate language to capture the dramatic changes in the material culture of the era.[71] But it was not simply new phenomena that produced this void, but also the impoverishment of existing concepts. As a result, scholars of the time found it difficult to grasp the nature of the changes they were experiencing. Henry Adams exemplified this intellectual confusion in his well-known response to the Exposition Universelle of 1900 as he stood in baffled amazement before the "occult mechanism" of the electric dynamo.[72]

The net result of these conceptual changes was to remove human agency from the discourse of industrial modernity at the turn of the twentieth century. In the never-ending parade of speeches, books, and articles about the material progress of the age, writers granted agency to only a few men, the individual creative geniuses supposedly responsible for specific inventions or discoveries.[73] During the twentieth century, this limited scope for human agency became a key attribute of the instrumental discourse of technology, especially when *technology* was defined as the application of science.

By 1910, use of the term *art* to refer to what would today be called *technology* was becoming antiquated. As the literary scholar Sidney Colvin noted in his entry on *art* in the eleventh (1910) edition of the *Encyclopaedia Britannica*, "the word Art, becoming appropriated to the fine arts," was no longer used for the "large number of industries and their products to which the generic term Art . . . properly applies."[74] But with *art* no longer available, another term was needed for the material processes at the heart of industrial civilization. In the first third of the twentieth century, *technology* assumed this meaning, in effect absorbing the abandoned content of the mechanical arts.

SIX

Technology in the Nineteenth Century: A Marginal Concept

There was no *technology* before the twentieth century. The term existed, of course, but it referred primarily to "the science of the arts," or more narrowly to treatises on the arts or descriptions of technical terminology.[1] *Technology* was a sixteenth-century neologism, appearing first in Latin to refer to a system of classifying the arts, both mechanical and liberal. The term spread to English in the seventeenth century, but with an important change: *technology* came to imply the science of the mechanical rather than the liberal arts.[2]

However, the term did not become common until the nineteenth century. In German-speaking countries, *die Technologie* became a minor scholarly field in technical universities. The English cognate also became more important in higher education, especially following the adoption of the term in 1860 for Boston's new school of science and engineering, the Massachusetts Institute of Technology. In the last third of the century, anthropologists and other social scientists occasionally used *technology* to refer to the study of human artifacts and production methods.[3]

Yet technology was not the focus of any sustained theoretical discourse in the nineteenth century, not even in German. I have found only five scholarly articles written in English from the era that discuss this concept in a sustained way.[4] Just three authors produced these five papers, and none mentioned the others, even though they were contemporaries. In fact, the entire nineteenth century, a period that witnessed the greatest transformations in material culture since settled agriculture, had no broad, unifying term for these changes comparable to *technology* in our own era.

In the first half of the twentieth century, scholars largely forgot these earlier discussions of technology. Only when technology became a major concept after World War II did these discussions assume retrospective significance. Yet the nineteenth-century use of the term did matter for its twentieth-century meanings, for two reasons. First, technology in the nineteenth century was understood as a field of science, which implied scholarly rather than artisanal knowledge. This connotation of *technology* as elite knowledge continued even as the meaning of *technology* was transformed in the twentieth century. Second, and perhaps more important, the presence of the term in the late nineteenth century created a false sense of familiarity that helped obscure its shift in meanings in the early twentieth.

Origins of the English Term *Technology*

The word *technology* has a Greek etymology; most accounts of its history start in the classical world. However, the term was barely present in classical Greek or Latin. Instead, *technology* entered English through a Latinized Greek neologism in the sixteenth century. Still, the classical roots of the term are important, because scholars repeatedly invoked them to define *technology*. The word is a compound of two common terms from Greek philosophy, *logos* and *techne*. *Logos* is a complex word that denotes both language and reason, while *techne*, as I discussed in chapter 2, is a broad term usually translated as "art," "skill," or "craft."[5]

Although both techne and logos were key concepts for Plato and Aristotle, neither philosopher had much interest in their relationship.[6] Nor did either philosopher feel compelled to join the terms into a new compound word. The sole exception is Aristotle's *Rhetoric*, a subject that the Greeks classified as a techne. Four times, Aristotle used a verb form of *technology* whose meaning is ambiguous. The nineteenth-century lexicographers Scott and Liddell defined it as "to bring [something] under the rules of art, to systematise." But in a recent translation, Joe Sachs renders the term as "speech-art making," arguing that it is actually a pun on the rhetorical manuals, "technê tôn logon," that

Aristotle criticized in the *Rhetoric*.[7] In any case, Aristotle's odd term appears to have had no influence on future writers.

Only later were the two terms *techne* and *logos* regularly combined into a noun, τεχνολογία (*technologia*), by Hellenistic and Byzantine authors. However, this compound noun contained a fundamental ambiguity. Did τεχνολογία refer to the logos of techne, that is, reason applied to the arts, or the techne of logos, that is, the arts of language? It was, in fact, understood as the latter, as technical aspects of language, mainly rhetoric and grammar.[8] Such usage continued in Byzantine texts at least until the twelfth century.[9]

Regardless of its meaning, the Latin equivalent of this Greek term, *technologia*, was unknown in classical and medieval Latin; Cicero used the term once, but in Greek.[10] The German scholar Wilfried Seibicke concluded that τεχνολογία had "no direct influence" on scholarly discourse during the Middle Ages and the early modern era. In other words, no continuous history of discourse links the Greek τεχνολογία with the current meanings of *technology*.[11]

Technologia became significant in Latin only during the Reformation, essentially as a neologism drawn from the Greek, but with a new meaning as the logos of techne—that is, science of the arts. It was first used in this new way by Peter Ramus, the influential sixteenth-century French philosopher, educational reformer, and convert to Protestantism. In his work, Ramus sought to reform the teaching of the liberal arts, which he saw as stultified by stale Scholasticism. *Technologia* referred to his system for restructuring the arts.[12] He understood art broadly as covering all forms of knowledge, rejecting the Aristotelian distinction between theoretical and practical disciplines. Technologia for Ramus was primarily an exercise in classification, based on his method of dividing concepts into dichotomous subcategories.[13]

Ramist ideas were developed with enthusiasm by Puritan theologians, in particular Alexander Richardson and William Ames.[14] In 1633, Ames produced a succinct summary of this system of the arts under the title *Technometria*, a term he used synonymously with *technologia*. In the writings of Puritans like Ames, the arts reflect God's wisdom, and therefore, as Perry Miller noted, "technologia was an assertion that the arts direct conduct to ends enunciated by God."[15] In principle, Ramist technologia encompassed all the arts, mechanical as well as liberal, although in practice it referred mainly to the liberal arts. Students at Harvard and Yale, for example, analyzed the liberal arts using Ramist principles in a whole series of *theses technologicae* from the mid-seventeenth century into the eighteenth century.[16] Ames, however, did explicitly include the "less dignified" mechanical arts in his system, defending them as both necessary and virtuous.[17]

Ramus was more influential as an educational reformer than as a philosopher. Outside seventeenth-century New England, Ramist technologia found few followers. Proponents of the new mechanical philosophy, among them Francis Bacon, had little use for Ramus.[18] Nevertheless, even as interest in Ramus faded, English-language dictionaries began listing a new word, *technology*. The definition of this term was consistent with the Ramist understanding of *technologia* as systemized knowledge of the arts, but with a key difference. The dictionary definitions focused on not the liberal but the mechanical arts.

Perhaps the earliest attestation of this usage is provided by the second edition (1661) of Thomas Blount's *Glossographia*, a dictionary of "Hard Words," generally of foreign origin, that had become common in "our refined English tongue." Although the first edition in 1656 did not have an entry for *technology*, the second edition identified the term as Greek and defined it as "a treating or description of Crafts, Arts or Workmanship." Similarly, John Kersey's 1706 edition of Edward Phillips's dictionary, *The New World of English Words*, defined *technology* as "a Description of Arts, especially the Mechanical."[19] This new definition also appeared in Latin in the work of the German Enlightenment scholar Christian Wolff, who in 1728 defined *technologia* as "science of arts and of works of art, or, science of what humans produce by the work of the organs of the body, mainly the hands."[20] Despite this shift from liberal to mechanical arts, *technology* in any of its forms remained rare through the end of the eighteenth century. Samuel Johnson's famous English dictionary of 1755 did not include it, nor did subsequent editions of that dictionary well into the nineteenth century.[21]

From Craft to Elite Knowledge: *Technologie* in German Cameralism

Technology first became a significant concept in German, not English. The term was introduced into German academic discourse by Johann Beckmann's book, *Anleitung zur Technologie (Guide to Technology)* (1777). Beckmann was a prominent figure in German cameralism, a set of practically oriented academic disciplines concerned primarily with providing guidance for state administrators.[22] In the *Anleitung*, Beckmann self-consciously developed the concept of *Technologie* as a discipline devoted to the systematic description of handicrafts and industrial arts.

Yet for Beckmann and his followers, Technologie constituted knowledge not for the artisan but rather for officials in the royal courts who had administrative responsibilities related to artisanal trades. Technologie was therefore a form of elite rather than artisanal knowledge.[23]

Much has been written on Beckmann's significance for the concept of Technologie, mostly in German.[24] He has been described as the "founder of the science of technology."[25] His concept of Technologie supposedly provided "the starting point for general technological education" through its concern with "crystallizing out the general principles of technology."[26] Indeed, Beckmann's *Anleitung* was a significant contribution to German academic discourse, firmly linking the concept of Technologie to craft work and industrial production. With the *Anleitung* and later works, Beckmann helped turn Technologie into a minor academic field within cameralism, as German universities appointed professors of Technologie and added courses in the new field.[27]

Yet cameralist *Technologie* remained very different from twentieth-century *technology*. Beckmann defined *Technologie* as "the science that teaches the processing of natural products or the knowledge of handicrafts."[28] In this definition, Technologie was a field of *Wissenschaft* (science), which meant not natural science but rather a systematic field of scholarly knowledge.[29] Beckmann conceptualized the material aspects of the crafts as an object of study, which he hoped to put on a systematic, theoretical basis.[30] Specifically, he sought to make Technologie into a science by creating a classificatory scheme equivalent to the Linnaean system for plants and animals.

Yet the significance of Beckmann's conceptual innovation should not be exaggerated. His work was in the tradition of the natural history of trades, a project first proposed by Francis Bacon and realized in part by the *Encyclopédie* of Diderot and d'Alembert. Beckmann, a natural historian by training, did try to classify the trades based on similarities in technical processes, but with little success. In the introduction of the *Anleitung*, he classified 324 arts and crafts into fifty-one groups, though he remained dissatisfied with the schema and did not use it in the body of the book. Instead, he focused on a smaller number of traditional crafts, from wool weaving to coinage, including trades such as beer brewing, starch making, and leather tanning.[31]

As Wolfhard Weber has argued, Beckmann had little interest in new machines or industrial processes. He ignored, for example, the Newcomen steam engine, first demonstrated in England in 1712.[32] When the *Anleitung* was published in 1777, steam engines were already operating in German states, and knowledge of them "was already fairly widespread."[33] But Beckmann gave short shrift even to well-established forms of industrial power such as water mills and windmills. He also paid little attention to German improvements in machinery, improvements that impressed even the French. In 1775, the French ironworks manager Pierre Clément Grignon described Germany as "the land of machines," noting that "in general the Germans diminish handwork considerably, by means of machines suited to all sorts of motions."[34] Although Grignon was surely exaggerating, his description of mechanical innovation was sharply at odds with the static craft practices described by Beckmann.

Despite Beckmann's desire for a more rigorous system of classification, the body of his *Anleitung* differed little from other descriptions of the arts and trades by his German contemporaries. From 1761 to 1779, for example, the Prussian natural historian Johann Samuel Halle published the six-volume *Werkstäte der heutigen Künste; oder, Die neue Kunsthistorie* (*Workshops of the Present-Day Arts; or, The New History of Art*). Halle's volumes covered a wider range of crafts in far more detail than Beckmann's work and included illustrations, which are notably lacking in the *Anleitung*. Halle also contributed to a related effort, begun by the leading cameralist scholar Johann Heinrich Gottlob von Justi, to produce a German translation of the multivolume *Déscriptions des arts et métiers*, an encyclopedia of the trades that the French Academy of Sciences began publishing in 1761. Von Justi's translations began appearing in 1762, eventually totaling twenty-one volumes.[35]

In 1782, five years after Beckmann's *Anleitung*, Halle published a one-volume abridgment of his *Werkstäte*, now renamed *Technologie; oder, Die mechanischen Künste* (*Technology; or, The Mechanical Arts*). In effect, Halle simply rebranded his older work with the new term popularized by Beckmann, while also including *mechanical arts* in the title to make clear the book's subject. These German works on the mechanical arts show that Beckmann's writings on Technologie were not particularly innovative but rather part of the continuing interest in craft production among the ruling elites of European states.

But works about the mechanical arts did differ in how they embodied social relations of knowledge. Bacon's project, as noted earlier, was always about creating knowledge for elites rather than artisans. In contrast, Diderot and d'Alembert did honestly seek to elevate the status of the mechanical arts through the *Encyclopédie*, even if they did not succeed in their goal. In this regard, Beckmann was heir to Bacon, while Halle's *Werkstäte* was more aligned with Diderot. Halle insisted that descriptions of the arts had to be based on direct observation of the "dirty workshops of the artists," not just on written accounts.[36] Like Diderot, he also complained bitterly of the vexations involved in getting information from artisans, describing them as discourteous "arcanists" who hid their "old-fashioned guild knowledge" out of self-interest and ill will. Yet Halle shared Diderot's leveling spirit, and he lamented the divide separating artisans and scholars. He insisted that "all the arts generate a common bond," one that linked craft and scholarly knowledge.[37] Halle ultimately blamed the mistrust between artisans and scholars on the low regard in which craftsmen were held. "People consider artists only as smoky machines of the countryside, refusing to live with them in respectful familiarity."[38]

In contrast to Halle, Beckmann was quite explicit about the elitist ends of his interest in craft knowledge. His Technologie was, after all, a field within *Cameralwissenschaft*, the science of state administration, and as such was meant for administrators, not craftsmen. The policy sciences of cameralism were designed to steer the private interests of the farmer, handworker, and merchant toward the interest of the state as a whole. According to Beckmann, Technologie was not makers' knowledge but rather knowledge for "those who are to organize, invest in, regulate, judge, govern, sustain, improve, and utilize" the arts. The relationship of the cameralist to the worker, in other words, was like that of the farm owner to the peasant, or the artillery officer to his men; the officer did not have to be a good shot himself.[39]

Beckmann's Technologie was thus thoroughly instrumentalist. It was knowledge to be used by elites to control the productive activity of artisans. Beckmann showed little concern about the moral status of the arts, the meaning of arts to the artisans, or the creative and expressive aspects of the mechanical arts. In contrast to Halle, who discussed the "art of the painter" in the first volume of his *Werkstäte*, Beckmann excluded the fine arts from the *Anleitung*.[40] Thus, the eighteenth century's separation of the mechanical from the fine arts was incorporated into the concept of Technologie. Almost every writer after Beckmann followed him in excluding the fine arts from the scope of technology.[41]

Beckmann indeed deserves credit for making the concept of Technologie significant in German technical education.[42] His followers, most important his Göttingen student Johann Poppe, built on Beckmann's descriptive efforts well into the nineteenth century.[43] But aside from helping spread the term *Technologie*, Beckmann's work did not involve any fundamental conceptual innovation. Instead, his approach to the arts and crafts remained firmly within the eighteenth-century encyclopedic tradition, a descriptive treatment that harkened back to Francis Bacon's project for a natural history of the arts.[44] In fact, Beckmann's contributions to the discourse of technology actually attracted little notice before World War II. Only in the postwar era, when *technology* emerged as a keyword, was his *Anleitung* granted retrospective significance.[45]

As a concept, cameralist *Technologie* proved to be a dead end, both in Germany and abroad. Cameralism itself was almost completely ignored by contemporary French- and English-language writers. In the second half of the nineteenth century, its influence waned even in Germany. And Beckmann's concept of *Technologie* declined along with cameralism. Its taxonomic structure was not suited to engineering education, particularly to the theory-based approach of the French polytechnic institutes.[46] Nor did this concept find a place in political economy, perhaps because cameralists focused on state administration rather than markets. And the term *technology* is absent from the works of the classical British political economists: from Adam Smith through John Stuart Mill, none used it in their theories.[47]

But even as cameralism became passé, the terminology of *Technologie* never completely disappeared. German universities, for example, continued to appoint chairs of *Technologie*. German chemists embraced the word, creating the field of *chemische Technologie*, or industrial chemistry, while engineers used the term more generally for fields dealing with the transformation of industrial materials.[48]

But though the term survived in German polytechnic schools, *Technologie* did not refer to technical instruction as a whole. Instead, it was just one division of the curriculum, corresponding roughly to what would later be called mechanical and industrial engineering. For example, the 1827 article "Polytechnik" in the Brockhaus encyclopedia listed *Technologie* as one of five main subjects of instruction, the others being chemical technology (*chemische Technik*), physics, mathematics, and a subject area roughly equivalent to civil engineering.[49] Similar classifications appear in publications of German-speaking technical schools, such as an 1848 pamphlet on the Hannover Polytechnic School by Karl Karmarsch, one of the school's founders and a key advocate of *Technologie* as a scholarly discipline. Yet *Technologie* was just one of twenty subjects of instruction, specifically *mechanische Technologie*. This field was, according to Karmarsch, primarily focused on production methods for transforming industrial materials into finished products, covering topics like metal- and woodworking, spinning and weaving, and papermaking.[50]

Bigelow's *Elements of Technology*: A False Origin Story

Despite the rise of *Technologie* as both a term and a concept in early nineteenth-century German, its English equivalent, *technology*, remained marginal in Anglo-American discourse. While the term did become more common, especially in the second half of the century, it remained closely tied to technical education. Aside from the five articles I noted above, no scholar used the term to make a substantive theoretical point or to do any significant boundary work.

Insofar as *technology* had a place in nineteenth-century English, it was mainly as a translated residue of the German cameralist tradition. Yet links to this tradition were generally indirect. Few cameralist works were translated into English, and those that were had little to say about technology. Beckmann's multivolume *A History of Inventions and Discoveries* was published in English in 1797, going through multiple editions well into the nineteenth century, but this work mentioned *technology* only in passing.[51] Also significant was Friedrich Knapp's *Lehrbuch der chemischen Technologie* (1847), a detailed, two-volume compendium of chemical processes used in industry. This book was almost immediately translated into English by two British chemists who had studied with the famous German scientist Justus Liebig.[52] Knapp drew explicitly from Beckmann's *Anleitung* (1777) for his definition of *Technologie*. As rendered in the English translation, "'Technology' . . . comprises in its literal signification the systematic definition ($\lambda o\gamma o\varsigma$) of the rational principles upon which all processes employed in the arts ($\tau\varepsilon\chi\nu\eta\varsigma$) are based."[53] Through the influence of Knapp's textbook and similar works, *technology* did appear occasionally in works written in English, for example in the phrase "technological science" used in the petroleum industry.[54] *Technology* also was used occasionally in its older meaning as a description of technical terminology.[55]

However, a different version of *technology*'s origin story was embraced by a generation of historians after World War II. These historians pointed to Jacob Bigelow's book, *Elements of Technology* (1829), as the starting point for the modern English meanings of the term.

In 1816 Bigelow, a physician and botanist, received an appointment at Harvard as the first Rumford Professor of the "application of the sciences to the useful arts." He was just thirty years old at the time, and for the next ten years he gave a series of lectures that surveyed the practical arts, including writing, printing and printmaking, machinery, and preservation of organic substances. Agriculture and medicine were notably absent, even though at this time both were included among the useful arts. Bigelow's *Elements of Technology* was based on these lectures, revised into a substantial five-hundred-page treatise with twenty-two plates of illustrations.

This book would most likely be forgotten if not for its title. It is a dry compendium of details compiled primarily from specialized treatises. Bigelow made no claim to any direct knowledge of the arts he described, and the level of description was too general to be of much use to a skilled artisan.[56] As Bigelow explained, his purpose was not to instruct artisans but rather to familiarize the college student with "the common technical terms which he meets with in a modern book of travels, or periodical work."[57]

In his preface, Bigelow tried to explain why he used the obscure term *technology* in the book's title. He said he chose this word to describe the "various topics" addressed in its pages, adopting "the general name of Technology, a word sufficiently expressive, which is found in some of the older dictionaries, and is beginning to be revived in the literature of practical men at the present day." He did not define this term explicitly, but implied that *technology* encompassed "an account . . . of the principles, processes, and nomenclatures of the more conspicuous arts, particular those which involved applications of science, and which may be considered useful."[58] This usage was fully consistent with the contemporary dictionary definition of *technology* as a description of or treatise about the arts.[59] After this rather vague discussion in the preface, however, Bigelow ignored the term for the rest of the book.

Bigelow never clearly explained why he worded the book title as he had. Though rare, *technology* was present in American and British periodicals of the early nineteenth century, including publications Bigelow was likely to have read. He could have encountered the term in its meaning as technical terminology in the medical literature, but that was not how he used it.[60]

Instead, Bigelow most likely borrowed the term from German cameralists, though without acknowledgment. His *Elements of Technology* was quite similar to cameralist surveys of the practical arts, particularly Beckmann's *Anleitung zur Technologie* and Poppe's *Handbuch der Technologie*.[61] Yet Bigelow did not cite Beckmann or other cameralist works, which suggests that he did not know German. Nevertheless, he may have encountered *technology* in its cameralist sense in reviews of German literature in English-language periodicals. For example, in the early nineteenth century the London *Monthly Review* had a regular column on German publications with a subsection that covered "technology and commerce." Other British and American periodicals of the day mentioned technology as a field of instruction in German technical education.[62] Given the similarity between his work and cameralist *Technologie*, Bigelow could very well have picked up the term from these English-language summaries.

Regardless of how he discovered the word, what does his book tell us about the significance of *technology* at the time? Very little, actually. But many historians have claimed otherwise. After World War II, when *technology* emerged as a powerful keyword in American culture, historians ascribed new significance to Bigelow's *Elements*. Among the first to do so was Hugo Meier, a pioneering historian of technology who discussed Bigelow in a dissertation completed in 1950 at the University of Wisconsin.[63] In a 1957 article, "Technology and Democracy," Meier identified the 1829 publication of Bigelow's *Elements* as the moment when "the new term entered into popular usage." Yet nowhere in Meier's quote-laden article did he remark that aside from Bigelow, none of his sources actually used the term.[64] *Technology* was a keyword of Meier's era, not of his historical subjects'.

Given the power of the term *technology* in the post–World War II era, Meier's assessment of Bigelow's significance had a deceptive plausibility. It was repeated so often that scholars stopped citing anyone but Bigelow himself.[65] Even the brilliant intellectual historian Perry Miller waxed poetic over Bigelow's contribution, as Howard Segal notes. Miller hailed the *Elements* as "a major document in American intellectual development." He lamented the neglect of Bigelow, whom he described as "a prophet more relevant to the later economy than either Emerson or Jefferson." He insisted that Bigelow had "in effect declared the independence of the nineteenth century from the eighteenth—of the practical, materialistic, hardheaded, utilitarian age from that of ideology and benevolence."[66] But Bigelow had done nothing of the sort. *Elements* was a tertiary work of compilation, a useful exercise no doubt but not a major shift in nineteenth-century American thought. But until Ruth Oldenziel questioned the Bigelow myth in 1999, only a few historians dissented from the received wisdom, and then only in passing.[67]

Bigelow's *Elements* did little to popularize the term *technology*. Most reviewers found the title odd. The *Journal of the Franklin Institute* took pains to note that "the work is not, as some might be led from its title to imagine, a mere *technological vocabulary*, containing definitions of words used in the arts."[68] Another reviewer complained that "[the] word Technology gives but an imperfect idea of the contents of this volume," and suggested that a better title would have made clear Bigelow's focus on "the scientific and practical principles of many of the useful, curious, and elegant arts."[69] A third reviewer, the inventor Donald Treadwell, also found *technology* "not so familiar in our language as could be desired." Treadwell, who would later succeed Bigelow as the Rumsford Professor at Harvard, did agree that "some word of the kind . . . has become necessary" to convey the topics in Bigelow's book, and he concluded that the term would "probably come into general use."[70]

But despite Treadwell's prediction, the term did not become common in the nineteenth century. Bigelow apparently recognized the marketing problem posed by the title of his book, because he renamed its third (1840) edition *The Useful Arts*, and dropped any mention of *technology* in the text itself. Google n-grams also fail to show a clear increase in frequency of the term between 1830 and 1870. *Useful arts* remained roughly an order of magnitude more common than *technology*, and *invention* over two orders more common.[71] Before the Civil War, *technology* clearly was an obscure term of no theoretical significance.

Despite the limited influence of Bigelow's turgid compendium, the book does shed light on changing attitudes toward the mechanical arts. Much like Beckmann's *Anleitung*, it was not written for artisans. Nor was it sufficiently detailed to be useful for owners and managers of the new manufacturing plants sprouting up all over New England.[72] Moreover, Bigelow's book did not quite fit into the long tradition of popular works in natural philosophy, which often described ingenious devices and industrial processes.[73] Instead, this text represented an appropriation of artisanal knowledge for the middle class, not to help in managing production but rather as an assertion of middle-class cultural authority over the useful arts. In this sense, Bigelow's choice of title did anticipate the future.[74]

MIT and the Concept of Technology

Even if Bigelow can't be credited with introducing *technology* in its modern meanings, there is a far stronger claimant for this honor: the Massachusetts Institute of Technology, chartered in 1861. MIT, although not the first school of higher technical education in the United States, was indeed the first to adopt the title "Institute of Technology."[75] Before MIT, a more common name for technical schools in the United States was "polytechnic," after the École polytechnique in Paris.

But the founding of MIT still did not represent the arrival of technology as a significant concept for understanding the modern world.

Planning for the institute had begun in 1847, but *technology* became part of its name only in 1860. Before then, educators had framed their plans for MIT in terms of a "polytechnic institute" or a "conservatory of art and science."

Yet the name change was apparently of no great significance; there is no evidence of any deliberation over the shift from *polytechnic* to *technology*. Historians of MIT have puzzled over this choice for decades. Neither published documents nor archival sources have revealed the reasons for the name change.[76] However, these documents do make it clear that *technology* did not serve as an organizing concept in discussions about the mission of MIT, with the exception of one celebratory speech given by Jacob Bigelow.

The founding of MIT needs to be situated within the broader history of higher technical education in the United States. The standard historiography focuses on a few pioneering institutions that created the foundations of American engineering education.[77] This story begins with the creation of the United States Military Academy at West Point in 1802, followed in 1817 by the establishment of a curriculum at West Point modeled on the École polytechnique. Despite its military purpose, West Point supplied more school-trained civil engineers in antebellum America than any other institution. (However, during this period most civil engineers still learned their profession through informal apprenticeships.) In the late 1840s Benjamin Franklin Greene, president of the Rensselaer Institute, established a three-year university curriculum for civil engineers, later extended to four years. This curriculum, derived from Greene's careful study of higher technical education in Europe, became a model for American engineering education. Greene later added *polytechnic* to Rensselaer's name.[78]

MIT emerged from this tradition. The first person to propose creating a polytechnic institute in Boston was William Barton Rogers, a professor of natural philosophy at the University of Virginia. Rogers developed this proposal in response to an 1846 letter from his brother Henry, who suggested adding a "School of Arts" to Lowell Institute in Boston. Both men had been advocating increased instruction in the sciences and "useful arts" in schools.[79] When Henry sent William his suggestion, William drafted a sort of manifesto titled "A Plan for a Polytechnic School in Boston."[80]

Rogers's plan is a revealing document. Most important, it reveals how utterly irrelevant the concept of technology was for the discourse of technical education in antebellum America. The word is entirely absent from the proposal. Instead, Rogers framed his goals as supporting the arts, manufactures, and invention. He proposed a two-stage curriculum to achieve these goals. The first stage was to provide a foundation in "general physics," which included the sciences of mechanics, electricity, heat, magnetism, and chemistry. The second stage was to be "entirely practical," emphasizing instruction in laboratories and workshops.[81]

This curriculum was based on Rogers's ideas about the relationship between science and the arts. He believed that practice was subordinate to theory. The "true and only practical object of a polytechnic school" was to teach not the details of particular trades but rather their "scientific principles." Since 1800, "progress . . . of every branch of the arts . . . is but the result of that general diffusion of a better knowledge of physical laws which has flowed from the researches and teachings of men specially devoted to natural science." Better knowledge of scientific principles would lead to fewer "utterly barren inventions, the laboured contrivances of [an] acute but undirected or misguided mind." He insisted on "the necessity of scientific principles" in construction, and he claimed, quite implausibly, that even in ceramics, "every step is but an application of some well-known scientific principle."[82]

Rogers enshrined this subordination of practice to theory in the very structure of the proposed school, with two "professors" in the "scientific department" to supervise the two "subordinate instructors," who taught practical subjects. He predicted that this polytechnic would grow "into a great institution comprehending the whole field of physical science and the arts," one that "would soon overtop the universities of the land . . . in all branches of positive knowledge."[83]

Such claims about the utility of natural science were hardly new, having been a staple of the British Industrial Revolution, though they were somewhat less common in the United States.[84] Yet when transformed into an educational program, William Rogers's "Plan for a Polytechnic School in Boston" mirrored America's growing divide between mental and manual labor.[85] While he claimed that his school would "elevate" the minds of the "operative classes," he wasn't aiming his curriculum at artisans or factory workers. (*Operatives* was a standard term for factory workers in the nineteenth century.) Instead, this was a curriculum meant to serve "the mechanician, chemist, manufacturer or engineer," along with the architect.[86] By "mechanician," Rogers meant not a humble artisan but someone more like a mechanical engineer. He intended his polytechnic, therefore, to serve the emerging middle classes by giving them both the practical skills and the cultural authority needed to assert control over skilled technical work.

William Rogers's original proposal for a polytechnic went nowhere. It took more than a decade for Rogers to get another chance to create his polytechnic in Boston, but by then it had become transformed into an "institute of technology."

He got this opportunity in 1860, as part of a debate related to the filling of Boston's Back Bay, a project that created both new income and new land for the State of Massachusetts. Rogers had resigned his chair at the University of Virginia and moved to Boston in 1853 to pursue independent research and writing.[87] He was peripherally involved in an early proposal that sought to reserve part of Back Bay for a Conservatory of Arts and Sciences. This proposed conservatory consisted of a set of museums devoted to a range of subjects, from agriculture to fine arts. When this proposal failed, its supporters called on Rogers to write a new memorial to the state legislature requesting land and funding. His new proposal, submitted in January 1860, continued to focus on museums, though with a stronger emphasis on useful arts. But Rogers included a new element: a suggestion that the conservatory eventually include a "comprehensive Polytechnic College" on the model of l'École centrale des arts et manufactures in Paris, a school dedicated to "la science industrielle." This aspect of the memorial was largely aspirational, though the aspirations were clearly Rogers's.[88]

Although this second proposal also foundered in the legislature, the undeterred Rogers led the effort to draft a third version. It was only in this version, completed in October 1860, that he embraced the idea of an "Institute of Technology." In this proposal, Rogers stressed the educational mission of the overall enterprise. The new institute was to have three major branches: "a Society of Arts, a Museum or Conservatory of Arts, and a School of Industrial Science and Art." For the Society of Arts, Rogers proposed a research institute that would publish a "*Journal of Industrial Science and Art*." For the museum complex, he stressed that its purpose would be for "practical instruction" rather than collecting a "multitude of objects."[89] Overall, the description of these first two branches had changed little from the January proposal.

But the third branch, a School of Industrial Science and Art, was a new element in the proposal, which in many ways represented a more refined version of Rogers's "Plan for a Polytechnic School in Boston" of 1846. The school was to embrace all the arts relevant to modern industry. Rogers sketched out departments focused on design, mathematics, physics, chemistry, and geology, but, surprisingly, none in engineering. The design department even included instruction in the fine arts, though its main purpose was to teach technical drawing and skills needed in "ornamental branches of manufactures." Aside from that department's practical focus, the other departments were to follow the two-stage process that Rogers had outlined in his 1846 plan: teaching fundamental principles first, and practical applications second. For example, the "first object" of the physics department was "to impart a thorough knowledge of the fundamental principles of the several branches of physics," and only then would students receive "more strictly practical instructions, as much as possible, under the guidance of these primary truths." Rogers insisted that it "cannot be doubted" that instruction in scientific principles would yield "practical advantage," both directly and also by inculcating habits of "close observation and reasoning."[90]

Despite Rogers's role in founding MIT, his proposal for a "School of Industrial Science and Art" was hardly visionary. It represented little more than a dogmatic assertion that instruction in natural philosophy and mathematics would yield practical utility. It contained nothing remotely resembling a draft curriculum. It revealed almost nothing about the content of the practical components of instruction. And despite its support from leading Boston industrialists, the proposal said nothing concrete about how the school would be of use to industry.[91]

What was most novel in the proposal was the name it gave to the overall organization. The suggested tripartite entity, consisting of the society, museum, and school, was to be known as the Massachusetts Institute of Technology. After the school opened in 1865, the society and museum soon faded in relative significance.[92] And MIT became the first school of higher education in the English-speaking world to include *technology* in its name.

Rogers apparently came up with the new name for the project during the summer of 1860, while drafting the new proposal for the state legislature.[93] Already in the January 1860 plan, he had proposed a department of "Mechanics, Manufactures, Commerce and *Technology* in general" while also predicting that with suitable support, this department would "rank with the more comprehensive *technological* museums of the Old World."[94] These occasional uses of this obscure word gave no hint that Rogers would soon include it in the name for the entire enterprise. Yet once he chose this novel name, it provoked absolutely no discussion—even during the extensive debates over the legislation that granted the proposed Institute of Technology its charter and land in April of 1861.[95]

The only sensible conclusion to be drawn from this evidence is nega-

tive: this use of *technology* had little significance for the historical actors themselves. Of course, Rogers's choice did shape the subsequent
use of the term, particularly as other technical schools emulated MIT's
name. But even then we should not exaggerate the significance of the
word. Polytechnic schools and institutes continued to proliferate without *polytechnic* becoming an important scholarly term. The *technology*
in MIT was little more than a moniker, a term sufficiently erudite and
foreign to convey intellectual authority. Perhaps Rogers had rejected
polytechnic for the entire three-part project because that term was used
only for instructional institutions. Rogers needed something both less
pedestrian than an "institute of arts and sciences" and sufficiently
vague to cover the entire enterprise. *Technology* was grand enough for
the purpose. As Leo Marx has suggested, *technology* was in effect an
empty category in the nineteenth century, but one that could be filled
with meaning by the cheerleaders of industrial modernity.[96]

Only one scholar seems to have cared about the name: Jacob Bigelow. Bigelow was not involved in the founding of MIT, although a
commencement speaker in 1890 spuriously attributed the choice of
its name to him.[97] Adding to the confusion, a wealthy textile inventor named Erastus Bigelow did play a role in establishing the institute;
however, he apparently was unrelated to Jacob. Only after its founding
in 1861 did Jacob Bigelow become involved with MIT, eventually serving as one of four vice presidents.[98]

Soon after instruction began at the institute in the fall of 1865,
Bigelow spoke before its Society of Arts. His presentation stressed the
significance of MIT's name, and pugnaciously defended the school's
curriculum against traditional education in classical languages and
literature. Bigelow opened his remarks by claiming that he had introduced *technology* into English usage with his *Elements of Technology*. Now, he continued, this word "gives name . . . to a vigorous and
popular institution" while also denoting "a great and commanding
department of scientific study in every quarter of the civilized world."
In other words, technology for Bigelow was, as the dictionaries of the
time made clear, a science. He insisted that this field "has advanced
with greater strides than any other agent of civilization," and was more
responsible for the "progress and happiness of our race" than anything
except Christianity.[99]

Bigelow's rhetoric is surprisingly modern. He equated technology
with applied science and credited both for all recent progress in the
industrial arts. But this apparent modernity is an artifact of his rheto-

ric. Bigelow elevated technology as both modern and useful in order to denigrate study of the classics as neither. Aside from his use of the term *technology*, his argument was a common one among advocates of new knowledge in the early industrial era. Bigelow actually said little of substance about technology, aside from waving it around as a talisman of useful knowledge. But he was not arguing for narrow utilitarianism. "Technology," he admitted, was but one of many "progressive" fields of knowledge that advanced the "dominion of mind over matter." The real problems were not the division between useful and useless knowledge, or the danger of pseudosciences such as "spiritualism, homeopathy, and mormonism." Instead, the true enemy was the "cumbrous burden of dead languages, kept alive through the dark ages and now stereotyped in England by the persistent conservatism of a privileged order." Technology for Bigelow was primarily a rhetorical cudgel to use against defenders of Greek and Latin.[100] Nevertheless, by explicitly linking technology to the concept of progress, Bigelow provided a preview of what would later become a powerful aspect of the concept.

But why did Rogers feel comfortable enough with *technology* to select this term in naming the institute? If the word had been truly odd, it would have elicited some comment, as happened with Bigelow's *Elements* when the book first appeared in 1829. In fact, the term had become somewhat more common since that time, thanks in part to the translation into English of the German word *Technologie*, one of two German terms that would shape the meaning of *technology*.

By the mid-nineteenth century, Technologie occupied a distinct place within higher technical education in Germany. As a subject in the polytechnic curriculum, it was not meant to elevate artisanal workers to managerial positions. Instead, it represented a middle-class appropriation of artisanal knowledge, training professionals to superintend industrial production. In this cameralist sense, Technologie fit easily into Rogers's vision for MIT. Even though there is no evidence that he had encountered the German term before naming the institute, he easily could have encountered it indirectly.

In the years before the founding of MIT, the first German-American intellectual exchanges were flourishing. American intellectuals found German scholarship attractive in part because it helped lessen their dependence on Britain, thus supporting their self-conscious project to create an American national culture. In Boston, influential Americans were forging strong ties to German scholarship. Among them were three men closely associated with Harvard: Edward Everett, George

Bancroft, and George Ticknor, all of whom had graduated from the University of Göttingen.[101] Rogers no doubt knew them personally, particularly Everett.[102]

Yet he probably learned nothing about the word *Technologie* from these Göttingen graduates, as their interests lay elsewhere—even though a generation before they studied in Germany, Johann Beckmann had founded Technologie as an academic discipline in Göttingen. Instead, Rogers most likely encountered the term in the work of George Wilson, the first scholar to hold the Regius Chair of Technology at Edinburgh University. Although professors of Technologie were widespread in German universities, Wilson's chair was the first in a major English-language institution of higher education. The chair was connected to the new Industrial Museum of Scotland, whose director was also to hold the technology chair and teach that subject at the university.[103]

It's not clear why the Senatus Academicus, the governing body at Edinburgh, chose to name the chair *technology*, but Wilson repeatedly felt compelled to explain this unfamiliar term.[104] In doing so, he elaborated what is probably the first serious discussion of the meaning of *technology* in English. In his 1855 inaugural address as Regius professor, he began by noting that *technology* "is so unfamiliar to English ears, and so inexpressive to English minds, that I must, at the very outset, explain what the branch of knowledge is which I am called upon to profess." He noted the term's German origins, which he credited to Beckmann, and the rise of "Chairs of Technology" in German universities. Drawing from dictionary definitions, Wilson described *technology* as "the Science of the Arts, or a Discourse or Dissertation on these," contending that its primary meaning should be as a science, not a treatise, in conformity with fields like theology, geology, and conchology.[105]

Wilson sought to make organizational and intellectual boundaries coincide in his new academic field. His first step was to insist on the status of technology as science, not practice, and thus a suitable subject for the university curriculum. "Technology . . . implies the Science; or Doctrine, or Philosophy, or Theory of the Arts. Its object is not Art itself, i.e., the *practice* of Art, but the principles which guide or underlie Art, and by conscious or unconscious obedience to which, the artist secures his ends." Despite the seemingly broad scope of this field, Wilson delineated its boundaries to avoid competing with more established disciplines. Technology, he asserted, excluded fine arts, covering only

"Utilitarian Arts," and not even all of these. He also excluded medicine from technology, and divided the remaining "industrial arts" between physical and chemical topics. Wilson sought to reassure his colleagues that he would "faithfully respect the rights of my brother professors" by not infringing on their intellectual territory. He then listed the subjects he planned to cover, which included, among others, practical chemistry and the "economic" applications of electricity, photography, metallurgy, building materials, and food (excluding agriculture, which was already covered by another professor).[106]

In delineating technology, Wilson appropriated key aspects of the cameralist version of the concept. For him, technology was an academic field, with specific boundaries defined by existing disciplines and courses of instruction. In scope, it covered much of what the term includes today, but with an explicit difference. *Technology* did not refer to the arts themselves, but to their systematic study and teaching.

George Wilson was not only Edinburgh's first professor of technology but also its last, at least for the nineteenth century. His sudden death in 1859, after four years in the position, led to the demise of the chair.[107] In fact, technology in general never caught on as a recognized subject in English-language technical education.[108]

Despite the truncated history of the Regius Chair of Technology, William Barton Rogers was almost certainly aware of Wilson's work. In 1855, well before the founding of MIT, he informed his brother Henry that he "eagerly" read every issue of the *Edinburgh New Philosophical Journal*.[109] In 1857, this journal published Wilson's second major address as Regius professor, one that further elaborated his concept of technology as a form of applied science.[110] In that same year, William and Henry Rogers visited Edinburgh University. In 1863, after the founding of MIT, William wrote to Henry, who was then in Glasgow, for information on the "Technology department at Edinburgh."[111] Such links make it quite clear that Rogers, given his interest in higher technical education, would have noticed Wilson's role in the new chair of technology.

Does my account solve the mystery behind MIT's name? Not really. Absent some undiscovered archival evidence, there is probably no definitive answer to how it was chosen. Rogers's choice of the uncommon term *technology* makes vague sense in its specific context, especially as a moniker for the institute's entire enterprise of society, museum, and school. The term was just familiar enough to avoid perplexity.

Ultimately, MIT's naming was significant mainly for its lack of significance. Historical events are sometimes like that. Even when they are full of sound and fury, such tales may signify nothing. Or more precisely, sometimes events of minor significance to the historical actors are granted retrospective significance by historians.

Nevertheless, Rogers's use of *technology* in MIT's name did make the term more common in the last third of the century. The *New York Times*, for example, published thirty articles with *technology* in the headline from 1870 through 1899. Of these, some twenty-two were about MIT, and over half of those concerned athletics ("Williams 12, Technology 0"). Five more articles dealt with Stevens Institute of Technology, which was founded in 1871 and was the first institution of higher technical education to emulate MIT's name. Aside from Stevens, few other technical colleges had *technology* as part of their name before the twentieth century.[112] Still, MIT and other schools of technology helped give the term a deceptive familiarity, even if few people knew what it meant.

It thus makes sense that *technology* remained closely linked to higher technical education for the remainder of the century. This link encouraged an elitist view of technology as a field of applied science whose subject matter was the industrial arts. Innovators in technical education were typically academics and elite physicians rather than practitioners of the industrial arts. Technical educators rarely defined their task as creating a unity of theory and practice. Instead, they sought to teach middle-class professionals how to apply theory to technical processes, often involving the supervision of manual workers.[113]

Technology as Industrial Arts in the Nineteenth Century: An Example of Misuse?

But what of *technology*'s other main twentieth-century meaning, its definition as industrial arts or material culture? If the first edition of the *Oxford English Dictionary* is to be believed, this meaning was already present in the mid-nineteenth century. The "T" volume of the *OED*, published in 1919, provided two main definitions of *technology*. The first was fully consistent with the term's nineteenth-century meanings: "A discourse or treatise on an art or arts; the scientific study of the practical or industrial arts." But the *OED* provided a second definition: *technology* as "the practical arts collectively."[114]

This is the earliest instance of *technology* defined as the industrial arts that I have found in an English-language dictionary. The *OED* gave two examples of this usage, one from 1859 and the other from 1864. Both were taken from descriptions of African travels by the British explorer Richard F. Burton. In the 1859 quotation, Burton described a kind of copal, a tree gum, that was exported to India and China, although "little valued in European technology." Elsewhere in the same text, Burton referred to the "larger asclepias" (milkweed), which the natives "ignore [despite] its use in Arab technology."[115]

The *OED*'s second example was taken from a particularly distasteful essay in which Burton disparaged Africans and defended slavery. *Technology* appears at the end of a paragraph worth quoting in full:

The negro has never invented an alphabet, a musical scale, or any other element of knowledge. Music and dancing, his passions, are, as arts, still in embryo. He cultivates oratory; and so do all barbarians. He is eternally singing, but he has no idea of poetry. His painting and statuary are, like his person, ungraceful and grotesque; whilst his art, like his mind, is arrested by the hand of Nature. His year is a rainy season; his moons have no names; and of an hour he has not the remotest conception. His technology consists of weaving, cutting canoes, making rude weapons, and in some places practising a rough metallurgy.[116]

Such racist sentiments were widespread among Europeans at the time, as Michael Adas has shown, but this appears to be the first time *technology* was explicitly connected to racial hierarchies.[117]

The *OED*'s editors appear to be correct in assigning the meaning of industrial arts to Burton's usage of *technology*, but such instances remained anomalous before the twentieth century. Even in Burton's own works, the word remained uncommon. The 1859 example appeared in the *Journal of the Royal Geographical Society*, yet in the fifty-year run of this journal (1831–80), the word *technology* appeared only in this one report by Burton.[118] *Technology* in the sense of the industrial arts was really a case of misuse, a slippage in meaning that conflated a field of study with the object of study. Similar examples abound in the nineteenth century, such as the use of *ornithology* to refer to birds directly rather than the study of birds.[119] But such misuse did not make *ornithology* into a synonym for *birds*, and it did not make *technology* equivalent to the industrial arts. By the time the "T" volume of the *OED* was prepared shortly before World War I, the industrial-arts definition of *technology* was already emerging, thus priming the dictionary's editors to project this definition onto earlier uses.[120]

Technology and *technological* did become more common in academic discourse of the later nineteenth century. There are a few other examples of slippage where the term was used to refer to the useful arts rather than the science of these arts.[121] As an adjective, *technological* was often used in the sense of *industrial* or *applied*, for example the phrase "technological science."[122] Nevertheless, the original meaning of *technology* as "the science of the arts" remained dominant through the end of the nineteenth century.

The continued usage of *technology* to mean "the science of the arts" is explicit in late-nineteenth century anthropology. In 1879, the famous explorer of the American West John Wesley Powell helped found the Anthropological Society of Washington, which quickly became the center of academic anthropology in the United States. In 1882, the Anthropological Society amended its constitution to add "Technology" as one of its main sections, along with "Somatology, Sociology, Philology, Philosophy, Psychology." This change in effect replaced *archaeology* with *technology*, probably in order to include the material culture of the present as well as the past. But *technology* in this context did not mean the industrial arts, or even the application of natural science to the arts. In his 1883 presidential address to the society, Powell defined each of the main branches of anthropology, including "technology, or the science of the arts."[123] For him, technology remained a branch of science, a fact he reaffirmed repeatedly for the remainder of the century.[124] This definition of *technology* remained largely unchallenged until the beginning of the twentieth century, when American social scientists began translating the German concept of Technik as "technology."

This chapter has been largely about negatives, focusing on changes in meaning that did not happen. Yet in the twentieth century, *technology* did acquire a new meaning, one that the *Oxford English Dictionary* defined correctly as "the practical arts collectively."[125] How did *technology*, a term that had for centuries denoted a type of science, become a substitute for *mechanical arts*? The answer to this question requires a detour through continental Europe, especially German-speaking countries, and specifically through the German engineering community.

SEVEN

Discourse of Technik: Engineers and Humanists

In German-speaking countries during the nineteenth century, a concept arose that replaced both the medieval category of mechanical arts and the cameralist discourse around Technologie. This concept was Technik. Starting in the mid-nineteenth century, German-speaking engineers adopted the word *Technik* as a core part of their professional identity. They used the term to claim all the arts of material production as the province of the engineer. Later in the century, this concept was adopted by humanist intellectuals, especially in the emerging social sciences. In the early twentieth century, English-speaking scholars drew from this academic understanding of *Technik* to reshape the English term *technology*, expanding the English term to cover the industrial arts themselves, not just the science of these arts.

The history of the concept of Technik is embedded in the tension between instrumental and cultural approaches to technology. This tension was implicit in older concepts such as mechanical arts and Technologie, but it became explicit in the discourse of Technik. German-speaking engineers produced theoretical writings about Technik as a defense of their social status against perceived hostility from humanist intellectuals. These engineers did so by developing an explicitly cultural understanding of technology, framed in terms of a relationship between Technik and *Kultur.*

The full history of the German concept of Technik is

far too rich to treat adequately here. Technik was the focus of a huge literature beginning in the late nineteenth century, but intellectual historians have written little in English about this literature, especially neglecting writings by engineers.[1] A few works have received attention far out of proportion to their significance, such as Heidegger's essay "Die Frage nach der Technik" (1954).[2] In this chapter, I focus on four works that are more representative of ideas about Technik: Marx's *Das Kapital*; a series of articles in German by the Russian engineer Peter Engelmeyer; an article by the German sociologist Werner Sombart; and Max Weber's discussion of Technik in his *Economy and Society*.

Technologie and Technik in Karl Marx's *Das Kapital*

Karl Marx published the first volume of *Das Kapital* in 1867. This book represents the most mature and systematic expression of his thought, prompting a vast literature of commentary and critiques, including studies of Marx's views on technology. But few scholars have recognized that *Das Kapital* marks an inflection point in the history of German concepts. In some ways, it represents the last gasp of the cameralist concept of Technologie. But Marx's book also provides evidence of a new discourse emerging around the concept of Technik.

In his mature work, Marx had plenty to say about topics that we now classify as technology.[3] But he did not have a theory of technology in general, because, like other political economists of his era, he had no general term to encompass the many aspects of technology relevant to economics, let alone its noneconomic aspects. When Marx invoked what we might describe as the technology of industrial capitalism, he had to use a list rather than a single term. For example, he noted that the productivity of labor depended on a variety of factors, including "the average amount of skill of the workmen, the state of science, and the degree of its practical application [*ihrer technologischen Anwendbarkeit*], the social organisation of production, the extent and capabilities of the means of production, and by physical conditions."[4] This jumble of ideas could all be included within a present-day economic concept of technology, except perhaps "physical conditions" (*Naturverhältnisse*).

Of course scholars can, as an interpretive exercise, construct Marxist theories of technology, mining Marx's often gnomic comments for insights into a general theory consistent with his thought.[5] But if technology was actually a central concept for Marx, he would have defined

it. In *Das Kapital*, Marx engaged in quite a bit of conceptual innovation, devoting many pages and sometimes entire chapters to his key
terms, such as *commodity* (*Ware*), *capital, labor power* (*Arbeitskraft*), *machinery*, and finally *value*, which included *use value, exchange value*, and
surplus value (*Gebrauchs-, Tausch-*, and *Mehrwert*). But with the exception of one footnote (discussed below), nowhere in Marx's corpus does
he articulate a general concept of technology, even one restricted to the
capitalist production process.

In his lack of a clear concept of technology, Marx was no different
from his predecessors, classical political economists like Adam Smith
and David Ricardo.[6] He was, in a real sense, the last of the classical
political economists. As Marx was writing *Das Kapital*, William Stanley Jevons was working on the beginnings of a radically new approach
to economics, which Jevons described as "a Calculus of Pleasure and
Pain," later known as marginal utility theory, the core of neoclassical
economics. Marginal utility theory dispensed with the heart of Marx's
approach, the labor theory of value, which had been a key part of political economy since Adam Smith.[7]

Yet Marx did use the term *Technologie* in its cameralist sense as a field
of science (*Wissenschaft*).[8] In his research for *Das Kapital*, he read extensively in the work of Beckmann and his followers. Their work, along
with that of British philosophers of industry such as Andrew Ure and
Charles Babbage, was central to Marx's understanding of modern large-
scale industry.[9] However, Technologie was not central to Marx's discussion of the material side of capitalist production, which he framed primarily in terms of "machinery," "modern industry [*große Industrie*],"
and "forces of production [*Produktivkräfte*]." He was, in fact, the last
major scholar in the nineteenth century to engage seriously with the
cameralist concept of Technologie, a concept that was already outdated
by the time he composed *Das Kapital*. Even Marx himself made a significant shift away from the term between the first and second editions
of *Kapital* (see below).

Nevertheless, Marx did contribute something important to the concept of Technologie by linking it directly to the political economy of
modern capitalist production. For him, *Technologie* referred to systematic knowledge of the production process in large-scale capitalist industry. In a well-known passage, he argued that "modern industry rent
the veil that concealed from men their own social process of production. . . . The principle which it pursued, of resolving each process into
its constituent movements, without any regard to their possible execution by the hand of man, created the new modern science of technol-

ogy [*die ganz moderne Wissenschaft der Technologie*]." According to Marx, this transformation of the labor process through technology involved the "technological application of science [*technologische Anwendung der Wissenschaft*]." He linked the productivity of labor to the developmental stage of science ("Entwicklungsstufe der Wissenschaft") and its technological application ("technologischen Anwendbarkeit").[10]

Marx's invocation of the word *Technologie* in these passages is compatible with cameralist usage, though distinguished by its explicit application to modern capitalist production. But his most detailed discussion of Technologie as a concept implied a much broader scope for technology in social theory. This discussion is contained in a lengthy footnote in the machinery chapter in *Das Kapital*. Marx began by commenting that "a critical history of technology would show how little any of the inventions of the eighteenth century are the work of a single individual." He lamented the lack of such a book, a clear criticism of the cameralist works he had mined for this chapter, among them J. H. M. Poppe's *Geschichte der Technologie* (1807–11) and Beckmann's *Beyträge zur Geschichte der Erfindungen* (1780–1805).[11] He then constructed an analogy to Darwin, who had portrayed "the history of Nature's Technology, i.e., . . . the formation of the organs of plants and animals." Marx called for a similar history of "the productive organs of man," which led him to his most famous statement on the nature of technology: "Technology reveals the active relation of man to nature, the direct process of the production of his life, and thereby it also lays bare the process of the production of the social relations of his life, and of the mental conceptions that flow from those relations."[12]

Marx's use of *technology* in this passage seems very close to its present-day meanings. He made the term applicable to the entire history of the human species, not just the capitalist era. Furthermore, he seemed to use *technology* to refer to the useful arts themselves rather than the principles of the arts. Such a reading would be misguided, however. *Technology* is not itself the "process of production" but rather "discloses [*enthüllt*]" this process. It is this "science of technology," the systematic knowledge of production, which uncovers the material basis of social relations. This analytical method, insisted Marx, was "the only materialistic, and therefore the only scientific one."[13]

Marx's understanding of Technologie in *Das Kapital* also had implications for his understanding of skill. For Marx, Technologie was not just the science of production but a bourgeois rather than proletarian science. Such a view was not in itself radical, as Technologie was already viewed as middle-class knowledge. But Marx's "new modern science of technology" specifically served capitalists by allowing them to shift skills from workers to machines, and thus destroy artisanal labor. In Marx's theory, capitalist industry used this science to reduce all forms of labor to abstract labor, that is, labor stripped of the special knowledge and skills used to produce specific use values. And even where skilled labor remained, it was reducible to unskilled labor in theory. Marx insisted that "skilled labour . . . represents a definite quantity of [unskilled] labour," with the ratio established "by a social process that goes on behind the backs of the producers," that is, through market competition.[14] For example, one hour of a skilled machinist's time has the same value as two hours from an unskilled laborer.

In this reductive understanding of skill, Marx was following Andrew Ure, who as noted earlier believed that the application of science to the factory could completely eliminate skilled workers from production.[15] Marx needed this reductive approach to skill not only for his labor theory of value but also for his theory of proletarian revolution. Because capitalist industry reduced skilled, concrete labor to abstract labor, it created the objective conditions for proletarian unity. This unity was a necessary precursor for the revolutionary overthrow of capitalism.[16]

As a consequence of these views, Marx had a strongly instrumental approach to technology. As Amy Wendling argues, this approach denied "the possibilities of real working-class knowledge and agency . . . from any work prior to capitalism's revolutionary demise."[17] Marx looked forward to the minimization of labor under communism, not its restoration as a meaningful, creative activity. His reductive notion of skill obscured the fact that new machines require new skills, both to build and to use them. Given this approach, Marx was unable to conceptualize machines under capitalism as anything but a means for the accumulation of capital. Such a view blinded him to the cultural functions of technology, as seen, for example, in machines such as fanciful automata, virtuoso displays of artisanal machine building that captured the imagination of the eighteenth century.[18]

Marx's use of *Technologie* failed to rescue the cameralist term, which continued to decline in the second half of the century. His famous footnote on Technologie was just one short paragraph in an eight-hundred-page treatise. Subsequent Marxists did not rush to develop a "critical history of technology," although some were influenced by Marx's insistence on the central role of technical processes in production. Much of the work of Max Weber and Werner Sombart, for example, can be read as a response to the perceived technological determinism of Marxist theory. Yet when these and other German-speaking scholars addressed

Marx's analysis of production, they did so using the term *Technik*, not *Technologie*.[19]

Marx himself used *Technik* a few times in *Kapital*. At one point, he referred to the "uninterrupted advance of science and technology [*ununterbrochenen Fluß der Wissenschaft und der Technik*]." Elsewhere, he mentioned the process by which "capitalist production, therefore, develops technology [*Technik*], . . . only by sapping the original sources of all wealth."[20] In these two cases, Samuel Moore and Edward Aveling's 1887 translation of *Capital* rendered *Technik* as "technology." In this context, *technology* meant the technical process of production itself rather than the science of this process. Using *technology* in this way was not standard English at the time, but apparently Moore and Aveling did not feel that there was a better term. This is the earliest instance I have found of such a translation, hinting at the future semantic influence of *Technik* on *technology*.[21]

Marx's use of *Technik* reflected a shift in usage that was just beginning in German-speaking countries. During the last third of the nineteenth century, a broad-based discourse of Technik almost completely replaced the discourse of Technologie. Although Marx did little to develop either of these two terms, his own terminology shifted significantly across the first four editions of volume 1 of *Das Kapital*. As a noun, neither *Technik* nor *Technologie* occurs often in any edition of *Das Kapital*, with *Technologie* occurring 7 or 8 times and *Technik* from 2 to 5. But the picture is quite different with the adjectival forms of these terms. Because *Technik* and *Technologie* were not central to Marx's argument, they appeared more often as adjectives than nouns.[22] *Technologisch* and its inflections occur 54 times in the first edition of 1867, while *technisch* and its inflections occur only 24 times. In the second edition of 1873, the situation is almost reversed, with just 21 instances for *technologisch* but 56 for *technisch*. The trend continued with the third edition of 1883, which was edited by Friedrich Engels based on Marx's notes: 14 instances for *technologisch* versus 71 for *technisch*. In other words, Marx himself was abandoning the largely outmoded term *technologisch* from the first to the second editions, a process continued by Engels in the third edition.

This clear shift in terminology is rather strange, as it was made without any acknowledgment. In many cases, Marx directly replaced *technologisch* with *technisch*, especially in references to *technologische Bedingungen, Grundlage*, or *Basis*.[23] These were all references to the technology of production. These references could have provided Marx with a starting point for theorizing more broadly about technical knowledge and skill. Instead, he invoked *technological* or *technical* aspects of production as if the terms needed no explanation. His terminological confusion might have been related to his long exile in London, where he lived from 1849 until his death. This would explain his use of the already dated term *technologisch* from the German cameralists, followed by his abandonment of the term in the second edition of *Das Kapital*.[24] Whichever term he used, Marx treated both *Technik* and *Technologie* instrumentally, as the means used by capitalists to increase the extraction of surplus value from the labor process.

The German Shift from *Technologie* to *Technik*

Marx did not make any significant contributions to the conceptual history of Technologie and Technik. Instead, he was following a trend toward the term *Technik* that was already under way. Like *Technologie*, *Technik* entered German through modern Latin. According to Wilfried Seibicke, the modern Latin *technica* was a late seventeenth-century neologism adapted from the ancient Greek *techne*. Scholars had been translating *techne* as *ars* since the Roman era, but *ars* carried centuries of interpretive baggage, which may have made the new term *technica* attractive. Although *technica* at first referred to technical terminology, by the early eighteenth century *technica* also took on the broader Greek meanings of *techne* as the practice and knowledge of the arts, both mechanical and fine.[25] In the late eighteenth century, the Latin *technica* was translated into a new German word, *Technik*.[26]

Technik remained uncommon in German before 1800, rising to the status of keyword only in the second half of the nineteenth century. And as noted in chapter 1, *Technik* has two main meanings, one generally translated by the English word *technique*, and the other by *technology*. Through the end of the eighteenth century, *Technik* had a broad meaning derived from *techne*. According to Seibicke, this meaning comprised every field of human activity, so that "under every conscious, repeated or repeatable action based on reason, there lies a Technik with a number of empirical-practical rules." In this sense, we could speak of the Technik of a painter or pianist.[27] This definition of *Technik* corresponds closely to its English cognate *technique*, a term that became common only in the last quarter of the nineteenth century.

Before then, however, *Technik* gained a second, narrower meaning, one that linked the term first to craft production and then to production in general. Seibicke termed this meaning "technische Technik," which was usually invoked without modification as *die Technik*. Used in this way, *Technik* gained an essential unity, referring to "the entirety of the means, instruments and methods of production in handicrafts, industry, manufactures and factories." As an abstract noun, *die Technik* encompassed all the arts of material production, conceived as a coherent whole. This usage was distinct from the older definition, *Technik* as art in the broad sense, "rules and procedures for achieving an end."[28]

Surprisingly, there was almost no relationship between the terms *Technik* and *Technologie*, even though both terms were linked to craft production. As the sociologist Guido Frison has argued, "The notion of *Technik* cannot be considered the heir of *Technologie*." *Technik* and *Technologie* were the focus of independent discourses, and they were almost never discussed together or compared.[29] Nevertheless, Seibicke argued that Beckmann's concept of Technologie did shape the meaning of *Technik*, encouraging the narrower definition that limited *Technik* to the mechanical or practical arts. Thus by the mid-nineteenth century, *Technik* had split into two distinct meanings: a broader meaning that encompassed art in its most general sense—the rules, procedures, and skills for achieving a specific goal—and a narrower meaning referring to the material aspects of industry and trade.[30]

Technik and German Engineers

Technik evolved into a central, hotly contested keyword in this narrower sense. A key step in the spread of the term was its appropriation by the new German engineering profession. Although *Technik* did cover the activities of craft workers, after 1850 engineers made the word central to their professional identity. (*Technologie*, in contrast, remained largely irrelevant to the self-conception of German engineers.)

This key role for *Technik* among engineers became clear with the founding of the first pan-German engineering organization, the Verein Deutscher Ingenieure (VDI). In the original 1856 constitution, the VDI defined its principal goal as the advancement of German Technik rather than promotion of engineers' interests. The organization defined membership almost completely in terms of Technik, with ordinary membership open to practicing *Techniker* (engineers and other technical professionals), teachers of Technik or technical sciences, and finally owners and managers of technical establishments.[31] Engineers became so identified with *Technik* that most German-English dictionaries in the twentieth century gave "engineering" as one translation of *Technik*.[32]

This strong connection between engineers and Technik helped spark a theoretical discourse on the nature and place of the concept in modern Germany. As Germany began its rapid industrialization following unification in 1871, engineers fought for more status within the German social hierarchy. For example, academic engineers in the *technische Hochschulen* sought the right to grant doctorates to their advanced students. Alois Riedler, one of the leading advocates of this position, admitted that such a title would be of little interest to practical engineers. But he insisted that doctoral degrees would have "social value" by giving engineering education the same standing as traditional university studies. Riedler's advocacy succeeded in 1900, when the Kaiser granted *technische Hochschulen* the right to grant doctoral degrees.[33]

But the emergence of a theoretical understanding of Technik in the late nineteenth century, which Carl Mitcham has termed the "engineering philosophy of technology," was more significant. This engineering philosophy served as a form of boundary work, implicitly and explicitly making claims about the scope and social position of *Techniker*. In particular, German-speaking engineers used this philosophy to defend their profession from the perceived and real slights of humanist intellectuals.[34] These debates continued through World War I and into the Weimar era.

The theoretical literature of Technik, although motivated by questions of professional status, was not narrowly focused. Instead, this literature engaged with fundamental questions of philosophy and social theory. Simply by engaging in philosophy, engineers were making a claim to social status. In German-speaking countries, elite professionals and humanist intellectuals used philosophical discourse as a marker of *Bildung*, that is, broad education in the principles of high culture.[35] Social elites used *Bildung* to distinguish themselves from lesser social strata. The elite engineers of the VDI, for example, spent considerable time arguing over whether engineering students should be required to learn Latin. Most of these engineering educators defended Latin, since classical learning was a necessary component of *Bildung*.[36] Similarly, when engineers made theoretical claims about Technik, grounded broadly in philosophy and history, they were at the same time providing evidence of *Bildung*, demonstrating their right to belong among the elite professions.[37]

A central theme in this engineering philosophy was the relationship between Technik and culture. Since the early nineteenth century, German scholars had posited an opposition between the concepts of civilization (*Zivilisation*) and culture (*Kultur*). Although this distinction was originally grounded in a contrast between French aristocratic manners and authentic German spirit, the meaning of these terms expanded as educated German elites embraced the opposition between them. As summarized by Fritz Ringer, *Zivilisation* "evoked the tangible amenities of earthly existence," including rationalization, technical progress, and economic growth, while *Kultur* "suggested spiritual concerns" and encompassed both fine art and pure science. In this view, Kultur and Bildung were close allies. But did Technik belong with Kultur, or was Zivilisation the more appropriate category? The answer depended on how Technik was conceptualized. If one defined Technik instrumentally, as "the most obvious manifestation of means-ends rationality," then it clearly belonged with Zivilisation, not Kultur.[38] But if Technik belonged with civilization, then it was excluded from culture, thus undermining technical professionals' claims to Bildung.

By 1900, the "culture question" pervaded the engineering discourse of Technik, as illustrated by a series of articles, "General Questions of Technik," by Peter Engelmeyer. He was a worldly, philosophically minded Russian engineer who graduated in 1881 from the Moscow Imperial Technical College with a degree in mechanical engineering. Like other elite engineers of his day, Engelmeyer traveled widely in Europe, working in both Germany and France. He published extensively on philosophical issues of interest to engineers, writing in German and French as well as in Russian.[39]

Engelmeyer's articles on Technik appeared from 1899 to 1900 in *Dinglers Polytechnisches Journal*, sandwiched between technical articles full of equations and line drawings of machinery. Engelmeyer surveyed a wide range of literature, much of it written by engineers, that examined the nature of modern Technik and its relationship to economics, culture, ethics, and law.[40] According to Hans-Joachim Braun, Engelmeyer was not particularly original, but he did have a broad grasp on the literature of Technik, expressing ideas that were "typical for his time."[41] Engelmeyer's articles thus provide a detailed overview of engineers' thinking about Technik at the start of the twentieth century.

Many of the themes in Engelmeyer's articles remain current in present-day technology studies. He insisted, for example, that Technik be examined in terms of its interactions with other "social factors." He also analyzed the relationship between technology and natural science and between technology and economics. He weighed the benefits of technical progress against its burdens, considering, for example, the role of mechanization in undermining individuality.[42] But his most central theme was the relationship between technology and culture, or more precisely, his belief that technology was an integral aspect of culture.

Engelmeyer's arguments about the nature of Technik were directly linked to his desire to raise the cultural standing of technical professions through philosophical discourse. He believed that the technical professions should occupy a social position that reflected the central role of Technik in human history. Such status could be achieved, he argued, by exploring Technik as a general concept, and not by focusing on individual technical fields. A general understanding would allow comparison of Technik to other branches of learning, such as science, arts, and ethics. Only through this theoretical exploration, he argued, "will we arrive at the correct assessment of the high cultural significance that has always been inherent in Technik."[43] To this end, Engelmeyer discussed the writings of other philosophically inclined engineers who shared his concerns about cultural status, among them Franz Reuleaux, Josef Popper-Lynkeus, and Alois Riedler.[44]

The cultural status of Technik was a key theme among German engineers at the end of the nineteenth century. They claimed that culture had both a material and an intellectual-spiritual (*geistig*) side, and that Technik was obviously the basis of the material side. The kinematics pioneer Franz Reuleaux commented along these lines in his 1885 lecture, "Kultur und Technik," insisting that modern, scientifically grounded Technik had become a powerful factor in culture (*Kulturfaktor*). Similarly, Engelmeyer described Technik as a "stream in the history of culture. Technik produces objective [*sachliche*] culture, the correlate of intellectual [*geistigen*] culture." The Austrian inventor-engineer Josef Popper-Lynkeus took this argument further, claiming in 1888 that it made no sense to speak of the cultural significance of art, science, or Technik, because "culture consists of nothing other than art, science and Technik." In other words, Technik was simply one part of culture. Many engineers stressed that Technik, like other elements of culture, had creative and spiritual aspects, making Technik similar to fine arts. Engelmeyer explicitly defined Technik as an art (*Kunst*) in the broad meaning of the term, that is, "any objectifying activity." Thus, "Technik as art is a productive activity [*schaffende Thätigkeit*]," an activity that is especially "creative [*schöpferische*]" during the initial stage of an invention.[45]

After 1900, the theoretical literature about Technik expanded well beyond the boundaries of the engineering community, growing into a flood of writing after World War I. Many of these post–World War I works focused on technology as a disruptive social force. The cultural role of Technik remained a central theme in these writings, not only among engineers, but also among philosophers, theologians, historians, and social scientists.[46] At the same time, the field grew more contentious. One participant, the X-ray inventor Friedrich Dessauer, later termed these Weimar-era debates the "controversy over technology [*Streit um die Technik*]." In part, this controversy set humanist critics of Technik against its engineering defenders, of whom Dessauer was one. Yet Dessauer himself admitted that the controversy was one-sided, since nontechnical scholars mostly ignored works by technical professionals.[47]

But the dividing lines were not so simple. Only the most extreme cultural pessimists went so far as to condemn Technik in general as the enemy of culture.[48] More common was an ambivalent approach that embraced some aspects of Technik while rejecting others.[49] Both liberals and cultural conservatives questioned the benefits of technology, although admittedly from different directions.[50] On the German right, humanists as well as engineers embraced Technik as an essential, authentic expression of German culture. The historian Jeffrey Herf has called such thinkers "reactionary modernists." But many apolitical, centrist, and left-leaning scholars also embraced Technik as a potentially positive cultural force. Like the reactionary modernists, these left-leaning scholars rejected the reduction of Technik to instrumental reason, accepting Technik as an integral part of Kultur. Unlike the reactionary modernists, however, left-leaning scholars also rejected the racist understanding of culture that was central to reactionary modernism.[51]

Technik and German Social Science

This German philosophical debate about Technik and Kultur was largely invisible to scholars writing in English.[52] Instead, these scholars learned about Technik mainly through the German social sciences. In the late nineteenth century, German social scientists began incorporating Technik as a concept in their work. It found a receptive home among scholars who grappled with the problems of industrial civilization, especially members of the younger historical school of German economics, such as Gustav Schmoller. By the early twentieth century, Technik had become a significant concept in the works of the founding fathers of German sociology, including Max Weber, Georg Simmel, and especially Werner Sombart.[53] In contrast to the engineer-philosophers, these pioneers in German sociology were less concerned about the precise nature of Technik; rather, they sought to understand it in relation to other social phenomena.

This social-scientific engagement with Technik took place in conversation with Marx, or more specifically as a critical response to the perceived determinism of contemporary Marxist theory.[54] By the early twentieth century, Marxist determinism had taken on a technological cast, which owed more to the later work of Friedrich Engels than to Marx's own writings. Marx took little notice of the emerging discourse of Technik before his death in 1883. But Engels did live long enough to briefly incorporate Technik into Marxist theory. He made his most significant statement about Technik in a well-known letter written in 1894 and published in 1895, the year he died. In this letter, Engels elaborated "our" (his and Marx's) theory of historical change, firmly incorporating Technik into an emerging Marxist orthodoxy of historical materialism. "Economic relations," he insisted, were the "determining basis" of history. These relations referred to both production and exchange, and "thus engage with the entire Technik of production and transport." "This Technik," he continued, "determines [*bestimmt*] . . . the mode of exchange, the distribution of production, . . . the division of classes, relations of domination and servitude, the state, politics, law, etc." In this statement, Engels made a classic argument for strong technological determinism, although he did allow some autonomy to politics, religion, and art by delaying determination until the "final instance."[55] Later Marxists drew from Engels's determinism to create technicist versions of Marxist theory that strongly influenced Soviet thought.[56]

One prominent German social scientist sought to counter this Marxist determinism by articulating an explicitly cultural understanding of Technik. This scholar was Werner Sombart. Sombart is widely regarded as one of the founding fathers of German sociology along with Max Weber, Ferdinand Tönnies, and perhaps Georg Simmel. Of these four, Sombart is probably the least remembered, in part because of his gradual shift from Marxism to National Socialism over the course of his career. Most of Sombart's key contributions, however, were made well before his embrace of National Socialism.[57]

Sombart's most theoretical discussion of Technik came in a 1911 paper, "Technik und Kultur." In it, he rejected the technological determinism that he saw as an integral element of Marx's materialist conception of history.[58] This article has received some attention from historians and science and technology studies scholars, but mainly as an example of antimodernist cultural conservatism.[59] Such rhetoric is certainly present, for example in Sombart's denigration of American-style hotels and American popular music, forms of culture he correctly linked to modern technology. But such intellectual snobbery was widespread at the time among the *Bildungsbürgertum*, the university-trained intelligentsia, and hardly represented a fundamental hostility to modern technology.[60] As Mikael Hård has argued, Sombart's article is more interesting for its methodological insights than as cultural critique.[61]

Sombart's "Technik und Kultur" originated as a presentation to the first congress of the German Sociological Society in 1910. A stenographer's transcript of the talk appeared in the published proceedings of the conference, along with responses by attendees, most notably Max Weber. In 1911, Sombart published a revised and expanded version of the talk under the same title in the *Archiv für Sozialwissenschaft und Sozialpolitik*, the most important journal for German social science in that era. In the revised version of the paper, Sombart described the initial talk as a "fiasco" that completely failed to make its key points clear. The discussants, he insisted, could have made their comments without having heard the presentation, and he dismissed Weber's response as "not of a fundamental nature."[62]

The revised paper, Sombart hoped, would bring clarity to the concepts of Technik and Kultur, and thus help dispel the misconceptions that had long plagued this discussion. In place of a simplistic schema of Kultur determined by Technik, Sombart sought a more sophisticated approach to the relationship between culture and technology. He structured the essay by examining first the meanings of *Technik*, then *Kultur*, and then the *und*, that is, the interactions between the two concepts. Sombart began with a discussion of the term *Technik*, noting first its broad meaning as a "system or complex of means suited to achieving a specific end." As he noted, even "the arts of love [*Ars amandi*] are also 'Technik' in this broad sense." Sombart's broad definition of *Technik* was clearly instrumental, similar to the English *technique*.[63]

But Sombart was more interested in narrower meanings of *Technik*, those that referred to the "material goods [*Sachgüter*] that people use to achieve a specific goal," as well as the methods for employing such goods, which he termed "instrumental" or "material" Technik. In present-day English, this meaning would generally be rendered as "technology." (*Instrumental* in this context referred to physical instruments and the procedures for using them.) But Sombart's focus was narrower still. *Technik*, he claimed, when used without modification, usually referred to "production Technik," the methods used to make material goods. Production Technik, Sombart wrote, was fundamental, the form of Technik on which all other material Technik depended, and the form of Technik most significant economically. It was this restricted meaning that Sombart intended to relate to Kultur.[64]

Yet even under this restricted meaning, Sombart still conceived of the concept broadly. It encompassed both knowledge (*Kennen*) and ability (*Können*), that is, a combination of theory and practice. And when people spoke of the Technik of an age, for example "modern Technik," the term referred not just to the sum of available production methods. Instead, the Technik of an age implied "the special spirit [*Geist*] of this Technik," that is, its general principles, or what one could call the "style" of the Technik, in the same sense as one speaks of a cultural style.[65]

Sombart then turned his attention to *Kultur*, which he felt was more difficult to define. He classified Kultur using a complex schema of overlapping categories, which I can treat only briefly here.[66] Most significantly, Kultur for Sombart was both material (*materiel*) and ideational (*ideel*). Material culture arose, not from the material artifacts themselves, but rather from their possession and use, that is, from consumption. For example, "the culture of purity finds its objective expression in the quantity of soap, toothbrushes, bathtubs, sponges, ointments, etc., available to a population." Ideational culture, in contrast, started from a substrate of material goods, but it also possessed an immaterial or intellectual (*geistig*) aspect. Ideational culture could be objectivized into formal institutions, such as the church or the state, or it could remain mental (*geistig*), for example as ideals or moral values. Culture for Sombart was also either objective or subjective, with objective culture residing in the symbolic significance of material objects, while subjective culture resided in individuals, in their education and training, physical as well as mental.[67]

With these definitions in hand, Sombart then turned to examine the "and" in his title, that is, the relationship between Technik and Kultur. He insisted that the relationship was bidirectional, and that analyzing each direction of causality was a distinct task. Although his paper focused on the influence of Technik on Kultur, he briefly sketched out an approach to the reciprocal problem, the influence of Kultur on Technik. Most generally, a cultural field could either hinder or promote technical progress. Yet Sombart did not view culture as merely affecting the pace of technological change; culture could also move Technik in specific directions, producing "qualitatively different technologies." In fact, he noted, any history of Technik had to take culture into account, examining how specific cultural forms produced the necessary conditions and driving forces for technological change.[68]

Sombart's main focus, however, was on Technik as a causal factor shaping Kultur. Analyzing this causal role led him to develop some of the most subtle points in the paper. He began by rejecting the "technological view of history," which posited Technik as "determining" and the rest of human history as "determined." Sombart ascribed this determinism to Marx. But as Torsten Meyer has argued, Sombart's interpretation was based on a misunderstanding of Marx's concept of productive forces (*Produktivkräfte*), which Sombart equated with Technik.[69] Yet Sombart was hardly tilting at windmills; such a determinist reading was in tune with Marxist theory of the time, as Engels's 1894 letter shows.[70]

Regardless of what Marx actually meant, Sombart's antideterminism was uncompromising. He argued that "Technik, like other components of culture, cannot be imagined as 'absolute,' that is, only self-determined and not at all determined. There is no Technik in a (social) vacuum; there is also no Technik that could have effects starting from an Archimedean point, external to human culture." In many ways, this analysis is quite similar to the critique of technological determinism that emerged among American historians of technology in the 1960s and 1970s.[71]

Sombart's analysis of the culture-technology relationship shares much with that of engineer-philosophers like Engelmeyer, most important Sombart's conceptualization of Technik as inherently cultural. In the quotation above, for example, he clearly classified Technik as a part of culture, not as something that interacts with culture from the outside. Nevertheless, Technik occupied a special place among the components of culture, being of "paramount significance," a universal element of all cultural phenomena, because "underlying every cultural act is a necessary material substrate."[72] This point too was shared by the engineer-philosophers, even though Sombart's paper made no reference to their work.

Despite these general principles, Sombart argued against building a grand theory of the influence of Technik on Kultur. He argued for limited theoretical claims based on careful case studies without, however, adopting the atheoretical approach of historians. The technological view of history, he claimed, had actually impeded proper appreciation of the role of Technik in culture. "Dogmatists," that is, Marxists, "did not believe it necessary to trouble themselves with details," while skeptics simply rejected any significant role for Technik in culture.[73]

But when it came to empirical examples, Sombart failed to heed his own advice. Some of his examples read like caricatures of conservative cultural critiques. He repeatedly defended traditional high culture against the inroads of technological modernity, denouncing, among other things, amusement parks, new forms of popular music, and cheap reproductions of artworks. Sombart also engaged in precisely the kind of grand generalizations that he had warned against. For instance, he made sweeping claims about Technik as a stimulus for women's rights. In another grandiose generalization, he argued that the system of handicrafts would be impossible under the rational and scientific Technik of capitalism.[74]

However, other examples were more balanced, even perspicacious. His analysis of technology and music, which he offered as "paradigmatic," examined separately the influence of Technik on production, reproduction, exchange, and consumption. Such an approach shares much in common with the concept of the circuit of culture now common in cultural studies.[75] His discussion of the relationship between science and Technik was also insightful. Sombart insisted that science depended on Technik as a condition of its existence. Technik was not only a necessary condition of science but also "frequently a determining influence" on the course of scientific research.[76] In any case, Sombart admitted that his empirical examples were provisional.[77] They functioned as "just so" stories, methodological exemplars whose empirical shortcomings do not undermine Sombart's conceptual contributions.

Most German social scientists did not share Sombart's explicitly cultural view of Technik, among them Max Weber, probably the most influential of all classical sociologists. His work remains central to the discipline, in stark contrast to Sombart's. Weber and Sombart were friends as well as rivals. Early in their careers, they worked together closely to create institutional structures for German sociology, including the journal *Archiv für Sozialwissenschaft*.[78] They both endorsed a causal role for "spirit [*Geist*]" in history, in opposition to Marxist historical materialism. Yet they also grew apart in fundamental ways, with Weber remaining a firm believer in liberal democracy until his untimely death in 1920, while Sombart's work increasingly reflected the outlook of a romantic reactionary.[79]

Weber was not terribly interested in technology, either conceptually or empirically.[80] But Technik was nevertheless significant to his conceptual structure. Weber was engaged in boundary work as part of his move from the field of national economy (*Volkswirtschaft*) to the nascent field of sociology after about 1900. His concept of Technik played a part in this work.

Weber's key move was to erect a sharp boundary between technical and economic types of action. He made this distinction in less than two pages of his posthumously published *Economy and Society*, probably the most famous work in the history of sociology.[81] Weber analyzed economics (*Wirtschaft*) and Technik as ideal types, abstracted from the complexity of actual instances. His discussion of Technik occurred after he had defined his four main types of action: value-rational (*wert-rational*), instrumentally rational (*zweckrational*), affective, and traditional. These four categories overlapped with his distinction between economic and technical action in ways that Weber never fully clarified, even with regard to rational action.[82]

For Weber, the key distinction was that Technik concerned the choice of means to a given end, while *Wirtschaft* dealt with the allocation of means among multiple ends. Thus, Technik excluded considerations of costs, because costs implied comparison of the utility of means for a variety of alternative ends. Weber illustrated this point by positing a "purely technical" choice between platinum or iron for a machine part. Assuming that both materials were available, the "technical" decision would be based on achieving the desired result while "minimizing" use of resources. In other words, it might be technically rational to make an engine oil pan out of platinum. But as soon as the designers took the relative cost of platinum and iron into account, the action also became an economic choice among different ends. Rational economic action would exclude costly platinum for the oil pan, because platinum could be better used to satisfy other ends.[83]

Weber based his ideal type of technical action on one of the two main meanings of *Technik*: its instrumental definition as skills and procedures used to achieve a given end. Such a definition encompassed both immaterial and material practices. This usage could reasonably be rendered in English as "technique." In a well-known passage, Weber made this meaning quite clear:

In this sense there are techniques of every conceivable type of action, techniques of prayer, of asceticism, of thought and research, of memorizing, of education, of exercising political or hierocratic domination, of administration, of making love [*erotische Technik*], of making war, of musical performances, of sculpture and painting, of arriving at legal decisions.[84]

But when Weber gave examples of the distinction between economic and technical action just a page later, he shifted from the instrumental to the industrial meaning of *Technik*. In fact, all of his hypothetical illustrations of the distinction between economy and technology were taken from the industrial arts rather than arts of prayer or eroticism. For example, in addition to the purely "technical" choice between platinum or iron for a machine part, Weber also posed a thought experiment of imagining a rational machine with no economic purpose, such as a device to make atmospheric air.[85]

Weber acknowledged that his distinction between economic and technical action had little relevance to the actual history of Technik. He admitted that "the economic determination of technological development" had been "the main emphasis at all times." He also insisted that "non-economic motives" such as artistic concerns, "the games and cogitations of impractical ideologists," and also "other-worldly interests and all sorts of fantasies" shaped the history of Technik. Sombart and later historians of technology all would have agreed with this list. Yet all this rich clashing of ends was in fact excluded from Weber's concept of Technik, which he limited to the narrowest form of instrumental rationality.[86]

In Weber's defense, his concept of Technik was an ideal type, that is, a selective appropriation of actual phenomena to produce a "unified analytical construct"—a "utopia" in the sense that the ideal type did not necessarily exist anywhere in its pure form.[87] Yet even in terms of ideal types, this analysis was problematic. As an ideal type, Weber's Technik functioned not to make Technik a part of either economics or sociology but rather to exclude it. Weber's Technik belonged to neither field because it dealt with neither comparison of ends in terms of utility nor consideration of ultimate ends, that is, values. Yet Technik as an actual empirical phenomenon was significant precisely because it engaged with questions of costs and values. By sundering these questions from his instrumental concept of Technik, Weber made it impossible to understand the empirical significance of Technik as industrial arts. This limitation represented a failure of the ideal type, since, as Weber insisted, the ideal type was not an end in itself but rather a means to obtain "knowledge of the cultural significance of concrete historical events and patterns."[88]

As I noted above, engineer-philosophers had long resisted various attempts to limit Technik to narrow forms of rationality, subordinate to either science or economics. Weber's narrow understanding of Technik was connected to a long history of instrumentalist thinking among humanist scholars. But it remains puzzling that he barely acknowledged the cultural side of Technik, given his thorough knowledge of Sombart's writings on the topic.[89] His lack of interest in Technik was perhaps a by-product of his animosity toward Marxist historical materialism, which both Marxists and their opponents often defined in technological terms.[90] Sombart shared Weber's opposition to historical materialism, but dealt with this problem not by ignoring Technik but by embracing the concept as an expression of human culture.[91]

In many ways, the German discourse of Technik anticipated much of the later twentieth-century discourse of technology. Engelmeyer and Sombart alone touched on almost every key issue that philosophers and historians of technology dealt with after World War II. These issues include the critique of technological determinism, the relationship between technology and natural science, the social shaping of technology, the status of technology as a cultural force, and the dehumanizing effects of modern technology. Some of these issues had been dealt with earlier, for example in the Scottish historian Thomas Carlyle's famous critique of the "mechanical age" in his essay "Signs of the Times."[92] But Carlyle's key term, *machinery*, was poorly conceptualized and inadequate to capture his critique of industrial civilization. In contrast, the German discourse of Technik brought together the whole range of key issues under a single, powerful rubric. And Engelmeyer and Sombart were hardly unique, as can be verified by the tall stack of philosophical works about Technik published between about 1890 and 1933.[93]

Germany was the center of this conceptual shift because of a fortuitous conjunction of circumstances. First was that nation's shockingly rapid industrialization after its unification in 1871, with engineers serving as the vanguard.[94] Second was its highly stratified professional structure that slighted the social aspirations of engineers.[95] And third was the intellectual context of *Bildung*, which linked social status to philosophical discourse, sparking theoretical inquiry into Technik among engineers and humanists—both of whom were beset by deep-seated status anxieties.

In the following two chapters, I will show how the German discourse of Technik contributed far more to the twentieth-century meanings of *technology* than did the original nineteenth-century term and its Continental cognates. Technik shaped the meanings of the English term when scholars began translating, or more accurately mistranslating, *Technik* as "technology" in the early twentieth century. This mistranslation occurred in part because of the "semantic void" that arose when the concepts of art and science narrowed in the nineteenth century, as discussed previously. As a result of this void, English-language scholarship on the industrial arts was far less sophisticated than the German discourse of Technik. With just a few exceptions, nothing as nuanced as Sombart's and Engelmeyer's understanding of Technik exists in English before the 1960s.[96] English terms like *applied science, invention,* or *industrial arts* simply lacked the conceptual breadth of *Technik.*[97]

EIGHT

Thorstein Veblen's Appropriation of *Technik*

In the early twentieth century, American social scientists created a new discourse of technology. These scholars, most important Thorstein Veblen, replaced the original meaning of *technology* as the science of the mechanical arts with a new meaning that denoted the industrial arts in general along with the material means of production. These changes were strongly shaped by borrowings from the German discourse of Technik. Oddly, none of these scholars seemed aware that their linguistic borrowings were transforming the English-language term.

Because of this borrowing, *technology* acquired the layers of meaning that I described in the introduction: applied science, industrial arts, and technique. This contradictory set of meanings continues to dominate its academic use. Depending on how these meanings were deployed, the concept of technology could support either an instrumental or a cultural view. And in part because of its internal contradictions, technology was able to serve as both a reifying concept, concealing human choice and relations of power, and a liberating concept, laying bare these same choices and power relationships. Veblen exemplifies the use of technology as a liberating concept, while Charles Beard reified technology, thereby concealing the human choices that create material culture.

The Second Industrial Revolution and the Problem of Art

At the turn of the twentieth century, the United States was at the height of what historians label the Second Industrial Revolution. This era, running from roughly 1870 to 1930, was even more transformative than the British Industrial Revolution about a century earlier. The first Industrial Revolution gave the world machine-made clothing, steam-powered factories, coke-smelted iron, river steamboats, and the beginnings of the steam railroad. Yet these transformations did not create a recognizably modern world, at least technologically. For the vast majority, even in Britain, early nineteenth-century domestic life was more akin to that of the seventeenth than the twentieth century. Most cooking was done on stoves fueled by coal or wood, which were hot and dirty. Indoor plumbing remained rare, and flush toilets nonexistent. Artificial light was dim and flickering, produced by candles, whale-oil lanterns, or the new coal-gas flames.[1]

In contrast, the Second Industrial Revolution occurred simultaneously across a number of industrial areas. The period witnessed global diffusion of the railroad and the electric telegraph, Bessemer and open-hearth steel production, synthetic dyes, the first pharmaceuticals, petroleum refining, the telephone, the phonograph, the electric light, the electric streetcar, the ocean steamship, the internal combustion engine, the sewing machine, the bicycle, interchangeable parts, photography, the steel-framed skyscraper, the elevator, the automobile, the airplane, the radio, and the moving assembly line, along with innovations in papermaking and printing that drastically reduced the cost of newspapers and books. A time-traveler leaving New York City in 1918 would have been more at home in 2018 than in 1818.[2]

Yet when observers sought to describe the Second Industrial Revolution, the English language remained surprisingly impoverished, at least for the period's higher-level concepts. There was, as I pointed out earlier, no general term for this dramatic transformation in material culture, what we would call modern technology. Art remained the most important concept for describing material culture, typically modified by *useful, mechanical,* or *industrial.* The modifiers were needed to distinguish these forms of art from fine art, which increasingly became the default meaning of *art* by the late nineteenth century. Other terms like *science, machinery, industry,* and *invention* were widespread, but none had the generality of *art.*

The persistence of art as a core concept for describing industrial modernity is clear in the emerging disciplines of anthropology, sociology, and economics, as well as among academically minded engineers. Engineers described their work in terms of uniting science and art. The German-born British engineer William Siemens used precisely this language in his well-known presidential address to the British Association for the Advancement of Science in 1882. Siemens's analysis was reminiscent of Diderot's entry on *art* in the *Encyclopédie*.[3] The sociologist Herbert Spencer used similar language in his essay "The Genesis of Science" (1854).[4] Likewise in anthropology, art remained the operative category for discussions of material culture and craft skills. This usage continued even when anthropologists employed *technology* in its nineteenth-century meaning. In 1883, the pioneering American anthropologist John Wesley Powell, for example, named "technology, or the science of the arts," as one of the five main divisions of anthropology. But when he discussed tools and machines, he dispensed with *technology* and instead framed his analysis using *art*.[5] And in the new marginalist economics, *art* too remained the general term for the phenomena that contributed to the productivity of labor.[6]

But if late nineteenth-century authors were willing to frame discourse about the Second Industrial Revolution in terms of "progress in the arts," why didn't they simply retain the category of art, especially with a suitable modifier? After all, even today the liberal arts remain an established concept, rarely confused with fine arts. In the second half of the nineteenth century, similar modifiers were already used to restrict the arts to what we would now call technology. First was *mechanical art*, which remained common despite its medieval origins. Second was *useful arts*, which arose in the eighteenth century as a new category in contrast to the fine arts.[7] This term is enshrined in the US Constitution, which grants Congress the authority to give "authors and inventors exclusive right" to their products in order "to promote the progress of science and useful arts."[8]

The third modifier was *industrial arts*. It was of nineteenth-century coinage, and had fewer of the archaic associations of *mechanical arts* and *useful arts*. The term *industrial* was rare before the mid-nineteenth century. Until then, *industry* usually referred to industriousness, not organizations of production.[9] The use of *industry* and *industrial* in its new sense was popularized by Thomas Carlyle, among others; Carlyle used these terms to refer to changes in British society that he despised.[10] The specific phrase "industrial arts" was most likely borrowed from the French *arts industriels*, where the term arose as a concept, like the

English useful arts, in opposition to the fine arts.[11] George Wilson invoked the industrial arts in his 1855 inaugural lecture as Regius Chair of Technology at Edinburgh University. According to Wilson, "The arts included in the domain of Technology, are the Industrial Arts, or Handicrafts."[12] In 1888, the US Commissioner of Patents published an impressive illustrated catalog titled *The Growth of Industrial Art*, which included everything from beehives to blast furnaces.[13] Into the 1920s, Thorstein Veblen referred often to the industrial arts, using the phrase "the state of the industrial arts" interchangeably with *technology*.[14]

Why, then, didn't writers embrace industrial arts as the key concept for the Second Industrial Revolution? And why didn't scholars translate the industrial meanings of *Technik* as "industrial arts"?[15] Explaining why something did not happen is always difficult for historians, and some consider it an illegitimate question. Yet arguments about historical cause always imply a counterfactual, that is, the claim that if a particular causal factor had not been present, then the outcome would have been different.[16] Following Leo Marx, I have argued that the "semantic void" in the late nineteenth century set the stage for the emergence of new meanings for *technology*. Yet *industrial arts* seems like it was the perfect candidate to fill this void, obviating the need to create new meanings for *technology*.

However, *industrial arts* had other meanings that made it problematic as a central concept for the material culture of industry. Most important, the term could not escape association with the fine arts. This association was less of a problem for *liberal arts*, which predated the category of fine arts by roughly a millennium. However, *industrial art*, when used in the singular, often referred to what we would now classify as industrial design or decorative arts, particularly in British English. For example, the South Kensington Museum, predecessor to both the Victoria and Albert and the Science Museum, published a number of books on industrial arts that focused almost entirely on decorative arts.[17] At the same time, *industrial arts* became associated with vocational training in manual trades, typically for children of the working classes. In the United States, such training was sometimes conducted in specialized schools. By the 1910s, the term *industrial arts* referred almost exclusively to teaching manual arts to schoolchildren.[18] In essence, the term was appropriated by well-established communities in the visual arts and education. In contrast, no group embraced *industrial arts* in its technological meaning. It became, in effect, unavailable for the role previously played by *mechanical arts*.

This undermining of the technological meaning of *industrial arts* helped create Leo Marx's semantic void, which set the stage for English-language scholars to import the meanings of *Technik* into the emerging social sciences of the early twentieth century.[19] English-language social scientists of the period lacked a general concept of technology comparable to that of Technik, and this lack limited their ability to generalize about the place of technology in modern civilization. At the same time, the waning of *art* as a term for industrial production interrupted the direct lineage linking *art* and *technology*, thus obscuring two thousand years of discourse about techne, ars, and art.

Appropriating *Technik*: The Translation Problem

The German discourse of Technik did not simply expand into this English-language semantic void. Language, like technology, is created by human choices, not objective laws. Anglophone scholars mostly ignored the German discourse of Technik. This ignorance is surprising, given the influence of German scholarship on American thought. After the Civil War, American academics turned increasingly to Germany for intellectual inspiration. Academics in the physical sciences led the way, but German influence was also strong in the social sciences, especially in the new fields of economics and sociology. An early twentieth-century survey found that 59 of 116 American economists and sociologists had studied in Germany, with 20 receiving German doctorates. Albion Small at Chicago, Richard T. Ely at Wisconsin, and Herbert Baxter Adams at Hopkins were just a few of the leading scholars who either studied in Germany or were deeply influenced by German social theory. This influence waned somewhat after World War I, but the links continued, especially in the work of theorists like Talcott Parsons. In the 1930s, refugees from National Socialism brought a new wave of German-speaking social scientists to America, helping maintain a place for German ideas in American social theory.[20]

These strong links ensured that American scholars would encounter the German concept of Technik. Yet Americans rarely remarked on this concept, even when reviewing or citing works with extensive discussions of Technik.[21] Instead, they haltingly borrowed elements of this discourse with little awareness of what they were doing.

This borrowing created a translation problem, since the technological meaning of *Technik* had no direct English-language equivalent. The most obvious translation was the English cognate *technique*. This term had been rare in English before the second half of the nineteenth century. When *technique* became more common later in the nineteenth century, it was most often used in describing medical procedures or skills in the execution of fine arts.[22] Thus, "technique" provided a reasonable though imperfect translation of the broad meaning of *Technik* as skills and methods suited to a particular end, whether poetry or surgery. But "technique" was never an accurate translation of the technological meaning of *Technik*. The English word was what translators call a "false friend," that is, a misleading cognate. When American scholars rendered *Technik* as "technique," without recognizing that they were in effect redefining the English word, they created a terminological muddle.

This confusion is clear in one of the earliest works of American theory inspired by the German discourse of Technik, Edwin R. A. Seligman's influential essay "The Economic Interpretation of History" (1901).[23] Seligman, the urbane son of a New York banker, was one of the founders of the Institutionalist school of economics. He made key contributions to public finance and the history of economic thought during his forty-year tenure as a professor of political economy at Columbia.[24] In the late nineteenth century, he studied in France and Germany, where he became familiar with the works of Marx and Engels; at this time, few of their works were available in English. Seligman's 1901 paper drew attention to European debates over the Marxist theory of historical materialism, a debate that, he noted, had "scarcely reached out shores." He claimed that the theory had merit. Echoing Marx and Engels, Seligman argued that the social conditions for the sustenance of life, that is, the economy, were responsible for "transformations in the structure of society," at least "in the last instance." Yet Seligman was no socialist, insisting that Marx's socialism "had no bearing on the truth or falsity of his philosophy of history."[25]

Seligman's understanding of the economic interpretation of history was decidedly technological. His terminology, however, involved a conflation of *technique* and *technology* that became widespread among American social scientists influenced by German theories of Technik. Abetting his confusion were his Marxist sources. Seligman quoted both Marx's use of the cameralist concept of Technologie from *Capital* and Engels's use of Technik. When quoting Engels, he translated the industrial meaning of *Technik* as "technique," a translation as odd in 1901 as it appears today. He drew from Engels to argue that Marx was not claiming that "purely *technical or technological* modes of production" were the determinants of history. "Even though it is claimed that changes in *technique* are the causes of social progress, we must be careful not to take too narrow a view of the term [i.e., technique]." Seligman noted that *technique* encompassed "relations between production and consumption" in general, leading him to reject the "technical interpretation of history" in favor of the "economic interpretation of history."[26] His conflation of *technique* and *technology*, and his abandonment of both for *economic*, illustrate the terminological difficulties in translating the German discourse of Technik into English.[27]

The mistranslation of *Technik* as "technique" remained common well into the twentieth century. For example, one of the first works of the German historical school of economics to be translated into English was Karl Bücher's *Entstehung der Volkswirtschaft*, published in 1901 as *Industrial Evolution*. The translation consistently rendered the technological meaning of *Technik* as "technique."[28] This practice even continued well into the twentieth century, for example in the 1986 English edition of Bertrand Gille's encyclopedic survey, *History of Techniques*.[29] Some Anglophone scholars adopted this usage for their own prose. One was Frederick Nussbaum, whose text, *A History of Economic Institutions* (1933), was an interpretation of Sombart's *Moderne Kapitalismus*. In this work, Nussbaum consistently used *technique* in the same sense as Sombart's *Technik*.[30] Similarly, Bertrand Russell repeatedly used *technique* where present-day scholars would use *technology*, for example in the phrase "scientific technique," which referred not to experimental methods but to the fruits of science in industry.[31] Russell continued to use *technique* in this way even after World War II.

This repeated mistranslation of *Technik* as "technique" could have transformed the English word, giving it the same two senses that exist in German and other Continental languages. But no such semantic change occurred. In fact, no English-language dictionary ever defined *technique* as a collective term for the practical or industrial arts. Instead, it was *technology* that carried the industrial meanings of *Technik* into English. This too was a mistranslation, at least until the meanings of the English word changed. *Technik* did not denote the "science of the industrial arts," the principal meaning of *technology* at the start of the twentieth century. Yet by the 1930s, *technology* had changed enough to make this term a suitable translation for *Technik*. Although many scholars had a role in this transformation, none was more important than the American social scientist Thorstein Veblen.

Thorstein Veblen: *Technology* as *Technik*

Both Leo Marx and Ruth Oldenziel have emphasized Veblen's central-ity to making *technology* a keyword of modern American culture. Old-enziel has focused on Veblen's later, more polemical works, especially *The Engineers and the Price System*, a collection of articles published in 1919.[32] However, the terms *technology* and *technological* became a key part of Veblen's conceptual armory much earlier, in the first years of the twentieth century. It was then that he began translating the Ger-man discourse of Technik using the terms *technology* and *technological*.[33]

Veblen remains difficult to classify. During his life he was usually considered an economist, but sociologists are more likely to claim his work today. He drew from many sources for his ideas, but tracing the origins of his thought is a difficult task. Veblen read voraciously in a variety of languages across many fields of social thought, synthesizing his own terminology in the process. Following the scholarly practices of the time, he usually alluded to rather than cited his sources, though he did refer to specific works often enough. His convoluted and often ironic prose can easily be misinterpreted. Scholars have connected Veblen to many intellectual traditions; his purported predecessors in-clude Charles Darwin, Herbert Spencer, Edward Bellamy, Henrik Ibsen, Charles Fourier, William James, Jacques Loeb, Franz Boas, Edward Ty-lor, Karl Marx, Immanuel Kant, Gustav Schmoller, and Werner Som-bart. Though Veblen's most fundamental allegiance was to Darwin, he also drew heavily from Marx and Sombart, despite significant disagree-ments with their theories. These German writers encouraged Veblen, at least indirectly, to incorporate the concept of technology into his theory of history.[34]

Technology was for Veblen a key concept in his theory of social evo-lution. He grouped technology with a set of beneficial forces in social evolution that were opposed by a contrary set of parasitic forces. On the beneficial side stood workmanship, industry, the machine process, and technological knowledge. On the parasitic side lurked predation, business enterprise, absentee ownership, and other pecuniary institu-tions. Veblen typically expressed this opposition as a conflict between business and industry, or between pecuniary and industrial institu-tions, but fundamentally this opposition centered on the distinction between wasteful and productive tendencies in human evolution. He viewed this opposition as universal in human cultures, arising from the conflict between the "instincts" of workmanship and predation.

(He used *instinct* unconventionally, insisting on its conscious, teleological character.) Veblen used this analysis to develop a critique of modern capitalism, arguing that the institutions of recent capitalism were incompatible with the peaceful development of modern industry for the benefit of the entire community.[35] Technology, in other words, first became a key concept in American social theory as part of a critique of capitalism.

If one detects a whiff in Veblen of the Marxist opposition between forces and relations of production, it was no accident. Like Marx, Veblen sought to explain the fundamental contradictions of modern capitalism by uncovering its historical dynamics. But his own terminology owed little to Marx's occasional use of *Technik* and *Technologie*. Even his most detailed discussions of Marx mention *technology* only in passing.[36]

Veblen's interest in the practical arts began well before he started using the term *technology* in the early 1900s. He had already incorporated technical factors into his theoretical framework in his book *Theory of the Leisure Class* (1899), the work that first earned him public acclaim.[37] But Veblen's initial term for the technical aspects of social evolution was not *technology* but *workmanship*, which he first examined in detail in 1898.[38] *Workmanship*, however, carried a strong sense of handicraft, which made it a poor term for the skills and knowledge that produced the great factories and new consumer products of the Second Industrial Revolution. Veblen needed a term broad enough to cover the practical arts from earliest agriculture to large-scale electrical systems. *Industrial arts* might have served this purpose, but this term carried other baggage, as I noted above. Veblen found the necessary breadth in the German term *Technik*.

Despite the difficulty of tracing Veblen's sources, ample evidence shows that he first encountered *Technik* in the works of the German historical school. This loose collection of scholars sought to uncover the laws of economic development through historical research and empirical studies, in contrast to the ahistorical abstractions of classical political economy and the newer marginalist economics.[39] Beginning in the 1890s, members of the historical school adopted *Technik* as a central theme in their largely empirical analysis of economic phenomena.

One of Veblen's earliest encounters with the German discourse of Technik began around 1892, when he was preparing his first published work. This work was *The Science of Finance*, a translation of a treatise on public finance by Gustav Cohn, a prominent figure in the German historical school. Charles Camic has examined this translation closely, and shows that the school had an enduring influence on Veblen's thought.[40]

In the book, Cohn connected the state's growing role in public finance to the increasing scale of industrial technology. He referred repeatedly to *Technik* throughout the book, in its meaning as both industrial arts and method. Veblen, however, struggled to translate the term, especially when used in its industrial sense. For example, he translated *Kriegstechnik*, military technology, once as "military art," and then four pages later as "military science." This confusion increased in Cohn's most detailed discussion of *Technik*, a section on technological progress and the state. In little more than a page, Veblen rendered *Technik* as "technical knowledge" (4 times), "technical efficiency" (3 times), "technology" (3 times), and "industrial knowledge" (1 time). These varying translations were hardly justified by the context; every time Cohn used *Technik* in this section, it was in its industrial sense. Even when Cohn repeated the standard phrase "Fortschritte der Technik," Veblen translated it variously as "progress in technical efficiency," "progress in technical knowledge," "progress in technology," and "improvements in technology," all in two consecutive pages.[41]

The problem wasn't with Veblen's understanding of German but rather with English. Clearly, Veblen was searching for suitable English words in 1892, but none quite fit the bill. Eight years later, when he returned to German writings on Technik, he was an established scholar, having achieved considerable recognition for *Theory of the Leisure Class*. But in contrast to his translation of Cohn, the more mature Veblen was willing to reshape the English language to fit his needs.

So after 1900, Veblen seized upon the still-obscure term *technology* as an equivalent for *Technik*. Even in its original meaning as the science of the arts, *technology* already had the necessary breadth, being applicable to prehistoric humans as well as the latest MIT graduates. But in appropriating *technology* Veblen also transformed it, moving it from a branch of science to a term that encompassed engineering knowledge and craft practices.

Veblen began linking *technology* to *Technik* in his reviews of books by Gustav Schmoller and Werner Sombart, both leading figures in the German historical school. In 1901, he wrote a long, laudatory review of Schmoller's *Grundriß der allgemeinen Volkswirtschaftslehre*. In it, Veblen focused on methodology, praising Schmoller for taking a Darwinian approach to economic theory that emphasized the historical genesis of economic forms.[42] Veblen singled out a forty-page chapter on Technik, in which Schmoller surveyed the development of technology from pre-historic times to the "modern West European-American machine age." He praised Schmoller's account of "modern machine industry," but objected to Schmoller's conservative moralizing on social questions, particularly "on the relation of technological knowledge to the advance of culture."[43]

But Veblen still remained inconsistent when translating *Technik*. In his review of Schmoller, he rendered the chapter title "Die Entwicklung der Technik in ihrer volkswirtschaftlichen Bedeutung" as the "Development of Technological Expedients and Its Economic Significance," equating *Technik* with "technological expedients." He also referred to "technological conditions" in a way that implied equivalence to Schmoller's *Technik*. In the same review, however, Veblen translated Schmoller's section title "Land, Leute und Technik" as "Land, Population, and the Industrial Arts." As I pointed out earlier, *industrial arts* was probably the closest English equivalent for *Technik* in its industrial meaning. "Technological expedients" was also a fairly accurate rendering of *Technik*, implying means that served industrial ends, but this phrase was a bit too polysyllabic even for Veblen. By translating *Technik* in these different ways, he implied a clear equivalence between technological things and industrial arts.[44]

After he reviewed Schmoller, *technology* became a new term in Veblen's conceptual armory. He used the term again the following year, when he critically reviewed a book on the Arts and Crafts movement by Oscar Triggs, his friend and colleague at the University of Chicago. In it, he condemned as "sophisticated archaism" the movement's rejection of machine production. Veblen invoked *technology* to support this critique. He insisted that art for the worker should be "in the spirit of the machine *technology*," and he argued for associating "[modern] art with the machine process and with the *technology* of that process." Although Veblen praised the aesthetic and functional goals of the Arts and Crafts movement, he insisted that these could be better reached "through the *technological expedients* of . . . the machine process" than through obsolete craft methods.[45]

Veblen decisively embraced *technology* in its new meaning in 1903, when he reviewed the first two volumes of Werner Sombart's *Der Moderne Kapitalismus*. Sombart made Technik a central concept in this work. But as noted in the last chapter, he was no technological determinist, insisting that Technik was not a "driving force" of economic development. For example, he pointed out that despite the importance of "the taming of steam," steam "remains only the driving force of a steam engine, without the ability to erect the engine and make it serve specific purposes." Rather than Technik, insisted Sombart, human wants and goals are the driving forces of economic change.[46]

Compared with his positive review of Schmoller, Veblen's review of Sombart was mixed. He praised Sombart's "modern, post-Darwinian spirit of scientific inquiry," an approach that sought causal explanations for the historical development of economic institutions. He also endorsed Sombart's "careful distinction" between business and industry, a distinction central to Veblen's own work. Nevertheless, he roundly criticized Sombart for his German-centered account of the origins of capitalism. The legal foundations of modern business, insisted Veblen, had their origins on English, not German, soil.[47]

This line of criticism he applied even more strongly to Sombart's analysis of modern industry, contained in the chapter titled "Die neuen Technik." Contra Sombart, Veblen argued for the English origins of modern industry, fully deploying his new concept of technology to support his claim. The English created not only the legal basis of capitalism, he wrote, but also

the material, the *technological* basis of business enterprise. The industrial revolution, which brought in the *technology* of the machine process and so laid the material foundation of modern business, is, of course, broadly an English fact—whatever fragmentary *technological* elements the English community may once have borrowed from southern Europe.

Veblen insisted that "English-speaking peoples" had retained their lead in the "modern machine industry" until quite recently, when other nations also came "into the first rank as creative factors in industrial *technology*." "English speech" had no monopoly on "thinking in terms of the machine process," but the epitome of such thinking remained with English speakers, "and this habit of mind is the spiritual ground of modern *technology*."[48]

Historians are unlikely to find clearer evidence of a direct link between the German discourse of Technik and a new American discourse of technology. Before Veblen, no American social scientist regularly used *technology* in remotely this way, as a broad synonym for the practices and principles of the industrial arts, the technical basis of modern industry.

Yet Veblen, always the careful semanticist, was not simply importing *Technik* into *technology*. Instead, he was constructing a set of related concepts for understanding social and economic change. What would now be termed modern technology Veblen referred to as the machine process. He clarified his definition of *machine process* in 1904 in his *The Theory of Business Enterprise*, drawing explicitly from Sombart. The machine process, he argued, referred not just to individual machines but to a system, "the whole concert of industrial operation . . . made up of interlocking detail processes." When Veblen referred to "the technology of the machine process," it suggested that *technology* was separated from *machine process* by a level of abstraction, so that technology comprised not the physical system itself but its principles, the knowledge and skills embodied in its operation. In this same book, Veblen began referring to *technology* as "the state of the industrial arts," a phrase he continued using throughout his remaining writings.[49] In effect, his usage subtly combined the older definition of *technology* as a field of knowledge with the concept of Technik as a set of material practices. However, his subtlety would be lost on later users of the term.

Veblen's Understanding of Technology

Over the next few years, Veblen developed his idea of technology into a sophisticated concept that drew from both the cultural and the instrumental aspects of Technik. Despite his veneer of scholarly detachment, powerful ethical values informed his theory of social evolution. Veblen had nothing but contempt for predatory institutions and values, even while accepting their central historical role. Instead, his sympathies lay with the productive factors in social evolution, and more important with the producers themselves. This affinity for producers shaped his understanding of technology, putting him more in the cultural than the instrumental camp. Some scholars see Veblen as a technocrat who advocated an elitist view of technical expertise. Yet he actually believed that technological knowledge rightfully belonged to the entire community of producers rather than the "vested interests."[50]

Two aspects of Veblen's concept of technology demonstrate this non-elitist approach. First, he conceptualized technology as the collective knowledge of the industrial arts shared by the entire community. Second, he developed one of the first explicit analyses of the relationship between science and technology, one that did not subordinate technology to science.

Veblen's approach to technological knowledge was grounded in two of his articles from 1898, written several years before he embraced the concept of technology. One of these developed his idea of the "instinct of workmanship." In this article, Veblen argued that humans were not innately averse to labor, as assumed in classical political economy. Instead, he maintained that humans actually possessed an "instinct of workmanship" derived from the essential role of labor and toolmaking in human survival. Although Veblen never explicitly defined *workmanship*, it referred primarily to "sophisticated human skill in manipulating material objects."[51] But workmanship, though it aided human survival, was not merely instrumental. It also carried a moral imperative, expressed in the condemnation of waste and inefficiency. "All men," Veblen insisted, "have this quasi-aesthetic sense of economic or industrial merit," to which "futility and inefficiency are distasteful." The prejudice against manual labor arose relatively later in human history when predatory elites who lived off the labor of others became dominant. According to Veblen, such elites valued military prowess over productive skills, viewing work that involved "tamely shaping inert materials to human use" as "unworthy" and "debasing."[52]

Also in 1898, Veblen published an article disputing another tenet of classical political economy: the idea that property was rooted in the productive labor of the individual. Instead, he insisted that the institution of property arose as a result of coercion by nonproductive predatory elites. Production was an inherently collective process, relying not just on cooperative labor but also on shared "traditions, tools, technical knowledge, and usages." Individuals are not "productive agents" in isolation, because "there can be no production without technical knowledge," and "no technical knowledge apart from an industrial community." Because of the collective nature of production, the idea that "ownership rests on the individually productive labor of the owner reduces itself to absurdity."[53]

After 1900, Veblen drew from these ideas about workmanship and technical knowledge to develop a non-elitist understanding of technology as a collective set of skills and knowledge. He elaborated this idea in detail in an article, "On the Nature of Capital" (1908). Veblen began by arguing that "technological knowledge" was integral to all human communities, even the most primitive. Primitive technological knowledge included language, the use of fire, the use of simple tools for cutting, and basic fiber arts. This knowledge constituted what Veblen termed the "immaterial equipment" of a community, as opposed to the material equipment of tools and machines. Technological knowledge was always collective, in that it exceeded the grasp of any single individual, and it was also cumulative, growing through experience transmitted by members of the group. Natural resources, machinery, and other types of physical capital became useful only through collective technological knowledge.[54] But when material capital or natural resources were in short supply, predatory elites could usurp the community's collective technological knowledge by controlling the means needed to utilize this knowledge, such as labor or land. In this way, predatory elites created "the basis of pecuniary dominion." In effect, when elites owned what Marx called the means of production, for Veblen it represented a theft of the community's collective technological knowledge.[55]

Veblen maintained a veneer of scholarly detachment in this analysis, but his choice of words left no doubt where his sympathies lay. He clearly saw technological knowledge as a source of virtue for the productive, non-elite members of society, Veblen's "common man." Even though he did not use *culture* in the same sense as the German *Kultur*, he made it clear that technological knowledge had cultural value.[56]

Veblen was not, however, endorsing Puritanism's aversion to idleness. Instead, he saw cultural value in idleness as well as work. This position is apparent in his article "The Place of Science in Modern Civilization" (1906). In it, Veblen provided one of the first discussions of the relationship between science and technology explicitly using these terms.[57] According to Veblen, scientific and technological knowledge arose from two autonomous instincts, idle curiosity and workmanship respectively. Idle curiosity became the basis of modern science, while workmanship drove the shift from handicrafts to the machine process. But the machine process itself shaped science by encouraging a shift in the way people viewed the natural world. In effect, Veblen was proposing a sociology of scientific knowledge. He argued that the nineteenth-century shift from craft to machine production inspired scientists to move from craft to machine models of causation. Thus, the utility of science in Veblen's time was the result of a "fortuitous . . . coincidence" that made possible "the employment of scientific knowledge for useful ends in technology, in the broad sense in which the term includes, beside the machine industry proper, such branches of practice as engineering, agriculture, medicine, sanitation, and economic reforms."[58] This seemingly tossed-off definition of *technology* demonstrated the breadth of Veblen's concept, which included not only medicine but, somewhat oddly, economic policies.

Thus, by the first decade of the twentieth century, Veblen had taken *technology* far from its nineteenth-century meaning as the "science of the industrial arts." His concept was surprisingly sophisticated, in many ways compatible with the cultural understanding of the concept that historians of technology have developed since the 1960s. Technology for Veblen included knowledge as well as practices. It was closely related to but independent of science. This independence from science made technology applicable to all of human history and prehistory, not just to the era of modern industry. In addition, Veblen applied technology to a wide array of activities, from domestication of animals to large-scale industrial systems. He emphasized technology in use, refusing to reduce it to invention. Insofar as he had a theory of technological change, he stressed gradual accretions of skill and knowledge rather than major breakthroughs. Finally, his understanding of technology was, in principle, neither deterministic nor progressive. Veblen saw nothing automatically beneficial in the progress of technological knowledge, particularly when used for military purposes or for socially pernicious commerce. "Technological proficiency" was itself neutral, neither "intrinsically serviceable [nor] disserviceable to mankind."[59]

Veblen's view of technology as neutral might imply that he belongs more in the instrumental than the cultural camp. As I pointed out in chapter 2, describing technology as morally neutral in effect defines it as serving ends outside itself, and thus lacking inherent virtue. Yet in most ways, Veblen fits better with the cultural approach. In particular, his understanding of technology emphasized human agency, not the determining effects of material forces. He argued that the historical role of capital goods, and by implication technology, "is a question of how the human agent deals with the means of life, not of how the forces of the environment deal with man."[60]

There was, nevertheless, a latent determinism and progressivism in Veblen's theory. He assumed that the autonomous development of "technological proficiency" was beneficial unless contaminated by predatory instincts. Growth in the human mastery of the material world served as the principal progressive motor of history. This progressivist, deterministic undercurrent in Veblen's understanding of technology grew more pronounced in his later writings, especially in his last book, *Absentee Ownership*.[61]

Veblen's ideas about technology reached full flower in his most controversial work, *Engineers and the Price System*, a book based on a collection of articles he had written for the leftist periodical the *Dial* in 1919. Writing in the wake of the Bolshevik Revolution, Veblen detailed conditions for an overthrow of the "vested interests" in the United States, with the caveat that any such overthrow was a "remote contingency." As leaders of this revolutionary overthrow, he anointed "production engineers," whose rule would be effected by a "Soviet of technicians."

Most of the book's controversy can be traced directly to this surprising role for engineers.[62]

Veblen scholars differ deeply in their interpretation of *Engineers and the Price System*. According to Rick Tilman's summary of this debate, some scholars view the book as a "serious program of reconstruction for the American economy," while others see it as Swiftian satire intended to expose the failures and inequities of modern capitalism.[63] But in terms of Veblen's concept of technology, *Engineers and the Price System* was thoroughly grounded in his prior works.

Veblen's emphasis on engineers was a logical extension of his earlier analysis, particularly as developed in "On the Nature of Capital" and *The Theory of Business Enterprise*. These works portrayed technology as a form of knowledge belonging to the entire community, a point that Veblen repeated in *Engineers and the Price System*. Traditionally, he argued, skilled craft workers were the vanguard of technology. But the "new technological order" required not only skilled workmen but also "a corps of highly trained and specially gifted experts" who, like the skilled workmen, based their expertise on the collectively held stock of technological knowledge. These "experts" did not constitute a clearly defined occupation; Veblen suggested that they "be called 'production engineers,' for want of a better term." He most certainly did not advocate giving control of industry to the existing engineering profession, which he viewed as thoroughly subservient to the vested interests, a subservience that made the engineer the "awestruck lieutenant of the captain of finance." Veblen's "Soviet of technicians" was to consist not of these "commercial" engineers but rather of nonpecuniary industrial experts who would be faithful stewards of technological knowledge, which was the collective property of the entire community. Veblen's proposed Soviet of technicians constituted, in effect, a utopian vision of the liberating potential of technology, providing Veblen with a vantage point for critiquing the existing technological order.[64]

Engineers and the Price System was Veblen's most controversial and widely read book. Ruth Oldenziel has argued that the book portrayed "engineers as the chief bearers of technical knowledge" and helped spread a concept of technology centered on engineers. I am convinced that this was not Veblen's intent. Nevertheless, scholars as astute as Daniel Bell have reached essentially the same conclusion as Oldenziel. Many contemporary readers surely missed much of the irony in Veblen's analysis, as well as his distinction between the existing engineering profession and the nonpecuniary industrial experts who would constitute his Soviet of technicians. Similarly, many readers may indeed have seen an engineer-centered concept of technology in the book. Regardless of Veblen's intent, *Engineers and the Price System* undoubtedly helped popularize the idea of technology in general, understood as "the state of the industrial arts." This definition was much closer to the German concept of Technik than to the nineteenth-century English-language concept of technology.[65]

Ultimately, Veblen subscribed to a populist concept of technology rooted in human agency, with skilled workers and production engineers at its vanguard. Technological knowledge belonged to the "community at large," not just technical elites and certainly not commercial interests.[66] Technology for Veblen was not a form of elite knowledge that gave middle-class engineers and managers authority over workers. But this populist understanding of technology did not survive World War I.

NINE

Veblen's Legacy: Culture versus Determinism

Veblen introduced a new meaning of *technology* into academic English, specifically technology as industrial Technik. This new meaning spread among English-speaking academics beginning in the early 1900s, often in ways that can be traced directly to Veblen. Yet even as the concept of technology spread, it lost much of the critical edge that Veblen had used in his analysis of capitalism. Most social scientists, even those directly influenced by Veblen, jettisoned his antideterminism and his emphasis on human agency. Instead, they embraced technology as an autonomous, largely beneficent agent of social change. In this way, technology became linked to the ideology of progress, a link that emphasized the instrumental over the cultural aspects of the concept. Ultimately, some liberal academics embraced a crude technological determinism that differed little from the position of orthodox Soviet Marxists.

At the same time, scholars continued the unconscious appropriation of the German discourse of Technik, even when they were thoroughly familiar with Veblen's work. This engagement with the German concept of Technik added to the confusion. Prominent scholars continued to use *technology, technique,* and *technics* in ways roughly equivalent to *Technik*.

But a number of social scientists picked up Veblen's Technik-inflected meaning of *technology* in the first decades of the twentieth century, among them the economist Herbert Davenport and the historian Charles Beard. In the early 1930s, Veblen's *technology* burst out of the academy and became an element in the briefly popular technocracy movement, which advocated an economic system run by scientific experts. Only one scholar truly embraced the cultural aspects of Veblen's *technology*: Lewis Mumford. But Mumford's debt to Veblen was overshadowed by his direct engagement with the German discourse of Technik, as is clear from his choice of *technics* rather than *technology* as a core concept in his work.

Veblen's Early Influence and the Rise of Technological Determinism

Veblen's influence is hard to gauge in general, let alone for his specific contributions to the concept of technology. Although one of the most important scholars of the early twentieth century, he remained outside the main currents of American social science. His stress on evolutionary themes and qualitative methods was at odds with the growing influence of social-scientific models drawn from the physical sciences, especially in economics. Furthermore, his iconoclastic stance, oft-interrupted academic career, and unconventional personal life deprived him of all but a few graduate students. Lewis Mumford, a great admirer of Veblen, described him as a "suspected heretic in the academic world"; even Veblen's own students rarely acknowledged his influence.[1]

A further barrier makes it hard to assess Veblen's specific influence on the concept of technology. At the time, scholars did not recognize that he was reshaping the meanings of the concept.[2] As a result, scholars who adopted Veblen's usage did not credit him as the source. Nevertheless, there are many direct connections between Veblen and later scholars who used *technology* in a similar way in their work.[3]

One of the earliest economists to employ the term in its new sense was Veblen's student, colleague, and later patron, Herbert Davenport. In 1904, Davenport published an article on the concept of capital. Echoing Veblen's distinction between pecuniary institutions and industrial institutions, Davenport distinguished technological capital from competitive capital. For Davenport, competitive capital was defined by market valuation, whereas technological capital referred to "all wealth held for the purpose of further production." He was not, however, entirely comfortable with using *technological* in this sense. In a footnote, he remarked that "etymologically speaking, there are manifest objections to this use of the term 'technological' as referring especially to capital regarded in the mechanical and industrial sense; but no better term seems to be at hand."[4] Here we see clear evidence of the tension between the old meanings of *technology* as a field of knowledge and its new meaning as industrial arts.

At least Davenport felt discomfort with the terminological confusion around *technology*. Later scholars tended to treat the concept as unproblematic, not recognizing that they were using it in a new way. An early example of naïve usage is a 1921 article by a young economist, Alvin H. Hansen. Hansen took issue with Edwin R. A. Seligman's 1901 essay "The Economic Interpretation of History." He attacked Seligman's characterization of Marx's theory of history as "economic," insisting instead that this theory was "technological." In contrast to Seligman's deep familiarity with the principal texts of European Marxism, Hansen's research was shallow and monolingual. In his analysis, he insisted on defining Marx's "forces of production" as technology, while identifying "relations of production" as economics. In effect, Hansen echoed Sombart in defining Marx as a technological determinist. But more interesting was his nearly interchangeable use of *technology* and *technique* along with *technological* and *technical*, another example of the unrecognized merger of *Technik* into *technology*.[5] In addition, by drawing a sharp distinction between *technology* and *economics*, Hansen helped separate the concept from critiques of capitalism. His narrow view of *technology* marked a decline in conceptual sophistication from Veblen's usage, and a shift toward a more narrowly instrumental view.

This shift from critique to apologetics was even clearer when *technology* became linked to a deterministic concept of material progress. At the vanguard of this conceptual change was the progressive historian Charles A. Beard. After joining the faculty at Columbia University in 1904, Beard embraced the economic interpretation of history developed by his colleague Seligman. Beard left Columbia in 1917 to protest wartime infringements on academic freedom, and helped found the New School for Social Research in 1919. He was soon joined by Veblen, whom Beard greatly admired.[6]

Like many Progressives, Beard's enthusiasm for reform was coupled with a profound faith in the progress of civilization. The economic view of history, stripped of its Marxist dialectic, offered promise that material progress could lay the basis for moral progress. In the late 1920s, Beard added technology to economics as a motive force in history. This shift to technology occurred at a time when his faith in economics as the principal motor of history was beginning to fade.[7]

Beard first granted technology this new role in his presidential address to the American Political Science Association in 1926. In discussing the transformation of American society since 1783, he singled out the steam engine and the spinning machine as key agents of change. Such inventions, he claimed, suggested important lessons for the future, "the ideas of indefinite progress—the continuous conquest of material environment by applied science."[8] As Leo Marx, John Kasson, David Nye, and others have documented in great detail, Beard's wonders of material progress had long been standard rhetoric in American culture.[9] From the early years of the American republic, orators, editorialists, and intellectuals embraced the technological marvels of their day as visible manifestations of progress. As an American historian, Beard knew this rhetoric intimately.

Yet Beard added something fundamentally new by explicitly linking the concept of technology to the idea of progress in a way that made technology itself the motive force of history. He did so in paradigmatic language that reverberates into the present:

Not one whit less inflexible [than time] is technology—also a modern and Western Leviathan. Like time, it devours the old. Ever fed by the irrepressible curiosity of the scientist and inventor, stimulated by the unfailing acquisitive passion—that passion which will outlive capitalism as we know it and all other systems now imagined by dreamers—technology marches in seven-league boots from one ruthless, revolutionary conquest to another, tearing down old factories and industries, flinging up new processes with terrifying rapidity, and offering for the first time in history the possibility of realizing the idea of progress so brilliantly sketched by Abbé de Saint-Pierre.

Under the "convulsive pressures of technology," he continued, all systems of thought would be transformed.[10]

Beard's rhetoric was a forceful expression of the theory of technological determinism.[11] Historians of technology have generally traced this theory to the nineteenth century.[12] But in terms of the history of concepts, that origin story seems flawed. Without a concept of technology, nineteenth-century writers could not have articulated a theory of technological determinism. Instead, they had to work with their own concepts, most important art. Yet I have found no prominent nineteenth-century scholar who conceived of the arts, whether industrial, mechanical, or useful, as an autonomous motor of history. As I pointed out in chapter 5, the concept of art implies an artist or artisan. Even with the rise of large-scale machine production, craft models still dominated nineteenth-century ideas about the practical arts.

When it came to technological determinism, the real question was not whether one aspect of culture or society determined other aspects.[13] The real question was about the role of human agency, about the relationship between human choices ("free will") and constraints that limited, or even eliminated, these choices, whether these came from God, a Hegelian *Weltgeist*, nature, economics, or technology. When framed in terms of human agency, technological determinism is revealed as an extreme form of the instrumentalist view, incompatible with the cultural view. Technological determinism denied humans any role in setting the goals of technology, let alone deciding how to achieve these goals. Technology, in this view, was reduced to not just means serving goals set by others, but means serving an immanent end, progress, an end set without human agency.

In developing this deterministic view of technology, Beard was clearly influenced by Veblen. In effect, he made manifest the latent determinism in Veblen's theory, especially as expressed in Veblen's *Absentee Ownership* (1923).[14] Beard's concept of technology lacked Veblen's subtlety, yet proved more enduring for its later history.[15]

There were four key aspects of Beard's technological determinism. First was the notion of autonomous technological change, metaphorically likened to the inalterable movement of time, driven by "curiosity" and "acquisitive passion" grounded in human nature itself. This formulation harkens back to Veblen's theory of instincts while effacing his notion of conflict between productive and pecuniary pursuits. Second, technology, not economics, became the key determining force in history, ruthlessly transforming not just material culture but also intellectual and spiritual life. Third, this autonomous, deterministic force was not to be lamented but rather embraced as an agent of beneficent progress. Finally, Beard divorced technology from capitalism, insisting that its influence did not depend on any specific economic system. Except perhaps for this last point, Beard's prose could serve as copy for present-day technological enthusiasts.[16] His concept of technology, with its firm faith in human progress, was better suited to defending the established order than critiquing it.

Beard was a few years ahead of his time in his passionate use of *technology*. The term was not widely adopted in the social sciences until the 1930s.[17] Like most of his contemporaries, he did little to theorize technology, and he failed to recognize that he was adopting a novel concept. He used *technology* inconsistently, favoring instead phrases like "machine civilization" and "science and the machine," using this latter phrase interchangeably with *technology*, describing "science and the machine" as the unstoppable driving force of history.[18] With Beard, we see not only the wedding of *technology* with progress, but also the blurring of the distinction between *science* and *technology* that would bedevil future use of the term.

Veblen's influence is also apparent in other pockets of academic thought. Although most mainstream economists rejected his evolutionary economic theory, his work remained popular among Institutional economists. The field of Institutional economics served as a sort of loyal opposition to the dominant neoclassical trend in American economics after World War I. In some ways, the Institutionalists were the heirs to the younger historical school of German economics, the school of Schmoller and Sombart that influenced Veblen's ideas about technology. Clarence Ayres, a prominent early Institutionalist, embraced Veblen's ideas on technology. Ayres even argued in the 1920s that technology, not science, was the driving force of modern civilization. Like Beard, he put a determinist stamp on Veblen's concept of technology, though his take was less strident. In any case, his influence on the discourse of technology remained limited, despite his long academic career at the University of Texas.[19]

The Technocracy Movement: Veblen's Problem Child

Beard was not the only writer to give a determinist interpretation to Veblen's concept of technology. Technological determinism also pervaded one of the most bizarre episodes of the Great Depression, the technocracy movement. Technocracy was originally an outgrowth of the academic debate over technological unemployment, but did more to bring technology to the attention of the public.[20] The movement arose during the darkest days of the Great Depression. From 1932 to 1933, Americans witnessed the accelerating collapse of the nation's economy, which reached its nadir in 1933. Meanwhile, the federal government remained paralyzed under the weak leadership of Herbert Hoover.[21]

Absent this context, technocracy would never have generated wide public attention. But it flared briefly and brightly like a just-lit match. In two months, from December 1932 through January 1933, the popular press was entranced by technocracy, with nearly two hundred articles appearing in major American newspapers, along with dozens of stories in national magazines. Coverage collapsed as quickly as it rose, but the movement endured at a slow burn for the remainder of the Depression.[22]

Led by a charlatan and fabulist named Howard Scott, technocracy was an intellectually vapid movement that called for replacing capitalism with an economic system based loosely on the science of thermodynamics and managed by scientific experts.[23] Scott claimed "important engineering posts" in the German chemical industry, a degree from the University of Berlin, and work as a consulting engineer for U.S. Steel; but according to a reporter from the New York *Herald Tribune* who checked him out, "practically all of this story of Scott's career is pure fiction." Rather, he was a denizen of Greenwich Village, entertaining acquaintances with tall tales of his picaresque adventures while supporting himself by delivering floor wax from a company in New Jersey.[24]

Of course, there was more to technocracy than Scott. He drew his ideas about basing the economy on energy flows from Frederick Soddy, a Nobel Prize–winning British chemist. Soddy sought to answer the question of why poverty persisted despite the progress of modern science and industry. His solution was to abolish credit and measure wealth in terms of energy rather than money. Economists widely dismissed Soddy's economic writings as amateurish and deeply misguided, but his ideas did impress some scholars outside economics. In effect, technocracy was the work of a charlatan relying on a crackpot. Technology, however, was not part of Soddy's conceptual repertoire.[25]

Technocrats repeatedly stressed their debts to Thorstein Veblen, and especially his *Engineers and the Price System*. Scott even claimed, quite implausibly, that Veblen owed key ideas in the book to him. But as I pointed out in chapter 8, many readers of Veblen's book missed its deep irony, taking seriously his suggestion for a Soviet of technicians, even though he explicitly stated that this was unrealistic. Furthermore, Veblen was quite clear that his "technicians," whom he also referred to as "production engineers," were not to be confused with the existing engineering profession. In a sense, he was proposing something like a Leninist dictatorship of the proletariat, but with an aristocracy of skilled technicians as the vanguard party. But rather than advocating rule by a technical elite, Veblen instead sought an economic system based on technology as a form of knowledge that belonged to the people collectively, not to any particular group. His "Soviet of technicians" was to serve as a steward of society's technological knowledge, not as a dictatorship of experts.

Despite these differences, the technocracy movement succeeded in casting Veblen as one of its founding fathers. A new edition of *Engineers and the Price System* was published, and sales soared. The book was far more widely discussed in the 1930s than it had been when it was first published, undoubtedly helping spread familiarity with Veblen's concept of technology.[26]

The technocrats often invoked technology as a central concept. Despite their purported debt to Veblen, they had a strongly instrumentalist and determinist vision of technology, quite similar to that presented in Charles Beard's 1926 presidential address to the American Political Science Association. This determinism is strikingly clear in one of the few works that Howard Scott published under his own name in technocracy's heyday. This piece, titled "Technology Smashes the Price System," appeared in the January 1933 issue of *Harper's Magazine*. The title was a clear reference to Veblen.[27]

In the article, Scott laid out a set of proposals for reforming the economy. He supported these proposals with a series of non-sequiturs rooted in his claim that conscious human labor is equivalent to a fixed quantity of mechanical energy. Money had no meaning, he argued, "in a country where 0.44 of a single pound of coal can do the work that an average man can do in eight hours." Not only did Scott conflate energy with human labor, but he also did not appear to understand basic thermodynamics, most important the second law, which limits the conversion of heat energy into work. Yet his ignorance did not stop him from making apocalyptic predictions of a coming economic collapse rooted in technological unemployment, itself the result of a massive increase in labor productivity produced by technology.[28]

In describing how this crisis would happen, Scott reified technology as a violent agent of social change. Technology revolutionizes, advances, sweeps away, and breaks through, making irrelevant all past ideologies and social systems, he insisted. Because "technology has laid its hands upon the building trades," prefabricated houses were just around the corner, to be "put together with a socket wrench." Housing was just one example of how "in every industry technology has swept away the human worker." In language quite similar to Beard's, Scott insisted that "with each step in technology the stride becomes greater and greater and more and more men are pushed aside." Not once did he ever suggest that technology was the product of human agency, or even that it had any cause whatsoever.[29]

Other technocrats were more sober in their rhetoric, but all shared a crude, deterministic understanding of technology. In 1933, most of the better-trained supporters of the movement split with Scott, among them Walter Rautenstrauch, a professor of industrial engineering at Columbia University, and Harold Loeb, a wealthy bohemian, novelist, and magazine editor. Like Scott, Loeb had no real technical education or experience. In January 1933, just before he split with Scott, he published *Life in a Technocracy*, which articulated the movement in considerably more detail than Scott ever had.[30]

In his book, Loeb did use the terms *technology* and *technological*, but they were hardly central, especially when compared with technocracy's key ideas of energy and power. His vision of technology was similarly instrumental and determinist, and he also subscribed to the model of technology as the application of science. The first sentence of the first chapter provided a surprisingly archaic definition of *technology* as "the science of the industrial arts." According to Loeb, this "new transforming science" of technology was a product of capitalism. Yet technology, the "offspring" of capitalism, now threatened to "destroy its parent." Capitalism was based on scarcity, but this scarcity was "being progressively destroyed by the application of science to production, known in its latest phase as technology." In these passages, we can clearly see Loeb's confusion of technology as a field of science with technology as the application of science. Using classic deterministic language, he concluded that ultimately, "technological processes will compel a social system congenial to their operation or they will ruin the state."[31]

As noted above, the technocrats did not really believe in rule by technicians in Veblen's sense. According to Veblen, such skilled experts depended on "that joint stock of technological knowledge carried forward . . . by the community at large."[32] His technicians were the bearers of technological knowledge, which, as he had made clear in earlier works, was distinct from science.[33] In contrast, technocrats viewed technology as the application of science, thus making technocracy into rule by science or scientists.[34] The technocrats' rhetoric also reflected the removal of human agency implied in the shift from industrial arts to technology, and the removal of artistry from the artisan. For technocrats, the artisan was replaced by the technician or technologist, who carried out the dictates of science. Of course, nothing of the sort really happened. Technicians are not value-free data-processing machines; they belong to ethical worlds that profoundly shape their technical judgments. Technocrats actually understood very little about the nature of technical judgment. This was true even of engineers in the movement, such as Rautenstrauch, who was an academic rather than a practitioner.[35]

Reaction to technocracy was almost universally negative. The Right viewed the proposals as communism, while the Left saw fascism. Leading scientists and engineers denounced technocracy as a fraud, while social scientists condemned the technocrats' almost total ignorance of the social sciences. One response came from the American Engineering Council, the executive body of the Federated American Engineering Societies, an organization that sought to represent the interests of all branches of the engineering profession.[36] In January 1933, the council passed a resolution that rejected technocracy and explicitly defended technology. According to the council, "There is nothing inherent in technical improvement which entails economic and social maladjustments. Indeed, technology offers the only possible basis for continuing material progress."[37] But technology remained a peripheral concept in most critiques of technocracy by scientists and engineers. For example, Karl T. Compton, president of MIT, penned a chapter titled "Technology's Answer to Technocracy" for a 1933 collection of essays critical of the technocrats. Compton absolved machinery and invention of any responsibility for unemployment, but technology was invoked only in the article's title, apparently in reference to Compton's institution. In the more than two hundred pages of critiques of technocracy in the collection, technology was barely mentioned.[38]

Lewis Mumford: Veblen's True Heir?

While the technocrats were spreading their determinist version of Veblen's concept of technology, Lewis Mumford was completing one of the foundational works for the history of technology, *Technics and Civilization*. Published in 1934, this work represented the first major attempt to construct a cultural history of technology in English. *Technics and Civilization* is an ambitious but deeply flawed work, filled with internal contradictions and interpretive excesses. Nevertheless, it remained among the most sophisticated analyses of technology in English into the 1970s.[39]

Technics and Civilization raised many key themes for the history of technology. In it, Mumford examined the relationship of technology to work, warfare, science, capitalism, fine arts, and nature. Most important, he insisted on understanding technology as the product of human agency and as "an element in human culture," setting forth his passionate endorsement of human agency early in the book:

Technics and civilization as a whole are the result of human choices and aptitudes and strivings, deliberate as well as unconscious, often irrational when apparently they are most objective and scientific: but even when they are uncontrollable they are not external. Choice manifests itself in society in small increments and moment-to-moment decisions as well as in loud dramatic struggles; and he who does not see choice in the development of the machine merely betrays his incapacity to observe cumulative effects until they are bunched together so closely that they seem completely external and impersonal. No matter how completely technics relies upon the objective procedures of the sciences, it does not form an independent system, like the universe: it exists as an element in human culture and it promises well or ill as the social groups that exploit it promise well or ill. The machine itself makes no demands and holds out no promises: it is the human spirit that makes demands and keeps promises.[40]

Mumford's emphasis on human choice contrasted starkly with the approach of scholars like Charles Beard, who posited technology as an autonomous motor of history.

In his research, Mumford delved deeply into the literature of technology across multiple languages. But three scholars exerted the strongest influence on his thinking: Veblen, Sombart, and Patrick Geddes. From Veblen, Mumford drew a concept of technology inflected with a critique of capitalism. From Sombart, he absorbed the understanding of Technik as an element of culture. From Geddes, he gained a sensitivity to geography and local context, but also a dash of geographical determinism.

Although Mumford used *technology* frequently in *Technics and Civilization*, the term played only a secondary role in the book. The most central concept of *Technics and Civilization* was the machine. In popular discourse of the time, *the machine* was the dominant term for what we would now call modern technology.[41] In one sense, the term signified the whole of modern technology, with reference to just one part of it. But the term also represented modern technology as a complex system that often acted independently of the will of its users.[42] In this sense, the machine came to characterize the entire era, the machine age.

Yet the machine was in many ways a strikingly inappropriate metaphor for the leading technologies of the era, such as electric power, chemical synthesis, and electronic communications.[43] Two decades after *Technics and Civilization*, Mumford noted the limitations of the machine as an organizing concept, blaming his relative neglect of agriculture and medicine to the "machine-limited ideology" of the machine.[44] But he was being too harsh on himself. He was in fact less constrained than most of his contemporaries by the limitations of the machine, because he also embraced a broader, more fundamental concept, namely technics.

Historians have puzzled over Mumford's choice of technics rather than technology as a key concept in this foundational work. In a 1970 letter, he explained that he used *technics* to refer to the industrial arts themselves because *technology* implied an "abstract, rational pursuit."[45] Mumford's post hoc explanation is not supported by the text of *Technics and Civilization*, however. Although *technics* is the more common term, Mumford made many references to *technology*. He defined neither term, nor did he distinguish them, even implicitly. For example, he mentioned "earlier forms of technology" in one sentence, and the "specific properties of technics" in the next. He defined "the machine" as shorthand for "the entire technological complex," one that "embrace[s] the knowledge and skills and arts . . . implicated in the new technics." He argued on one page that each stage of "the machine" was characterized by a distinct "technological complex," which on the next page became a discussion of the "technical complex."[46] If there is a distinction in the book between *technics* and *technology*, it is very subtle.

Mumford's use of *technology* was almost certainly shaped by Veblen. The two men met in 1919 when they briefly worked together at the magazine the *Dial*. By then, Mumford was already a great admirer of Veblen. In his memoirs, he noted that he had read all of Veblen's books, which "stirred" him in the same way as the works of Geddes. He especially admired Veblen's penchant to ignore disciplinary boundaries, noting that Veblen, like Geddes, "refused to recognize the no-trespass signs that smaller minds erected around their chosen fields of specialization."[47] In *Technics and Civilization*, Mumford referred to Veblen frequently, drawing from his distinction between pecuniary business enterprise and productive industry, and praising his insight that "capitalism and technics, . . . so far from being identical, are often at war."[48] Yet it is difficult to find specific links between Veblen's abstract analysis of technology and Mumford's own understanding of technology and technics.

If Mumford's use of *technology* came from Veblen, from where did he get *technics*? Some scholars have assumed that it came from Mumford's mentor, the Scottish social scientist Patrick Geddes,[49] whose influence on Mumford has often been stressed.[50] Mumford borrowed some key ideas from him, most important his division of the industrial era into paleotechnic and neotechnic periods. But for Geddes, paleotechnic and neotechnic were just periods within the industrial age, not fundamental concepts embracing the industrial arts in general.[51] He almost never used the terms *technics* or *technology* by themselves, although he occasionally used the adjective *technic* more or less as a synonym for *industrial*.[52] It therefore seems unlikely that Geddes directly inspired Mumford's use of *technics* as a core concept.

Furthermore, *Technics and Civilization* represented, in many ways, Mumford's break with Geddes. Rosalind Williams has argued that the death of Geddes shortly before Mumford was to travel to Europe in the summer of 1932 had a "liberating effect" on Mumford. She describes how Mumford transformed the ideas he took from Geddes for *Technics*. Yet even as transformed by Mumford, these ideas retained a strong element of geographical determinism—for instance the "valley section," which linked patterns of work and settlement to locations within a valley cross section. Williams notes that "Mumford repeatedly borrowed abstractions from Geddes . . . which he then reified as historical actors in his own dramatic structures."[53] By drawing from Geddes, Mumford introduced significant contradictions into *Technics*. Many of Geddes's ideas were simply incompatible with the book's central idea, namely that "technics and civilization as a whole are the result of human choices and aptitudes and strivings."[54]

It seems likely, then, that Mumford's use of *technics* came not from Geddes but rather from the German discourse of Technik. Mumford's word choice was not as odd as it might appear today. *Technics*, though an obscure word in English, was perhaps the most appropriate term in the early 1930s for translating the industrial meanings of *Technik*. Noah Webster's 1828 American dictionary, for example, defined *technics* as "the doctrine of arts in general; such branches of learning as respect the arts."[55] By the late nineteenth century, *technics* had become an alternative to *technique* for translating the industrial meanings of *Technik*. As early as 1884, a German-English dictionary had suggested *technics* as one such translation.[56] Books and articles translated from Continental languages often used *technics* in this sense, among them Oswald Spengler's slim volume, *Der Mensch und die Technik*, translated into English as *Man and Technics* in 1932.[57] Mumford reviewed this book in the *New Yorker* that same year.[58]

Yet in actuality, Mumford's embrace of *technics* occurred rather late in the process of drafting the book that became *Technics and Civiliza-*

tion. As Williams has discussed, his original conception for his book was very different from the one he eventually wrote. The original project, titled "Form and Personality," was to cover not just "the machine" but also art, architecture, cities, and regions. In August 1930, Mumford published the first fruit of this project, an article in *Scribner's Magazine* titled "The Drama of the Machines." In this essay, he sought to assess the cultural value of the machine, acknowledging its "perversions" while appreciating its potential "spiritual contributions to our culture."[59]

"The Drama of the Machines" contained only a hint of the more general concept of technics that would come to frame *Technics and Civilization*. Rather, in the longer drafts from which the article was drawn, the unifying concept was provided by the arts, a concept that Mumford used in its older meaning to cover both practical and fine arts. Around 1930, he did not yet recognize that either *technology* or *technics* could supply a core concept for discussing the human transformation of the material world.[60]

The decisive point in transforming "Form and Personality" was Mumford's four-month research trip to Europe in mid-1932. During this trip, he encountered the full breadth of the Continental discourse of Technik. He made at least two visits to the Deutsches Museum in Munich. There, aided by the museum's director, Oskar von Miller, he probably delved more deeply into the literature of Technik than had any previous American scholar.[61] In a letter to a friend, Mumford anticipated his return to the United States in August, at which time he could "count over my loot," which included "fifty books whose existence I had not even suspected."[62]

In the months after his return from Europe, Mumford gradually decided to limit his book to technology and to publish the other parts of the project as separate volumes. Not until February 1933 did the draft chapters become recognizable as *Technics and Civilization*.[63] The title itself was a decisive step, anointing technics as a fundamental concept for the book. Precisely when Mumford chose this title is not clear, as he usually just referred to "the book" in his correspondence, but by the summer of 1933 he began mentioning the title.[64] By then, he had adopted technics as a key concept that mirrored the German concept of Technik.

The fruits of Mumford's encounter with the literature of Technik were reflected in the annotated bibliography accompanying *Technics and Civilization*. In it, he cited dozens of philosophical and historical works on Technik that had been almost entirely ignored by American academics. He included books by engineer-philosophers, skeptical humanists, social scientists, and early German historians of technology such as Franz Maria Feldhaus. In particular, Mumford acknowledged his debt to Werner Sombart. He likened Sombart's *Moderne Kapitalismus* to the Mississippi River, while describing his own book as "the railway train that occasionally approaches its banks."[65]

Yet the precise contributions of the discourse of Technik to Mumford's book are difficult to disentangle. *Technics and Civilization* is a synthetic work without footnotes, and Mumford rarely cited specific sources. It seems doubtful that he carefully read the roughly eighty German books included in his bibliography, and no research notes survive from his European trip. Nevertheless, the general influence of the discourse of Technik is readily apparent in the text. Mumford used *technics* in a humanistic way that at the time was quite novel for a scholar writing in English, but that wouldn't have been at all unusual in German. He stressed connections between technics and culture that were common in the German literature, insisting that "technics . . . exists as an element in human culture." Like Sombart, Mumford portrayed technics as both shaped by culture and a shaper of culture. Like Engelmeyer and other German engineers, he rejected romantic critics who viewed technics as the antithesis of culture. And like most German writers, he portrayed technics as a product of the human mind and spirit, that is, *Geist*. "The machine is just as much a creature of thought as the poem," he claimed.[66]

Technics remained Mumford's preferred term into the 1960s, though he also continued to use *technology*.[67] In 1958, he justified his use of *technics* in a letter to the newly formed Society for the History of Technology. As a founding member, Mumford recommended naming SHOT's proposed journal *Technics and Culture*. He explained that

there would be a great gain in differentiating between Technics and Technology, as the Germans have done. Technics is an accurate and useful term for denoting the whole field; and though technology has been used for this purpose in English since 1859, it would be better reserved for the systematic investigation of methods and processes. I realize that this usage is not yet general; but it is time that a start was made.[68]

In other words, *technics* for Mumford had always been a direct translation of *die Technik*. However, he never articulated this fact in print, which helps explain why *technics* was not widely adopted.[69] In fact,

Mumford contributed directly to a conflation of *technics* and *technology*, not just in *Technics and Civilization* but also in later works.[70]

In many ways, Mumford's technics represented a full-throated appropriation of the German discourse of Technik, especially those aspects emphasizing the cultural relevance of Technik. In following the German discourse, in particular Sombart, he also remained true to the cultural view embodied in Veblen's concept of technology. But Mumford was not an academic. He was a public intellectual and popular author, making a living through writing. As an outsider to the academic world, he had limited influence on scholarly discourse.

Veblen's precise contribution to the concept of technology remains unclear. He directly influenced a number of writers who embraced the industrial-arts meaning of the term, even though most of them abandoned the critique of capitalism inherent in his version of the concept. Other scholars no doubt picked up the term from reading Veblen's work, but without leaving evidence of their debt to Veblen. Irrespective of Veblen's legacy, *technology* gradually became a more common term in the social sciences before World War II, though primarily in a determinist and instrumentalist sense.

TEN

Technology in the Social Sciences before World War II

From the turn of the twentieth century until World War II, the concept of technology gradually diffused throughout the social sciences. This diffusion did not, however, lead to a cohesive understanding of the term *technology*. As academic boundaries between the social sciences hardened, several largely independent discourses of technology emerged, primarily in economics, sociology, and anthropology. By the early 1930s, the concept of technology had largely shed its nineteenth-century meaning as the science of the industrial arts. By World War II, *technology* had become a general term in the social sciences that referred to the material means of production, transportation, and communication.

Despite occasional declarations about technology's central role in human history, it remained far from a core concept in any branch of social science.[1] Academics rarely reflected deeply on technology as a concept in social theory, continuing the historical neglect of human productive activities by elite scholars. Some social scientists defined technology as outside the boundaries of their disciplines. Others embraced an instrumental or determinist vision of technology. Almost without exception, these scholars never acknowledged that an alternative view was possible, one that rejected determinism, questioned instrumentalism, and acknowledged technology as an integral component of culture. Social scientists almost completely ignored Veblen's belief in technology as the collective property of the common man, along with Mumford's optimistic faith in technics as the product of human values.

Between about 1900 and 1940, social scientists began using the word *technology* in a number of ways that were more or less independent of Veblen. In the first two decades of the twentieth century, sociologists and psychologists sometimes used *technology* to distinguish between the theoretical and the practical aspects of a social science, while economists began using the term as part of the debate over technological unemployment that began just before the financial crisis of 1929. Among leading sociologists, Talcott Parsons embraced a definition of *technology* as instrumental rationality, drawing from Max Weber's theory of social action. And in the 1930s, the sociologist William Ogburn became the most prominent scholar before World War II to make technology an explicit field of study. However, Ogburn tended to reduce *technology* to *invention*, and he promoted a crudely deterministic understanding of the effects of technology on society.

Boundary Work: Technology as Applied Social Science

Veblen's concept of technology, with its clear connection to the German discourse of Technik, was not the only one circulating among social scientists in the early twentieth century. Another meaning of *technology* was carried over from its nineteenth-century definition as the science of the arts: the idea of technology as the practical aspects of a science. This usage was part of a larger discourse of technology as applied science, which is discussed in the next chapter.

A number of social scientists embraced *technology* in the sense of practical science in the early twentieth century. In this era, many academics adopted a pure-science ethos in defense of professional autonomy, particularly in the growing research universities, which were often funded by conservative businessmen or meddlesome state legislatures.[2] In connection with this pure-science ideal, a number of prominent social scientists sought to bolster their pure-science credentials by distinguishing their field of science from technology. *Technology*, when used in the sense of practical or applied science, was usually viewed as subordinate or inferior to pure science. This view of technology as practical science emerged parallel to Veblen's redefinition of *technology* as industrial arts. These two definitions were incompatible, yet scholars did not appear to notice.

The idea of technology as practical science is clear in the phrase "social technology." A number of sociologists adopted this phrase in the first two decades of the twentieth century to refer to the practical side of the social sciences. The phrase first appeared in the work of Albion Small, one of the founders of the Chicago school of sociology, along with his colleague C. R. Henderson. Henderson and Small treated *social technology, applied sociology,* and *practical sociology* more or less as synonyms.[3] Small contrasted *social technology* with *general sociology.* He defined *general sociology* as concerned with abstract principles of "human associations" rather than practical problems.[4] Henderson, however, refused to subordinate social technology to general sociology. He insisted on an interactive relationship between theoretical sociology and social technology, both of which he classified as sciences: "the debt and obligation are reciprocal, and progress in each science depends on parallel progress of the other." As Maarten Derksen and Tjardie Wierenga point out, Henderson had a strong commitment to the project of social reform, which informed his framing of the relationship between social theory and practice.[5]

After World War I, American sociologists increasingly abandoned their commitment to social reform in favor of disinterested social science. Small's influential successor at Chicago, Robert Park, endorsed a value-free sociology that renounced "all ameliorist goals, which could only obstruct the dispassionate and systematic production of knowledge."[6] With the waning of the reformist impulse, academic sociologists abandoned Henderson's nuanced, interactive concept of social technology. Instead, they embraced a subordination of theory to practice modeled on their flawed understanding of the physical sciences. As Robert Park and Ernest Burgess stated in a widely used textbook of the period, "Science, natural science, is a research for causes, that is to say, for mechanisms, which in turn find application in technical devices, organization and machinery, in which mankind asserts its control over physical nature and eventually over man himself." The phrase "social technology" quickly declined among sociologists, replaced by a mechanistic model of applied social science sometimes referred to as social engineering.[7]

Yet *technology* in the sense of applied science did not disappear from the social sciences. A few academic social scientists took advantage of the ambiguity in the applied-science meaning of the term to make two arguments at once in support of disciplinary boundaries. First, they argued that their disciplines were, or should be, pure science, dedicated to the pursuit of truth and divorced from application, from technology. Second, they insisted that technology depended on the discoveries of pure science for its general principles, with technology left to deal with specifics. This kind of dual argument had been common since the second half of the nineteenth century among professionalizing natural scientists,[8] but it was rarely linked to technology before the twentieth century.

The Cornell psychologist Edward Titchener provides a striking example of these twinned arguments in a 1914 essay in *Popular Science Monthly*. This magazine, the predecessor to *Popular Science*, often published articles by prominent scholars aimed at general audiences. Titchener sought to free academic psychology from its dependence on what would now be called clinical psychology. He began his essay by complaining about demands that psychology devote itself to practical problems, claiming that such demands were widespread "in many other fields of scientific work." Such "hostilities" between practice and theory, however, were largely the result of a misunderstanding, "the neglect of a very elementary distinction," that between science and technology. According to Titchener, the fundamental distinction was about aims or state of mind. Science was distinguished from technology by its "disinterested attitude." The "germ of science" arose when people began valuing some aspect of knowledge as "intrinsically desirable." Titchener insisted that "scientific activity aims at no goal." Scientists do their work "self-directed upon an endless task." To demand that a scientist "interest himself in 'practical ends,' is simply to bid him cease from scientific activity."[9]

Technology, in contrast, was "defined by its end or goal." Titchener claimed that he was using *technology* in a broader sense than its usual meaning. In this broader sense, *technology* covered those "activities . . . ordinarily and misleadingly referred to as 'applied science,'" activities such as engineering, medicine, industrial chemistry, and eugenics. For Titchener, the distinction between the field of science and that of technology was absolute and definitional. But even though these fields were mutually exclusive, they were "closely related." Although technology depended on unscientific concepts such as "common sense" and "existing technologies," it also "draws continually on science." In turn, it helped advance science, in part by uncovering "some defect in theoretical knowledge." But technology depended more on science than science did on technology. Technology could not in itself be relied on to advance theoretical knowledge, as it remained "limited by its end," the solution of a particular problem.[10]

There was little new in Titchener's arguments about the purity of science, aside from the way he linked these arguments to technology.[11] He excluded technology from the sciences, because he regarded it as dependent on pure science, noting that "the technologist, for the very sake of his technology, needs the stimulus, the criticism and the assistance, of the man of science." Titchener clearly sought to establish a hierarchy, even though he claimed that he was not disparaging technology. His view stood in contrast to Henderson's, which drew from the older meaning of *technology* as a field of science, thus putting each on a similar epistemological plane. Titchener followed the British scientist Thomas Huxley in denying the existence of "applied science" at all, instead treating all application of knowledge as technology. As it had for Huxley, this denial helped him construct an intellectual boundary to protect academic scientists against the intrusion of practical problems.[12]

But why did social scientists use technology to do this boundary work? Huxley, John Tyndall, Henry Rowland, and other nineteenth-century pure-science advocates had gotten along perfectly well without the concept.[13] Henderson and Titchener began using *technology* in the sense of applied science during the same period when Veblen was importing the industrial-arts meaning of *Technik* into *technology*. Although I have found no direct link between the discourse of Technik and the social-science discourse of technology as applied science, circumstantial evidence suggests that such a link does exist. Both Small and Titchener did their postgraduate work in Germany. Small even married a German woman he met during his studies.[14] Henderson was thoroughly familiar with German social sciences and cited German sources extensively.[15]

This familiarity with German social science undoubtedly connected these scholars with the German concept of Technik. There is, in fact, a direct German cognate of *social technology*, written as both *soziale Technik* and *Sozialtechnik*. The term first appeared in the title of an 1881 pamphlet by the prominent Göttingen chemist Julius Post, who received some minor attention from the German press for his proposed system of guaranteed work for the unemployed. The phrase became somewhat more common over the next two decades, appearing in fields such as jurisprudence and pedagogy, and also in the widely read journal founded by Weber and Sombart, the *Archiv für Sozialwis-*

senschaft und Sozialpolitik.[16] Regardless of where Small and Henderson obtained their concept of social technology, they were clearly using it in a way similar to the instrumental meaning of *Technik*, as a set of methods for achieving practical ends. *Technology* in this instrumental sense referred to neither the science of the arts nor the industrial arts themselves.

But the idea of technology as applied social science was just a minor episode in the genealogy of *technology*, demonstrating the continued muddled meanings of the term. Over in the field of economics, probably the most clearly defined discipline in the social sciences, *technology* was rarely invoked before the 1930s. What changed the language of economics was the emergence of a discourse on technological unemployment in the late 1920s.

The Absence of *Technology* in Nineteenth-Century Economic Theory

Before the late 1920s, *technology* was a rare term among mainstream economists. As Guido Frison has pointed out, classical political economists almost never used the term *technology* or its cognates.[17] The term is absent from the work of Adam Smith, David Ricardo, and Thomas Malthus, and rarely used by later economists of the nineteenth century, including the fathers of marginal utility theory, William Stanley Jevons, Carl Menger, and Léon Walras. But this absence was not just about the term but also about the concept the modern term refers to. Classical economic theory had no concept dedicated to the role of technical knowledge, practices, and artifacts in production. Instead, its scholars used a variety of overlapping terms, such as *improvement, invention, industrialism, machinery, arts, applied science,* and others. None of these concepts was adequate on its own to stand for all economically relevant technological change.

Thus, when the scholars of classical political economy turned to theories of production, they were hindered by a historical legacy that ignored, disparaged, and subordinated the mechanical arts. This legacy made it difficult for them to grasp the key role of technical skill, and hence technology, in the rise of modern industry. The famous Austrian-American economist Joseph Schumpeter addressed this neglect of technology in his unfinished history of economic theory. Referring to the "pessimist" school of Malthus, Ricardo, and James Mill, Schumpeter noted that they wrote "at the threshold of the most spectacular economic development ever witnessed." But these theorists predicted "nothing but cramped economies, struggling with an ever-decreasing success for their daily bread." In this scenario, "technological improvement" would "fail to counteract the fateful law of decreasing returns."[18] Some historians of economic thought have challenged the harshness of Schumpeter's judgment, among them Maxine Berg.[19] Yet in the end, Schumpeter's assessment remains valid. In an era of widespread and often spectacular technological change, change that confounded predictions of stagnation, none of the leading British political economists grasped the key role of technology in economic growth, regardless of the terms they used.

This neglect continued in more modern forms of economic theory. As noted in chapter 7, Karl Marx's limited understanding of skill also undermined his analysis of the production process. But at least Marx paid serious attention to the technology of production. The economists who pioneered the marginalist revolution in the last third of the nineteenth century had little concrete interest in production, and that revolution gave birth to neoclassical economics, which remains the dominant form of economic theory today.

The founders of the mathematical theory of marginal utility were of course aware of technological change, but it remained peripheral to their analysis. This is the case with William Stanley Jevons, who pioneered the use of calculus in economic theory in his book, *The Theory of Political Economy* (1871). For example, he noted that his theory was in agreement with the classical economic doctrine on the tendency of the rate of profit to fall. "Unless there be constant progress in the arts, the rate of interest must soon sink toward zero." In other words, technological change, "progress in the arts," would counteract the trend toward economic stagnation. Yet Jevons did nothing to develop this point.[20]

The Swiss economist Léon Walras, another key pioneer of marginal utility theory, also grappled with the problem of incorporating technology into economics. In his theory, Walras proposed a set of equations with "coefficients of fabrication" that specified the proportion of inputs (labor, land, and capital) required for each product of an enterprise. These coefficients varied with the prices of inputs and with capital investments. But more fundamentally, changes in what he called "productive services," the introduction of new services and abandonment of the old, would change "the very nature of the coefficients of fabrication." This type of change Walras termed "technical progress [*progrès technique*]."[21]

German-speaking economists developed similar marginalist theories that applied the concept of Technik to the technical aspects of pro-duction.[22] Yet Technik was explicitly treated as external to economic theory, as a factor that impinged on the economic system from the outside. For example, in Eugen von Böhm-Bawerk's highly influential *The Positive Theory of Capital*, first published in 1889, he referred several times to *Produktionstechnik*, production technology, and he repeatedly used the adjective *technische* to distinguish economic from technologi-cal factors. He argued that the complex steps needed to produce most goods were "a purely technical fact," but that "to explain questions of technique [Fragen der Technik] does not fall within the economist's sphere."[23]

This quotation is taken from the 1891 English translation of Böhm-Bawerk's book. It shows that when German economic theory was trans-lated into English, *Technik* was usually rendered as "technique," a prac-tice that continued well into the twentieth century, despite Veblen's alternative translation of *Technik* as "technology." When Alfred Mar-shall, the great synthesizer of marginalist theory, published a book on industry in 1919, he used *technique* almost exclusively instead of *tech-nology*, typically in the phrase "industrial technique."[24] Even American Institutional economists, who were all influenced by Veblen, often used *technique* and *technology* interchangeably.[25]

The Economic Debate over Technological Unemployment

After the stock market crash of 1929, the debate over technological unemployment migrated from academia into public discourse, and economists' preference for *technique* over *technology* began to change as well. Amy Bix describes this debate in detail; it involved economists, engineers, business leaders, labor unions, and the federal government, and continued into the 1950s.[26] The debate itself was not new, as writ-ers on the topic admitted. Concern about the replacement of "men by machines" went back well before the British Industrial Revolution.[27] But the use of *technological* in this context was new, as indicated by its frequent enclosure in quotation marks.[28]

This new usage began before the collapse of the American stock market in late October 1929. Industrial production was booming in the 1920s, both in value and in volume, yet industrial employment appeared to have barely budged overall. In many growing industries, such as railroads and steel, employment actually fell significantly. Economists started worrying about falling employment in 1928, but they lacked reliable data, as the federal government did not conduct national surveys of unemployment before 1940.[29]

There were, however, employment figures available for a number of key industries and for industrial production as a whole. These figures were already troubling economists in 1928 and early 1929. In a typical analysis, the labor economist Sumner Slichter pointed out that industrial employment was lower in 1927 than in 1920 by roughly 2.3 million workers, even though production had increased substantially in mining, manufacturing, and transportation, implying major increases in output per worker. The loss in industrial employment was partly compensated by employment growth in the service sector, which included new occupations like bootlegging, but "no one knows precisely how large" this increase was. Slichter concluded that the number of unemployed people had probably grown since 1923, but the actual figure was highly uncertain. Estimates of this increase ranged from zero to three million.[30]

Not until 1928 was the term *technological* added to *unemployment*. Slichter, for example, did not use it in his 1929 paper, even though the paper was presumably the published version of his 1928 talk titled "Technological Unemployment."[31] The phrase was still somewhat controversial, and also quite political. When the Socialist Party candidate Norman Thomas launched his presidential campaign in August 1928, he demanded a "remedy for what has been called 'technological unemployment,'" which he defined as "unemployment due to increased efficiency in the use of machinery." Thomas insisted that the "benefits of machinery" should result in "increased leisure for the many" rather than "increased profits for the few."[32]

Widespread attention to the new phrase also arose from government-sponsored investigations of economic problems. In early 1929, the University of Wisconsin labor economist John R. Commons used the phrase in a congressional hearing on unemployment, where he distinguished "what we are now learning to call the technological unemployment" from seasonal and cyclical unemployment. He and other witnesses at the hearings linked this problem to rapid increases in industrial efficiency in the 1920s.[33] Commons's comments on technological unemployment drew further attention when the hearings were published in the spring of 1929.[34] A few months later, the US Department of Commerce released the report of president Herbert Hoover's Committee on Recent Economic Change. It too highlighted the problem of technological unemployment, drawing additional attention in academic journals and the mainstream press.[35] After the October 1929 stock market crash, the use of *technological unemployment* became even more common in the popular as well as the academic press.[36]

One would think that the phrase "technological unemployment" would have helped make technology into a major concept in Depression-era discourse. Yet for the most part, no such change occurred, at least not at first. One reason was the continued confusion between *technique* and *technology*, both of which were still being used in the same way as *Technik* in its industrial sense. The early debate about technological unemployment was framed not around technology but rather in terms of inventions, machinery, and, most significantly, technical change.[37] For example, in 1936 the Works Progress Administration launched a program to study the problem of technological unemployment that was called the National Research Project on Reemployment Opportunities and Recent Changes in *Industrial Technique* (emphasis added). As the decade progressed, the terminology gradually shifted to *technology* and *technological*, but not completely. The first report from the National Research Project repeatedly used the phrases "industrial technique" and "production technique" alongside "technological change." Yet just three years later, in 1940, a summary of the project's findings was framed overwhelmingly in terms of technological change and technological progress.[38] But economists never fully abandoned their use of *technique* as a synonym for *technology*, particularly in literature on the "choice of technique."[39]

Despite the ongoing terminological confusion, the debate over technological unemployment did help spread awareness of technology as a key concept of industrial modernity, but participants in the debate did little to develop this concept. For the most part, economists produced a decidedly determinist and instrumental understanding of technology. Because neoclassical economists treated technology as an external force that the economy responded to, they rarely examined the causes of technological change.[40] Even when discussing specific technologies in detail, their analyses remained descriptive.

Consider the final report from "Technology and the Concentration of Economic Power," a series of congressional hearings. These hearings were conducted in 1940 by the Temporary National Economic Committee of the US Senate, which framed the issue of technological change in terms of the "impact of technology on modern society." The final report of this committee, titled *Technology in Our Economy*, never addressed the causes of technological change, even when describing these changes in detail. John M. Blair, the economist who drafted the report's analytical section, repeatedly described technology as an unmoved mover, an external instrument that transformed the economy. Blair explicitly defined *technology* as applied science, "the application of the analytical methods of science to the industrial arts." Yet he understood technology narrowly as a tool for "the displacement of human effort," that is, for enhancing labor productivity. He almost exclusively discussed technology in terms of its effects, impact, and influences, and thus reified it as an economic actor, a force that, "by bringing forth new industries, causes the economy to expand." He also claimed that technology encouraged economic concentration, "steadily increas[ing] the power at the command of giant concerns."[41] In Blair's analysis, technology was both instrumental and determinist, a narrow tool for reducing labor costs and an external force shaping the economy from the outside.

Talcott Parsons: Embracing Max Weber

So much for the economists, who failed to play a leading role in developing technology as a concept in the social sciences before World War II. Sociologists, in contrast, contributed more to this development, though they did little to reduce the muddle of meanings. Two very different sociologists illustrate how confusion about the concept of technology continued: Talcott Parsons and William Ogburn.

Talcott Parsons was the most important American theoretical sociologist of the twentieth century. His concept of technology was directly shaped by his deep engagement with European social theory, especially that of Max Weber. Yet more than any other English-language sociologist before World War II, Parsons embraced an explicitly instrumental definition of *Technik*, one that helped marginalize technology within the social sciences.[42]

Parsons began his immersion in European theory during his graduate work at Heidelberg University, where he arrived in 1925 just a few years after Weber's death. While there, he studied philosophy, economics, and sociology with an impressive group of scholars, including Karl Mannheim, Karl Jaspers, Max Weber's younger brother Alfred, and the innovative Austro-Marxist Emil Lederer. Under the tutelage of Edgar Salin, a young political economist, Parsons produced a dissertation on capitalism that focused on the ideas of Marx, Sombart, and Weber. He defended the dissertation in the summer of 1927, just before leaving for a position at Harvard University.[43]

In 1928, Parsons drew from his dissertation for his first major academic publication, a two-part article on Sombart and Weber titled "'Capitalism' in Recent German Literature." The first part focused on Sombart's *Moderne Kapitalismus*. In it, Parsons consistently and repeatedly translated *Technik* as "technique."[44] He described Sombart's division of technique into three historical stages: first, traditional "medieval technique," which then gave way to a "rational" technique in the early modern era, followed by "modern technique [which] is scientific." This union of science and *Technik* had profound consequences, which Parsons captured in a particularly memorable phrase of Sombart's: "To the elimination of God [*Entgöttlichung*] from the conception of nature corresponds the elimination of man [*Entmenschlichung*] from technique."[45]

In the article, Parsons's assessment of Sombart was quite positive; he was more critical of Weber, in particular his method of ideal types.[46] But soon after his arrival at Harvard, he quickly switched his allegiance to Weber, whose ideas became the cornerstone of his theory of social action. In 1930, Parsons published his translation of Weber's *Protestant Ethic and the Spirit of Capitalism*, which remained the standard English version of the book for seventy years. In his translation, he continued the practice of rendering *Technik* as "technique," even when it was clearly referring to the industrial arts.[47]

In mistranslating *Technik* as "technique," Parsons was following the common practice of social scientists of his day. Yet by the late 1920s, many of his peers had shifted to *technology* as a general term for the industrial arts, often inspired by Veblen's usage.[48] As a result of this new meaning of *technology*, scholars began using both *technique* and *technology* to translate *Technik*, depending on the context. One of these scholars was a friend of Parsons, the neoclassical economist Frank Knight, who produced the first major translation of Weber into English. Knight had begun using *technology* in its Veblen-influenced sense in the early 1920s. In Knight's 1927 translation of Weber's *General Economic History*, he typically rendered *Technik* as "technology" when used in the sense of the industrial arts, though he sometimes translated this sense of *Technik* as "technique," for example in the phrase "industrial technique."[49]

During the 1930s, Parsons gradually accepted technology as a significant concept in his social theory. He altered his translation practices accordingly, and began using *technology* to render the industrial meaning of *Technik*. But Parsons went a step further. As part of his development of a general social theory of action, he also began using *technology* to translate the second major German meaning of *Technik*, that is,

Technik as any method for achieving a particular goal. By translating this meaning of *Technik* as *technology*, Parsons in effect redefined *technology* as the narrowest form of instrumental reason, precisely mirroring Max Weber's definition of *Technik*.

This shift in Parsons's terminology occurred in the context of German debates over the boundaries between Technik and *Wirtschaft* (technology and economy). He drew explicitly from Weber's distinction between the two (see chapter 7).[50] Weber defined *Technik* as the narrowest form of instrumental rationality (*Zweckrationalität*), a rationality that dealt only with the choice of means to a single end. In contrast, *Wirtschaft* dealt with the allocation of means among multiple ends. Thus, *Technik* in Weber's sense excluded considerations of costs, because costs implied comparison between means for a variety of alternative ends.[51] Weber admitted that costs were indeed central to industrial Technik, that is, technology. Yet he elided the distinction between industrial and instrumental Technik, illustrating his argument exclusively with examples taken from the industrial arts.

Parsons followed Weber's lead in conflating the two meanings of *Technik*, merging both into the term *technology*. In the mid-1930s, he began developing ideas for his first book, *The Structure of Social Action*. He drew from Weber's instrumental definition of *Technik* as part of his theory, shifting his terminology from *technique* to *technology*. Parsons's change first occurred in a 1934 article in which he defended the British neoclassical economist Lionel Robbins against a critique by Ralph Souter, a young Institutional economist.[52] Souter claimed that neither *economics* nor *technology* could be defined by an abstract formalism that disregarded the concrete ends of human action.[53] He disputed the marginalist divorce between economics and technology, arguing that "the meticulously constructed antithesis between 'economics,' with its 'plurality of ends,' and 'technology,' with its 'single end,' now dissolves into thin air."[54]

Parsons rejected Souter's critique of the distinction. He did so by embracing the instrumental definition of *Technik*, but now translated in terms of *the technological*. Parsons argued that "the simplest means-end relationship," which was the choice of means for a single end, "in a very broad sense . . . may be called the 'technological.'" This use of *technological*, he noted, "does not imply 'physical' or 'material' as is so often the case in defining the concept."[55] In other words, Parsons's definition included immaterial techniques as well, a clear echo of Weber's discussion of Technik in *Economy and Society*, but rendered now as "technology." By shifting from *technique* to *technology*, Parsons in effect layered the instrumental definition of *Technik* onto existing meanings of *technology*.[56]

These various ideas about *technology* all converged in Parsons's first book, *The Structure of Social Action*, now a classic of sociological theory.[57] The book represented a maturation rather than a rupture in his work. In it, Parsons made his instrumental definition of *technology* an integral part of his theory. He incorporated themes from his previous decade of writings, among them the idea of technology as the rational choice of means for a single given end, which he referred to as the "unit act." Parsons defined this "elementary atom" of action as "technological." Technological action in this sense was governed by the principle of "efficiency," which Parsons defined vaguely as "attainment of an end with a minimum of 'sacrifices.'" He insisted, however, that his "concept of technology" went beyond mechanical or energy efficiency, also including, for example, "the technology of mystical contemplation."[58]

Parsons thus made it clear that he was defining *technology* not in industrial or even material terms, but rather as the methods—or more accurately the choice of methods—for achieving a definite end.[59] In fact, he repeatedly referred to *technological* and *technology* in this instrumental sense throughout *The Structure of Social Action*.[60] Yet in this book he did little to develop a theory of technological action, even arguing against the possibility of such a thing. Parsons claimed that an "independent analytic science" of technology was impossible. Technological action (that is to say, the unit act) formed "the common basis of all the sciences of action." Therefore, he claimed, it could be studied only by a set of disciplines defined by the "concrete ends" of action, "such as industrial, military, scientific, erotic, ritual, ascetic, contemplative, artistic, etc." Parsons termed these (nonexistent) disciplines "the technologies." However useful these disciplines might be "concretely," he concluded, "they add relatively little to the systematic analytical theory of action."[61]

This argument is bizarre for more than just its weak, reductive logic. In effect, Parsons was excluding technology from his disciplinary division of labor in the social sciences, except as an empirical supplement to the true sciences of action. This exclusion prevented him from theorizing about technical action in general, since the sciences of action took the solution of technological problems as given.[62] But more important, his analysis also made it impossible to understand technology as a set of material practices. It was *technology* in this material sense that was most significant for the social sciences, especially those that grappled with industrial modernity.

Parsons cemented this instrumentalist definition of *technology* with his translation of the first volume of Weber's *Economy and Society*, published in 1947, a project begun by Alexander M. Henderson but completed by Parsons. In a footnote to Weber's discussion of Technik, Parsons addressed the difficulty in translating *Technik* into English. He correctly noted that *Technik* could be translated as both "technique" and "technology," depending on context. Nevertheless, he continued to translate *Technik* as "technology" even when Weber was clearly referring to instrumental action, which would have been more accurately rendered as "technique."[63]

This extension of *technology* beyond its industrial meanings represented another redefinition of the term. This new definition made explicit the instrumentalism that had always been present in how elite scholars understood craft skill and technical practices. Parsons's work thus contributed a third meaning to the definitional muddle, *technology* as a narrow form of instrumental reason, divorced from culture and unworthy of theoretical analysis. The meaning was added to the more common definitions of *technology* as applied science and *technology* as industrial arts.

Parsons's direct influence on the discourse of technology is hard to assess. Because he explicitly denied technology a place in his sociological theory, he was rarely cited in discussions of the concept. After World War II, a minority of scholars began to use *technology* to mean instrumental reason or methods. They could have picked up this usage through reading *The Structure of Social Action* or, more likely, through Parsons's translation of *Economy and Society*, which was standard reading for any aspiring sociologist.[64] At least one early historian of technology drew from the instrumental definition of *technology* to extend the concept to nonmaterial techniques, with clear allusions to Weber.[65] Somewhat later, organizational sociologists also adopted this practice by using *technology* to refer to bureaucratic procedures, that is, organizational techniques.[66]

Regardless of Parsons's direct influence, his appropriation of the instrumental meaning of *Technik* exemplifies the confusion that attends the translation of concepts with multiple meanings. The conflation of instrumental and industrial meanings of *Technik* persisted in European scholarship after World War II, especially in philosophy. That scholarship continued to shape English-language discourses of technology, exemplified by the works of European philosophers like Herbert Marcuse and Jacques Ellul.[67]

William F. Ogburn and the Sociology of Technology

By the late 1930s, American social scientists had largely settled on us-ing *technology* to refer to the material practices of industrial civilization. But the multiple meanings of *technology* continued to coexist, often in the same work. *Technology* was sometimes used in Veblen's sense as the state of the industrial arts, particularly among economists, who were concerned with the effects of technologies in industry. But economists also described *technology* as the application of natural science, a defini-tion that absolved them of any need to explain technological change within economic theory. And lurking in the background was Parsons's definition of *technology* as instrumental rationality.

As the term *technology* spread in the social sciences during the 1930s, particularly in sociology, it was increasingly linked to the con-cept of invention. This link emerged from a group of scholars that Da-vid McGee has termed the "early sociologists of invention."[68] McGee includes a number of people in this group, but only one, William F. Og-burn, fully embraced the term *technology* before 1940. More than any academic in the 1930s, it was Ogburn who established the sociology of technology. But in founding the field, he also cemented a deterministic understanding of technology centered on inventions and the applica-tion of natural science.

William Ogburn is perhaps the most forgotten major American so-ciologist of the twentieth century. He served as president of the Amer-ican Sociological Society in 1929, when he was viewed as the "high priest of scientific sociology."[69] Although he made significant contri-butions in the use of quantitative data over his long academic career, Ogburn is most often remembered for his cultural lag theory of inven-tion. He developed this theory in *Social Change with Respect to Culture and Original Nature* (1922). Much of the book was an argument against biological determinism. To that end, Ogburn deployed the concept of culture, drawing from Edward B. Tylor's 1871 definition as "that com-plex whole which includes knowledge, belief, art, morals, law, custom, and any other capabilities and habits acquired by man as a member of society."[70] Ogburn embraced the word *culture* to counter social sci-entists who modeled their theories on biological sciences, especially Darwin's theory of evolution. This intellectual shift, as Charles Camic and others note, provided sociologists with a way "to demarcate an au-tonomous realm . . . irreducible to the biological," a realm that would "constitute the proper sphere of the social sciences."[71]

Tylor's multifaceted definition of *culture* became the basis for Ogburn's theory of cultural lag. According to Ogburn, whenever changes in one sphere of culture outpaced changes in related spheres, the result was a cultural lag. In principle, lags could occur between any areas of culture, but in practice the greatest lags occurred when "material culture" changed more rapidly than "nonmaterial culture." When this happened, "strains" and "maladjustments" arose, creating social problems.[72]

Correcting these maladjustments required an active process of cultural adaptation, which could take decades. In one of the few detailed examples in the book, Ogburn described workmen's compensation laws as an adaptive response to rising rates of industrial accidents in the nineteenth century. Before these laws were enacted, the common law made it very difficult for injured workers to recover damages from their employers. According to Ogburn, the legal system lagged behind the industrial system, requiring almost half a century before states passed new workmen's compensation laws that provided reasonable protections for workers injured on the job. He treated the adaptive process as a mechanistic response, ignoring the bitter Progressive Era struggles that were required to enact these laws. Yet in his analysis, the adaptive response was not fully determined by the change in material culture. He pointed out that a variety of alternative adaptive responses were possible, among them "factory inspection, machinery safeguards, rest periods, . . . and perhaps prohibition of the sale of intoxicating beverages."[73]

In *Social Change*, the word *technology* was most notable for its absence. Ogburn used the term only about four times in the book, along with *technique* in the sense of industrial arts, as in the "technique of industry." Rather than the concept of technology, the governing concepts were material culture and invention. Although Ogburn noted that Tylor's definition of *culture* did not stress its material side, he insisted that "the use of material things is a very important part of the culture of any people." As the sociologist Toby Huff argues, Ogburn understood *material culture* primarily as artifacts.[74] This definition excluded material practices and knowledge. Since technological knowledge was at the heart of Veblen's concept of technology, it is not surprising that Ogburn made few connections to Veblen's writings on technology.

For Ogburn, autonomous change in material culture was largely the product of inventions. He was adamant that inventions were not primarily the work of great men. Instead, particular inventions were, "if not inevitable, . . . certainly to a high degree probable" at specific stages of the "growth of culture."[75]

There was nothing particularly novel or insightful in Ogburn's theory of invention, which remained quite underdeveloped. More sophisticated theories were developed by scholars such as the anthropologist Otis T. Mason and the sociologist S. Colum Gilfillan, who was Ogburn's own protégé.[76] And even though Ogburn seemed to make bold claims, he also hedged. Everything in *Social Change* implied the priority of material over nonmaterial culture, yet Ogburn nevertheless admitted that "some changes in non-material culture" could come first and induce adaptive change in material culture.[77] Despite such caveats, he clearly privileged the role of technical inventions in social change, but without adequately explaining the origins of the inventions themselves.

During the 1930s, technology increasingly became a core concept in Ogburn's theory of cultural change. He took the meaning of this concept for granted, taking advantage of its imprecision to conflate technology with invention. This conflation, so common today, was actually rare before the 1930s.[78] With increased attention given to *technology* in the early 1930s, it was easy for Ogburn to simply add *technology* to his terminology without rethinking his categories. In 1931, he published a popular article highlighting "technological progress" along with "inventions and scientific discoveries" as agents of change. As an example, he predicted that "rapidly growing technology" could eliminate poverty when combined with a declining population.[79]

Ogburn conflated technology and invention more explicitly in his work for President Hoover's Research Committee on Social Trends, which Hoover established just after assuming office in 1929. Ogburn was not only a key committee member but also the research director for the entire effort. In this position he oversaw "forty prominent investigators" and "an army of trained assistants."[80] The study was published in 1933 under the title *Recent Social Trends in the United States*, with twenty-nine chapters totaling over sixteen hundred pages.

The terms *technology* and *technological* were used freely throughout *Recent Social Trends*. Ogburn wrote the chapter titled "The Influence of Invention and Discovery," assisted by Gilfillan. At the start of the chapter, Ogburn proclaimed that

science and technology are the most dynamic elements of our material culture. Through technology men transform the physical environment, so that men, natural resources and inventions and discoveries are the primary factors which determine the wealth, standard of living and well being of a people.[81]

In this quotation, Ogburn clearly treated *technology* as equivalent to "inventions and discoveries." Elsewhere in the chapter, he equated *inventions* with *technological progress*. In the introductory section, "scientific discoveries and inventions" were similarly linked to "changing technology" as a major source of social problems. "More and more inventions are made every year, and there is no reason to think that technological developments will ever stop," proclaimed the unnamed section author, almost certainly Ogburn.[82]

By 1938, Ogburn's reductive embrace of technology as invention was complete, as demonstrated by "Technology and Sociology," an article he published in *Social Forces*. In his most theoretical discussion of technology yet, he practically merged it with invention, science, and the industrial arts generally. In a wonderfully muddled definition, he noted that "the word technology will be used to include applied science, and will be interchanged with invention and scientific discovery, which are causes of change in technology." In other words, he intended to use *technology, invention*, and *scientific discovery* interchangeably, even though *technology* was itself shaped by the latter two categories. At the same time, he also linked technology to material culture broadly, using the concept to encompass agricultural tools and the domestication of plants and animals. In effect, Ogburn combined in *technology* the ideas of applied science, invention, and technology-in-use. By merging applied science and technology, he was able to present technology as largely self-determining. And by negating the distinction between invention and technology-in-use, he could claim that technological change mechanically determined social change. Thus, he formulated the classic statement of technological determinism.[83]

Using this framework, Ogburn concocted examples of connections between inventions and social change. These examples read like "a parody of technological determinism," as the historian Rudi Volti puts it.[84] Echoing Charles Beard, Ogburn claimed that "technology develops, is let loose on society, sweeping all before it."[85] In one example, he portrayed elevators as a cause of increased employment for married women. "The invention of the elevator leads to an increase in the number of homes in apartment houses which leads to an increased density of population, which tends to lower the birth rate, which sets free more of a married woman's time, which finally increases the number of women employed outside the home." Similarly, in an article on the future consequences of invention, Ogburn argued that contraception would lead to scarcity of children and universal higher education, so that "plumbers will discuss Aristotle . . . as well as members of the professions."[86]

Ogburn's shortcomings as a theorist of technology are obvious in retrospect, but by the end of the 1930s his invention-centered sociology of technology had become highly influential, in part through his connection with the New Deal. The New Deal largely bypassed sociologists, who in the 1920s had tried to make sociology seem scientific by divorcing it from social problems. Even though Ogburn championed this divorce, he also supported the New Deal and built connections with the Roosevelt administration.[87] These connections helped him spread his understanding of technology not just in sociology but also in the popular press.

In late 1935, Ogburn drafted an ambitious proposal for another large government-sponsored study, "The Influences of Technology and Science on Social Change with Particular Reference to Planning." He sent the proposal to Frederic Delano, President Roosevelt's uncle, who chaired the Advisory Committee of the National Resource Committee, a prominent New Deal organization that included key members of the president's cabinet. The proposal was accepted, and Ogburn became chair of the Subcommittee on Technology, along with two other prominent academics, the biologist John C. Merriam, president of the Carnegie Institute of Washington, and Edward C. Elliott, president of Purdue University.[88]

As the director of the study, Ogburn encouraged contributors to adopt his invention-centered concept of technology, urging them to select technologies on "the frontier of change." His instructions conflated *invention* and *technology* pretty directly. For the chapters on specific technical fields, Ogburn instructed authors to "deal with those important technologies that are likely to be socially significant during the near future. These will be inventions (a) that are in existence but undeveloped, (b) that are in use to only a limited extent, and perhaps (c) that will probably be made." In other words, "important technologies" are equated with "inventions."[89]

In June 1937, Ogburn's subcommittee published its report in one large-format volume titled *Technological Trends and National Policy, including the Social Implications of New Inventions*. The bulk of the report consisted of separately authored chapters addressing "technology in various fields," which were mainly tedious compendia of recent developments in key industrial sectors. Ogburn and S. Colum Gilfillan restated their theories of invention and social change, while Merriam and Elliot contributed short, platitudinous essays on the relationship of science to technology.[90] In Ogburn's sections, there were some changes from his earlier work. Most important, Ogburn gave science a much more prominent role than he had in the 1920s, as he belatedly embraced the arguments for industrial research spread by the National Research Council (see chapter 11). He attributed an "increasing part of . . . technological advances" to "science and research," noting that "scientific discoveries in applied science and invention do have important social consequences."[91]

In addition, as Benoît Godin argues, Ogburn also refined the idea of invention as a process consisting of sequential steps. He was not the first to propose such sequences, but he went further. Rather than ending each of his sequences with the success of an invention, Ogburn went on to describe the invention's social effects. "In many important cases," he argued, "the change occurred first in the technology, which changed the economic institutions, which in turn changed the social and governmental organizations, which finally changed the social beliefs and philosophies." Although Ogburn did not claim that this sequence always held, he nevertheless imagined the process in mechanistic terms, using the metaphor of collisions produced by a billiard ball.[92]

Technological Trends did much to enshrine Ogburn's muddy concept of technology, which was thoroughly instrumental, stripped of inherent moral value or craft skill. For Ogburn, technology centered on invention and depended on science while possessing tremendous power to transform society, politics, and morality. People had little choice in the direction of change, even through collective action; they could only retard the inevitable adoption of new technologies. In many ways, this vision of technology still dominates today.[93]

Still, Ogburn's *Technological Trends* marked the arrival of *technology*, in the sense of industrial arts, into the mainstream. The book was widely reviewed and discussed in both the scholarly and the popular press, and the foreword, which summarized the book's findings and recommendations, was reprinted in its entirety in *Science*.[94] Several months before the book appeared, Ogburn asked Waldemar Kaempffert, science editor for the *New York Times*, to run a story on the project. Ogburn was rewarded with a detailed front-page article in the paper's Sunday edition. This was followed two days later by another glowing piece, with the headline "America's Future Studied in Light of Progressive Application of Technology; Inventions Point to Changed World." The semantic bond between *technology* and *invention* could not be clearer.[95]

Ogburn and his followers went on to pioneer both scholarly and popular studies on the social effects of new technologies.[96] In 1941, Laura and McKee Rosen published a popular book titled *Technology and*

Society. McKee Rosen, who had a PhD in political science from the University of London, had worked under Ogburn on *Technological Trends.* Ogburn wrote a lengthy introduction to the McKees' book that recapitulated his invention-centric concept of technology and its social effects.[97] Ogburn himself coauthored a 1940 sociology textbook that included a chapter titled "The Social Effects of Inventions." Although *technology* was used throughout the book, the index entry under "technology" directed the reader to "See inventions."[98]

By the early 1940s, *technology* had become an established term in American social sciences. Economists adopted the concept in response to the debate over technological unemployment. In sociology, Ogburn's invention-centered approach dominated. In anthropology, where the term had been abandoned in the 1890s, the evolutionary anthropologist Leslie White and the Marxist archaeologist V. Gordon Childe both began using *technology* in the sense of industrial arts in the early 1940s.[99] A few political scientists, perhaps spurred by Charles Beard, considered the relationship between technology and government.[100] Scholars in the new field of industrial relations also began using the term.[101]

Even though *technology* was used widely among social scientists by the late 1930s, as a concept it continued to be neglected, in part because it was not central to any scholarly discipline or profession. *Technology* was not yet an academic keyword, and thus did not matter enough for anyone to fight over.[102] Nevertheless, by 1940 technology was at the threshold of becoming an indispensable category, one useful for understanding not just modernity but all of human history.

Yet something major was still missing: a deeper understanding of the relationship between science and technology. The idea of an association between these two concepts was new to the twentieth century. Before then, the discourse of art and science dealt with this relationship. But with the narrowing of the concept of art, this discourse became obsolete. Instead, as Veblen showed, a transformed concept of technology could provide new ways to think about connections between the natural sciences and material practices. As the next chapter will show, most historians of science accepted an interactive model of the science-technology relationship, while physical scientists were more likely to adopt a concept of technology as the application of science.

ELEVEN

Science and Technology between the World Wars

Technology has been closely linked to *science* since at least the mid-eighteenth century, when Christian Wolff defined *technologia* as the "science of arts and of works of art." Yet because *technology* was defined as a branch of science, there was no talk of a relationship between the two terms before the twentieth century. A sustained discourse about the relationship between science and technology did not emerge until after World War I. This discourse drew from the association of technology with higher technical education, an association that helped support a definition of *technology* as applied science, especially among scientists and engineers. Yet the concept of applied science remained deeply ambiguous. It could refer to a field of practical knowledge, or it could imply the applications of academic science. Meanwhile, a parallel discourse about science and technology continued in German and other Continental languages, a discourse grounded in the industrial meanings of *Technik*. In the 1930s, historians of science drew from this Continental discourse to develop their own approach to the science-technology relationship. These scholars were responding in part to Soviet Marxists who applied a materialist view to the history of science in the early 1930s.

In a very real sense, this twentieth-century discourse about science and technology continued centuries of tension between scholars and craftsmen, between people who shape the world of ideas and people who shape the

material world. Like the early modern discourse about natural philosophy and art, the science-technology discourse was also divided into
roughly two camps. On one side were elite scholars, typically natural
scientists but also social scientists, who tended to subordinate technology to science and practice to theory. For scientists, this subordination
was often aimed at securing their social status by defining the boundaries of science as a profession.

On the other side were practitioners of various stripes, primarily
inventors, engineers, and industrialists, though rarely skilled workers.
Only a few practitioners denigrated formal scientific knowledge in favor
of hands-on experience and practical wisdom; Thomas Edison, for example, was famously skeptical of university-trained scientists, at least in
public.[1] Most practitioners presented a more nuanced understanding of
the science-technology relationship, particularly when compared with
the simplistic and ideologically charged claims of spokesmen for science.

However, elite scientists and practitioners were not really that divided in their concrete understanding of the science-technology relationship. Most practitioners occupied a middle ground, similar to the
zone of exchange between early modern scholars and craftsmen that
the historian Pamela Long describes.[2] Many natural scientists were also
comfortable in this middle ground, especially experimentalists and scientists in industry.[3] From there, both scientists and engineers could accept an interactive model of the science-technology relationship, one
that acknowledged the importance of natural science to technology,
but only as one of several key factors. At the same time, this interactive
model also allowed for the reciprocal influence of technology on science. In this conception, science and technology were seen as interacting, autonomous fields of knowledge and practice, each with its own
set of norms and goals.

It was only after World War II that most academics scientists (and
some academic engineers) endorsed a subordination model of the
science-technology relationship. This model led to a definition of *technology* as applied science, or more specifically, as the application of science. This definition, which first appeared in dictionaries just before
the war, became widespread during the Cold War.[4]

Historiography of the Science-Technology Relationship

Over the past fifty years, works by historians about the relationship
of science and technology have both helped and hindered historical

understanding. Many of these works have questioned the definition of *technology* as applied science. This research has demonstrated that much technological change owes nothing to institutionalized science. And even when scientific principles are essential to the success of a technology, the transformation of an abstract principle into a successful technology is never trivial. Instead, the path from scientific theory to technological practice requires creativity as well as persistence, the production of new knowledge along with new devices.[5]

But the rejection of the applied-science definition of *technology* served another purpose, once again helping shape professional boundaries. Beginning in the early 1960s, historians of technology used the critique of technology as applied science to assert the autonomy of their field from that of the history of science. At the same time, this argument helped support the social status of the engineering profession by rejecting its subordination to the natural sciences. Many early historians of technology came from engineering backgrounds or taught in engineering schools. During the Cold War, engineers suffered deep anxieties about their social status, especially in light of the high cultural prestige of natural scientists. It was particularly galling for engineers when natural scientists received credit for the engineering successes of, for example, the US space program.[6]

Yet there was a downside to this focus on disproving the applied-science definition of *technology*. Historians of technology tended, like most professionals doing boundary work, to reify key concepts, ignoring ambiguity and changes in meaning. In particular, they often interpreted the applied-science definition of technology in the narrowest way, as a claim that all the practical achievements of modern technology, from automobiles to radios, were reducible to the application of natural science, particularly scientific theory. In doing so, historians were arguing against a straw man. Advocates of academic science did often claim credit for the fruits of technological progress, typically when seeking funding from nonscientists. Yet they rarely claimed that technology could be reduced to the application of scientific principles.[7]

This historical context has made the science-technology relationship a vexing and inconclusive realm of scholarship. The discourse of this relationship should be heir to more than two millennia of debate about the relationship between art and science. Yet as I pointed out in chapter 5, the narrowing of the concepts of art and science in the nineteenth century produced a rupture between this older discursive tradition and a new twentieth-century discourse of science and technology. This rupture in effect erased any wisdom that had accumulated on the topic. Twentieth-century scholars had to reconceptualize this relationship from the ground up, in almost complete ignorance of the changing meanings of the concepts they used. The result was and continues to be a conceptual muddle.

Real progress in understanding the science-technology relationship requires two things. First, scholars need to acknowledge that the meanings of *science* and *technology* are neither fixed nor unified, and second, scholars need to come to terms with how their own social position has shaped and continues to shape these concepts and their relationship.

My argument here is not new. In 1976, the historian of technology Otto Mayr published a devastating critique of scholarship on the science-technology relationship.[8] He cast doubt on attempts to develop a general model of the science-technology relationship in light of the diversity revealed by case studies. Mayr urged historians to focus instead on the discourse about this relationship. He counseled historians to "recognize that the concepts of science and technology themselves are subject to historical change; that different epochs and cultures had different names for them, interpreted their relationship differently, and, as a result, took different practical actions."[9] Since the 1980s, historians of science and technology have largely embraced the negative side of Mayr's critique, abandoning attempts to generalize about the science-technology relationship. Yet until recently, historians have neglected Mayr's positive recommendation that they focus on the concepts used by historical actors themselves as a way of understanding what we now consider the science-technology relationship.[10]

Nevertheless, an ideal of "pure science" still interferes with analysis of the science-technology relationship.[11] This ideal has aristocratic roots.[12] The insistence on purity has long been a defense of privilege, a demand for autonomy from the social and political forces of the moment, an insulation from engagement, all of which conceal existing relationships of power. In recent years, some academics have tried to give the ideal of pure knowledge a progressive cast, using it as a foil against corporate capitalism and the neoliberal policies that have become increasingly dominant since the end of the Cold War.[13] Such arguments have worthy ends, but ultimately prove ineffective, because they obscure the economic and social structures required to support the autonomy of academic researchers.

Trends in the historiography of science and technology point to productive ways to go beyond this impasse. Beginning with innovative studies of experimental practice in the 1980s, historians of science have embraced the materiality of science and the diversity of scientific practices.[14] These trends have borne fruit in diverse scholarship that connects scientific concepts and material practices without essentializing the concepts of science and technology.

Most of this new work quite self-consciously avoids generalizations about the science-technology relationship. In the essay collection *The Mindful Hand*, editors Roberts, Schaffer, and Dear explicitly instructed their contributors to avoid using terms like *science* and *technology*. This injunction makes a great deal of sense, as it helps the contributors avoid applying anachronistic categories to the past. Yet understanding the past always requires a translation into the language of the present. In part, the contributors to *The Mindful Hand* merely avoid the ideologically charged pairing of *science and technology* by substituting less common and somewhat archaic concepts, such as "inquiry and invention" or "contemplation and manipulation."[15] This strategy of enforced strangeness can help historians grasp the past on its own terms, but it leaves unfinished the process of translating it into the present. In a sense, this chapter tries to complete this task of translation by showing how a new discourse about the science-technology relationship arose to replace the older discourse of art and science.

The Ambiguities of *Applied Science* before World War I

Explaining the new science-technology discourse requires a return to the nineteenth-century debates about applied science introduced in chapter 5. These debates rarely mentioned *technology*, but were instead framed around the claim that the natural sciences were essential to progress in the industrial arts. Much of this discourse emerged from middle-class "men of science" who sought to assert authority over the knowledge of artisans. These scientists claimed that they knew better than skilled craftsmen about how work should be performed.

But this summary slights deep ambiguities in the meaning of *applied science* that continued into the twentieth century, ambiguities that allowed applied science to support conflicting models of the science-technology relationship, both the subordination and the interaction models. In his classic article on the concept of applied science, Ronald Kline has identified four meanings of the term that were common among American engineers in the early twentieth century. First was application of the theories of science to the industrial arts. Second was application of the scientific method to practical problems, for example through systematic experimentation. Third was applied science as "a relatively autonomous body of knowledge," that is, autonomous from professional scientists. Fourth, some engineers defined *applied science* in terms of research and teaching, primarily in higher education. By using these ambiguous meanings, status-conscious engineering spokesmen were able to do a type of double boundary work using *applied science*. They drew from the high status of science to raise engineers above skilled workers while also staking a claim to professional autonomy from science.[16] The result was a kind of conceptual jujitsu, as engineers, especially academic engineers, sought to benefit from the cultural prestige of science without subordinating themselves to scientists.[17]

Ultimately, these ambiguities rested on the question of whether applied science was subordinate to or autonomous from "pure" (typically academic) science. The historian Graeme Gooday argues that *applied science* in Britain originally referred to "an entirely autonomous domain of practical knowledge." According to Gooday, Charles Babbage, the great advocate for British science in the first half of the nineteenth century, viewed pure and applied science as part of a "nonhierarchical division of labor." And Andrew Ure, also writing in the 1830s, was even more insistent on the autonomy of applied science. Ure's "science of the factory" had little in common with academic theory.[18]

But, explains Gooday, roughly half a century later, another great propagandist for British science, Thomas Huxley, rejected the autonomy of applied science. In an oft-quoted passage, Huxley "wish[ed] that this phrase, 'applied science,' had never been invented." Applied science was not distinct from pure science, he insisted. Instead, "applied science is nothing but the application of pure science to particular classes of problems." This famous quotation was often invoked to deny autonomy to applied science, as Huxley had intended. Yet as Gooday shows, it was also used to argue for research of immediate practical utility, the complete opposite of Huxley's intent. After all, if there was no real distinction between pure and applied science, as Huxley claimed, then science devoted to industrial problems was no different from the disinterested speculations of a pure scientist. This argument appeared, for example, in the first annual report of Britain's Department of Scientific and Industrial Research in 1917, which explicitly referenced Huxley.[19]

These multiple, conflicting interpretations of *applied science* continued well into the twentieth century, during the same period in which the meaning of *technology* in scholarly discourse was also changing. In fact, the very idea of a relationship between science and technology was premised on a change in the meaning of *technology* itself, its Veblen-inspired redefinition as Technik.

As long as *technology* was defined as the science of the industrial arts, it made no sense to discuss its relationship with science, a relationship of part to whole. Talk of a science-technology relationship would have been no more logical than that of a science-chemistry or a science-botany relationship. Scholars could and did discuss the nature of chemical, botanical, and technological science, but without questioning the fact that these fields were sciences. It was only after 1900, when Veblen and other social scientists transferred the meanings of *Technik* to *technology*, that it was possible to view science and technology as autonomous fields that could enter into some kind of relationship, even if that relationship was one of subordination. Hence the irony: only by redefining *technology* as industrial arts could the concept be equated with the application of science.

Thus, *technology* was almost never defined as applied science before the twentieth century, although it was still linked to the discourse of applied science. These links are evident throughout the nineteenth century, appearing as early as Jacob Bigelow's *Elements of Technology* (1829). As noted in chapter 6, Bigelow actually said very little about *technology* in this book, and he never defined the term. But the title of the book did link *technology* to the application of science: *Elements of Technology: Taken Chiefly from a Course of Lectures Delivered at Cambridge, on the Application of the Sciences to the Useful Arts*. In the introductory section, he repeatedly invoked the application of science. He rhapsodized on the "abundant rewards" produced by the "practical applications of science," while also praising the "application of philosophy to the arts" for making "the world as it is today."[20] But *technology* was not essential to Bigelow's discourse of application, as demonstrated by his 1840 revision of the *Elements*, which eliminated all references to *technology*, even in the title.[21]

Other nineteenth-century sources show similar links to the rhetoric of applied science. The 1848 translation of Friedrich Knapp's *Chemical Technology* significantly altered the title to emphasize application. In English, the full title was *Chemical Technology; or, Chemistry, Applied to the Arts and to Manufactures*. A more literal translation would have been *Textbook of Chemical Technology for Instruction and Self-Study*. The altered subtitle helped explain an unfamiliar term to English speakers; *technology* concerned the application of natural science to the industrial arts. But this connection remained limited to the title; nothing in the text linked *technology* to applied science.[22]

These connections were suggestive, but to actually define *technology* as the application of science would have been inconsistent with the term's nineteenth-century meaning as the science of the arts. When *technology* was added to discussions of applied science, the result was often confusion, as is clearly evident in a speech given by the US Secretary of War, Joel Poinsett, in 1840. Poinsett was president of the National Institution for the Promotion of Science, a short-lived organization that sought to create the first national scientific society in the United States. The National Institution was formed in large part to claim the bequest of James Smithson, the Englishman who left half a million dollars to the United States to create "an establishment for the increase and diffusion of knowledge."[23]

In his speech, Poinsett sketched out the framework for an organization worthy of Smithson's bequest, an "Institution for the Promotion of Science and the Useful Arts." This organization was to have eight branches, ranging from astronomy to fine arts, with one branch devoted to "the application of Science to the useful Arts." In describing this branch, Poinsett invoked *technology*, which he defined as "that science which teaches this application," that is, the application of science to the arts. In effect, he defined *technology* as a science of the application of science. He seemed oblivious to the awkwardness of his definition, even though he was fairly well informed about the concept of technology, making allusions to Bigelow's *Elements* and noting the teaching of technology in German universities.[24]

Only in the first third of the twentieth century did the ambiguous meanings of *applied science* spill over explicitly into the meaning of *technology*. The applied-science definition of *technology* appealed to members of new professions who were insecure about their social status, especially social scientists and engineers. Engineers in particular typically understood *applied science* as autonomous, although their rhetoric was inconsistent. This confusion is visible in one of the first extended discussions of *technology* as applied science, Henry Taylor Bovey's lecture titled "The Fundamental Conceptions Which Enter into Technology," presented at an international conference held in conjunction with the 1904 St. Louis World's Fair. Bovey was a distinguished British academic and engineer, and at the time Dean of the Faculty of Applied Science at McGill University in Canada.[25]

In a rambling but fascinating talk, Bovey articulated an interactive view of the science-technology relationship, speaking from the standpoint of a "technologist." Like pretty much every engineer who spoke about applied science, he stressed that successful practice required subservience to the laws of nature. This argument has been a truism since the time of Francis Bacon, whose widely quoted aphorism states that "we cannot command nature except by obeying her."[26] Yet as Kline's analysis of *applied science* has shown, subservience to the laws of nature does not necessitate subservience to a particular professional group, namely scientists.

Nothing in Bovey's talk was particularly novel. What was new, however, was the explicit connection between *technology* and *applied science*, which he treated as different names for the same area of knowledge. The demands of "practical necessity," he suggested, helped create "the science which has received the descriptive title of applied science and the general title of technology." Bovey accepted the common argument that pure science was pursued without any reference to utility, but that when scientific laws were linked to utility, "we call it applied science or technology." Overall, he concluded, "I see no essential difference between the use of the two terms 'applied science' and 'technology.'" The one exception was "the science of medicine, . . . an applied science which is never described as technology."[27]

Bovey did not believe that engineers or other "technologists" were subordinate to scientists. Instead, he deployed all four of Kline's categories of applied science to defend the autonomy of technology, which, he insisted, clearly deserved to "rank amongst the sciences." When he defined *technology* as a science, Bovey invoked the German term *Wissenschaft*, traditionally translated as "science," "which seems to embrace ordered knowledge of every kind." In contrast, the English term *science*, he noted, was equivalent to what the Germans called exact science.[28] By using the broader definition of *science* as systematic knowledge, Bovey was able to define *technology* as an autonomous body of knowledge, a definition broad enough to include even the natural history of the trades, a key component of cameralist *Technologie*.

But Bovey went further, also pointing out that new scientific laws could emerge from the act of applying science. Such laws could arise, for example, in studies of the properties of steam or electricity. Thus, the engineering profession, and by implication technology in general, contained "within themselves a pure science." In other words, technology could uncover the laws of nature in mathematical form.[29]

Finally, Bovey tried to blur the lines between *fine art* and *technology*. He argued that beauty and utility were compatible virtues while claiming that the imagination was as necessary "in the highest departments of technology" as it was in the fine arts. He also insisted that technology was a source of virtue, producing "moral strength, truth, and manliness."[30] Such claims to moral virtue were at the same time assertions of professional status, much like the moral virtues claimed by elite artisans in the early modern era, or by German *Techniker* beginning in the 1880s.[31] Such views are central components of what I have termed the cultural view of technology.

Bovey's linking of *technology* and *applied science* was relatively novel in 1904.[32] If this link had been a commonplace, he wouldn't have needed to equate the concepts explicitly, nor would he have qualified his statement by using the first person ("I see no essential difference") rather than just stating this equivalence as a fact. In any case, engineers rarely made technology the focus of their writings before World War II. Bovey's topic was chosen not by him but by the organizers of the St. Louis congress, as part of a volume on the "utilitarian sciences" grouped under medicine and technology.[33]

Technology among Natural Scientists between the World Wars

The definition of *technology* as applied science required that science and technology be linked in a new way. This link was expressed in the phrase "science and technology." This phrase first emerged from the nineteenth-century discourse of applied science, but did not catch on before the 1920s. Like "applied science," "science and technology" was deeply ambiguous. Although spokesmen for the natural sciences were almost unanimous in their praise of pure science, they almost never explicitly subordinated technology to science before World War II.

The phrase "science and technology" remained rare before the 1920s, being far less common than phrases linking *art* and *science*. As a first-order approximation, Google n-grams provide a reasonable estimate of the frequency of particular words and phrases. Comparing the n-gram "science and technology" with "arts and sciences" dramatically demonstrates the decline of one discourse and the rise of another (see figure 3). "Science and technology" barely registers on the graph before the 1920s. Most instances of the phrase before 1920 are actually false positives, the result of errors in the Google Books database. In contrast, "arts and sciences," despite a decline from the mid-1820s, remained more frequent than "science and technology" until 1956. Interestingly, in conjunctions of *art* and *science*, *art* was just as likely to precede *science* as the reverse, while the phrase "technology and science" has always been relatively rare, suggesting an unequal relationship between the two terms.

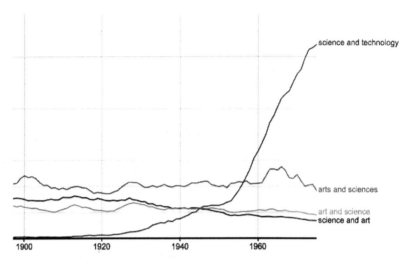

FIGURE 3 Google n-gram plot. Google Books Ngram Viewer, accessed 9/9/2017, https://books.google.com/ngrams/graph?content=science+and+technology%2Cscience+and+art%2Cart+and+science%2Carts+and+sciences&year_start=1895&year_end=1970&corpus=15&smoothing=2. Many instances of "arts and science" are nineteenth-century survivals, such as the "arts and sciences colleges" in many universities. But the n-grams are case sensitive, and therefore should exclude capitalized phrases such as journal titles and names of institutions.

However, n-grams tell us little about what these phrases meant, especially across time and in varying contexts. Even in context, "science and technology" remained deeply ambiguous; writers almost never defined what they meant when they used the phrase.[34] It first appeared in discussions of technical education, where it implied the teaching of both pure and applied science. But the phrase could also mean "science and its industrial application," and even this meaning was ambiguous—*science* in this context could refer to its broader, nineteenth-century meaning as organized knowledge, as well as to its narrower definition as formal knowledge in the natural sciences.[35]

In mid-nineteenth-century Britain, education reformers sought to raise the level of scientific knowledge among skilled workers, in part to help British manufacturers compete more effectively against Continental industries. The British government supported these efforts sparingly, funneling a limited amount of aid through a government agency called the Science and Art Department. These funds were intended to stimulate "instruction in science, especially among the industrial classes."[36] In 1857, the Science and Art Department was relocated to the new South Kensington complex of schools and museums in London, placing the department at the center of debates about technical education.

In 1859, a young military engineer named John Donnelly joined the Science and Art Department, where he became a tireless promoter of practical science education.[37] Frustrated in his attempt to obtain funding for a college of science at South Kensington, Donnelly turned instead to the Royal Society of Arts, a venerable institution founded in the eighteenth century to promote "arts, manufactures, and commerce."[38] In 1871, he proposed that the Society of Arts establish new examinations to encourage the "skilled artisan" to obtain more technical education. These exams covered "the *science and technology* of the various arts and manufactures of this country" in order to assess the "practical knowledge and skill required in the application of the scientific principles involved in each art or manufacture."[39] The exams represent the first use of this phrase in a repeated and consistent way. They had a very practical focus, covering both principles and practices in traditional trades. The first set of exams was given in 1873, and covered five subjects: "Cotton, Paper, Silk, Steel, and Carriage-building." New industries at the vanguard of emerging science-intensive industries, such as dyestuffs or telegraphy, were not included.[40]

By themselves, these science and technology exams did little to transform British technical education. But the early work of the Department of Science and Art eventually did bear fruit. In 1908, Great Britain enshrined the phrase "science and technology" in a new school dedicated to higher technical education, the Imperial College of Science and Technology.[41] Yet the founding of the Imperial College was more a culmination than an inauguration of the phrase.

In the United States, the phrase "science and technology" first found an institutional home as the result of new connections forged between the scientific community and the federal government during World War I. During the war, the National Academy of Sciences created a new organization, the National Research Council, to help harness academic science for the war effort. Ronald Kline has shown how the leaders of the NRC argued for the practical benefits of pure science, while also urging industry to establish industrial research laboratories to provide jobs for scientists. When the NRC adopted a formal organizational structure in 1919, it created a series of "divisions in science and technology," Several of these divisions combined both practical and academic fields. There was a separate division of engineering, but a combined division of "chemistry and chemical technology," and of "biology and agriculture."[42] The structure of the divisions implied a strong practical focus for the NRC. Such a focus may seem to contradict the pure-science propaganda of the council's leaders, but it actually makes sense. As David Edgerton has pointed out, most natural scientists do not focus on research. The incessant NRC propaganda for pure science was actually a plea for more funding for academic scientists, especially those not engaged in work of immediate practical importance.[43]

Yet until World War II, American natural scientists and engineers rarely defined *technology* as either applied science or the application of science, and rarely even discussed the relationship between science and technology in any depth.[44] For example, when Karl Compton replaced Samuel Stratton as president of MIT in 1930, each man briefly addressed the science-technology relationship in their speeches at the inauguration ceremony. Both Compton and Stratton were physicists. Both men acknowledged the debts that technology owed to "fundamental" research in the sciences, and they insisted that progress in science helped generate progress in technology. But neither man used the opportunity to reduce technology to applied science or the application of science, nor did they suggest that progress in science automatically led to new technologies. Even though Compton would shift MIT's focus away from problems of immediate practical benefit, he did not justify this shift by defining *technology* as applied science.[45]

The great experimental physicist Ernest O. Lawrence provided one of the rare detailed discussions of the science-technology relationship before World War II, in a 1937 commencement address to the Stevens Institute of Technology. In this talk, Lawrence jumped freely between technology and applied science without ever actually equating the two. Instead, he described the relationship between science and technology as interactive, "the great partnership of modern times." Similarly, pure science and applied science were linked by a "mutual dependence" that benefited "our material welfare." As one might expect, Lawrence's talk contained standard rhetoric about the practical benefits of pure science. But while he insisted on the "essential role played by pure science in industrial development," he also acknowledged that "technology is just as important for science as science is for technology."[46]

Lawrence illustrated the importance of technology for science with the example of Albert A. Michelson's famous interferometer experiments, which helped prove the invariance of the speed of light. Lawrence noted that it was "the perfection of glass technology [that] made possible the Michelson interferometer." The result was "an experiment that formed the basis of Einstein's theory," thus transforming not only physics but "our entire philosophy." Although historians of science reject this understanding of the relationship between Einstein's theory and Michelson's experiments, Lawrence did raise a valid point: insofar as experimental science pushes the boundaries of material practice, science depends fundamentally on progress in technology. It's not surprising that Lawrence would hold such views. His invention of the cyclotron made him a pioneer in big physics, and in some ways he was as much engineer as physicist. However, in terms of the conceptual history of technology, his commencement address shows that natural scientists, when actually reflecting on the subject, often endorsed an interactive model of the science-technology relationship that relied on the definition of *technology* as industrial arts, not applied science.[47]

In those few cases before World War II where *technology* was explicitly defined as applied science, it was primarily as an autonomous body of practical knowledge. Most writers who addressed this topic in any detail, whether they were natural scientists, social scientists, or engineers, embraced an interactive model of the science-technology relationship. *Technology* was rarely defined reductively as simply the application of science.

Historians of Science and the Concept of Technology

Sustained discussion of the science-technology relationship first emerged in the 1930s, but not from natural scientists, social scientists, or engineers. Instead, it was historians of science who led the way. Before the 1930s, these historians had relatively little to say about technology. That changed when Anglo-American scholars encountered Soviet Marxist theories about the science-technology relationship, theories that stressed the priority of technology over science.

This engagement was sparked in 1931 by a series of controversial papers presented by Soviet delegates to the second International Congress of the History of Science and Technology in London. The Soviet delegates weren't on the printed program, so their talks were presented in a separate, hastily organized session. Their quickly translated papers were published in a volume titled *Science at the Cross Roads*, and their session became known as the Cross Roads conference. Many of the papers discussed the relationship between science and technology, sparking considerable debate among Anglophone historians of science, who soon joined their Soviet colleagues in grappling explicitly with the science-technology relationship.[48]

At the time, *technology* was already a familiar term to historians of science. Fifteen years before the Cross Roads conference, the chemist and historian George Sarton had used the word in his proposal for establishing the history of science as a scholarly field. A tireless promoter of the field, Sarton founded both the History of Science Society and the journal *Isis*. His proposal was published in 1916, shortly after he had fled German-occupied Belgium in a peasant cart.[49]

In the proposal, Sarton examined four topics related to the history of science. One of these was the history of technology. He clearly conceptualized technology as industrial arts, and he insisted that its relationship with science was interactive. "The history of science is constantly interwoven with the history of technology," he wrote, "and . . . it is impossible to separate one from the other." The influence of technology on science went far beyond instrumentation; Sarton noted that technology even created new problems for scientists to study: "Industrial requirements are always putting new questions to science, and in this way they guide, so to say, its evolution." He even acknowledged that "commercial needs" shaped the development of mathematics, as in the spread of Arabic numerals and decimal fractions.[50]

Sarton's openness to the history of technology seems to be at odds with his later dismissive quip that he was interested in "thinkers, not tinkers."[51] Whatever the context of this statement, he was not hostile toward the history of technology early in his career. From the start of the *Isis* annual bibliography in 1913, he included technology as a category (initially *technologie* in French). This category was largely populated by works in German, as the Germans already had a number of journals devoted to the history of technology.[52] Nevertheless, before the Cross Roads conference, Anglophone historians of science paid little attention to the science-technology relationship.

This relative neglect changed decisively with the visit of the Soviet delegation to the International Congress in 1931. Although both science and technology were ostensibly topics of the congress, the history of science remained the principal focus. In contrast, the Soviet delegates argued that a close, interactive relationship existed between science and technology. Their talks, as translated by the Soviet embassy, used *technology* and *technique* more or less interchangeably, mainly in the sense of industrial arts. Although the original papers do not survive, these almost certainly would have used the Russian cognate of *technique*, техника. (The distinction between the Russian cognates of *technique* and *technology* was similar to that between the German terms *Technik* and *Technologie*, creating a similar problem for translations into English.)[53]

The Soviet delegates created quite a stir, given the exclusion of the Soviet Union from European intellectual life after the October Revolution of 1917. The British tabloids attacked the Labour government for giving a visa to the head of the delegation, Nikolai Bukharin, a prominent Bolshevik leader the tabloids considered to be a dangerous propagandist.[54] Bukharin, however, had already been outmaneuvered by Stalin and been pushed to the margins of power; he was executed during the great purges just seven years later.[55]

A sophisticated thinker who read widely in multiple languages, Bukharin still ranked among the leading Soviet theorists in 1931. He was most famous for *Historical Materialism* (1921), his attempt to reconstruct Marxism as a systematic social science.[56] In this book, Bukharin built on the mechanistic, technicist tendencies of the later Engels. The Russian text repeatedly employed техника, which was rendered as "technology" in the 1925 English translation. Bukharin identified technology with "the system of social instruments of labor." He often used the phrase "social technology" to emphasize the social context of these instruments, noting that a steamship at the bottom of the ocean was not social technology. Bukharin viewed technology as fundamental, "a point of departure in an analysis of social changes." Although his theory avoided crude economic determinism, his discussion of technology could be explicitly deterministic. For example, he insisted that the "system of social technology also determines the system of labor relations between persons."[57] Bukharin's understanding of technology was definitely instrumental; nothing in his approach suggested that technology carried any moral values or allowed humans to express themselves creatively.

In *Historical Materialism*, Bukharin also analyzed the relationship between technology and science,[58] arguing that science was a complex system that included material instruments and social organization in addition to ideas. Although the connections between science and productive forces (technology and labor) were complex, ultimately, he wrote, "every science is born from practice." Even as the sciences became more sophisticated, "their direct or indirect dependence on the stage of the productive forces" continued. Using Engels's phrase "in the last analysis" to soften the determinism, Bukharin insisted that "the content of science is determined in the last analysis by the technical and economic phase of society." This determination could, however, be quite vague, as when a "mode of thought" reflected a particular class structure.[59]

Despite some variations, the papers presented by the Soviet delegates at the International Congress supported Bukharin's argument that the natural sciences were rooted in economic and technical practices. Several of the papers went beyond Bukharin's reductive analysis to advocate an interactive model of the science-technology relationship. This was clear, for example, in the contribution by the prominent physicist A. F. Joffe, "Physics and Technology." Joffe argued that workers in "the realm of technique" freely admit their debts to science, but that "the pure scientist usually forgets" the practical origins of problems in physics. The economist Modest Rubinstein insisted on the "interdependence of science and technology" in connection with the social relations of capitalism. He argued that modern monopoly capitalism had become an impediment to the further progress of science and technology, an impediment that Soviet socialism would remove.[60]

But the paper presented by the physicist Boris Hessen generated the most controversy. In Soviet physics, Hessen was a strong supporter of Einstein's theory of relativity, a fact that he only alluded to in his paper. Instead, he sketched out a Marxist approach to the history of science. He focused his paper on showing that Isaac Newton's *Principia*, a paragon of pure science, was not simply the product of Newton's undeniable genius. The *Principia*, Hessen argued, was also rooted in the economic, technical, and ideological context of its time.

Hessen's paper had more impact than the others in part because it was the most concrete in its historical details and the most specific in its claims. Like Bukharin, Hessen argued for the primacy of practice over ideas. He claimed that economics and technology (or technique, depending on the translation) provided topics for Newton's mechanical theories, posing theoretical problems related to mining, pumping, navigation, shipbuilding, ballistics, and the calendar. He tended to describe the connection between economic requirements and scientific topics in terms of functional requirements—that is, what the capitalist social order needed. "In order to develop its industry," Hessen argued, "the bourgeoisie required a science that would investigate the properties of material bodies and the manifestations of the forces of nature." But he was not advocating a crude economic determinism. He also argued that Newton's theories were shaped by the ideological context of Newton's time, ideologies that were themselves a product of class struggle. Ultimately, Hessen explicitly endorsed an interactive model of the science-technology relationship. Under modern capitalism, "the immense development of technology was a powerful stimulus to the development of science, and the rapidly developing science in turn fertilized the new technology."[61]

Hessen's paper for the International Congress created quite a splash in the English-speaking world, and the ripples spread through the rest of the decade. Soviet participation at the event was a news story even in the United States.[62] Academic journals also covered the congress and the Soviet delegates' contribution. The tone of these articles was typically skeptical but rarely dismissive. And almost every report touched on the issue of the science-technology relationship.[63]

In the academic community, Hessen's paper spawned a number of direct and indirect responses that continued into the Cold War. Many scholars reacted with hostility toward Hessen's Marxist framing. According to the historians Gideon Freudenthal and Peter McLaughlin, this hostility led scholars to systematically mischaracterize Hessen's thesis. Hostile scholars portrayed Hessen as making a crudely deterministic argument that economic and technical needs alone determine the content of science.[64] This interpretation is clear in one of the first academic responses to Hessen and the other Cross Roads papers, by the British chemist and metallurgist Cecil Desch. In "Pure and Applied Science," a lengthy article published in *Science*, Desch summarized the "Soviet theory" as an argument that "all scientific research had its origins in practical needs."[65] Hessen had, of course, explicitly rejected this interpretation in his paper. Nevertheless, even Hessen's critics accepted his framing of the problem in terms of a relationship between science and technology.

The most detailed reply to Hessen was written by George N. Clark, a professor of history at Oxford. Clark had given one of the opening addresses of the session, and he was indirectly attacked by Hessen for supporting a great-man theory of history. According to Freudenthal and McLaughlin, Clark rejected an argument Hessen never made, that "the real personal motive of all the scientists in seventeenth-century England was utilitarian." Instead, Clark listed a whole range of factors besides economics "which have worked upon science from the outside." Much of his argument actually supported Hessen's contextual analysis of seventeenth-century science. His most fundamental critique, however, was that Hessen rejected as a motive "the disinterested desire to know." But nothing in Hessen's argument hinged on the personal motivation of the scientist, as Robert Merton and Joseph Needham soon pointed out. Even though Clark, as an economic historian, focused his core arguments on economic factors, he repeatedly returned to the interactions between science and technology, reinforcing Hessen's framing of the problem, while also promoting *technology*, rather than *technique*, as the preferred term.[66]

Perhaps no scholar took Hessen's work more seriously than Robert Merton. As a graduate student in sociology at Harvard University in the mid-1930s, Merton began working on the relationship between science and society, and this work helped create the sociology of science as a recognizable field. He cited Hessen in one of his first papers, "Science and Military Technique," when he was just twenty-five years old. In this paper, Merton took it as a given that "the foci of scientific interest are determined by social forces as well as by the immanent logic of science." He used the phrases "military technique" and "military technology" interchangeably in the paper, which mirrors the usage in the original translation of Hessen's paper.[67]

Merton embraced Hessen more fully in his doctoral thesis, published in 1938 under the title "Science, Technology and Society in Seventeenth Century England" in the journal *Osiris*. This work is most famous for the "Merton thesis," the argument that Puritanism played a crucial role in seventeenth-century British science. But the Merton thesis constitutes less than half of the dissertation. Just as much space (three chapters) was devoted to a detailed analysis of the influence of economic and technological problems on seventeenth-century British science, research that was directly inspired by Hessen's paper. Merton concluded, in a rough quantitative assessment, that just under 60 percent of research items considered at meetings of the Royal Society were related, directly or indirectly, to practical problems. Throughout this analysis, he repeatedly and consistently framed his argument in terms of connections between science and technology, avoiding his earlier use of *technique* in the sense of industrial arts.[68] Ultimately, Merton's thesis helped ensure that the old problem of the relationship between philosophy and the arts was henceforth framed as a relationship between science and technology.[69]

The reaction against Hessen, as Freudenthal and McLaughlin argued, was more about his Marxist terminology than the substance of his argument.[70] In fact, before Hessen it was not particularly controversial to argue that a particular technical problem drew scientists to their subject matter. Sarton made precisely this point in his 1916 essay, even though he later dismissed Hessen's work. In 1928, the Princeton physicist Augustus Trowbridge also acknowledged the debt that pure science owed to practical needs, although he framed the question in terms of engineering rather than technology.[71] Yet after World War II, especially with the advent of the Cold War, most non-Marxist scholars abandoned work on the technological and economic shaping of science, as the topic became ideologically suspect. Although the issue of the science-technology relationship did not disappear, most historians of science united to defend scientific purity and freedom, in part by denying the influence of technology.[72]

Before World War II, scientists did sometimes define *technology* as applied science, but this definition was ambiguous, often referring to *applied science* in its meaning as an autonomous body of scientific knowledge. This sense is present in one of the earliest dictionaries to define *technology* as applied science, the 1939 printing of the second edition of *Webster's New International Dictionary*. The third definition in the entry described technology as "any practical art utilizing scientific knowledge, as horticulture or medicine; applied science contrasted with pure science." By setting up *applied science* in opposition to *pure science*, the dictionary presented applied science, and by extension technology, as distinct types of science. However, the first part of the definition, technology as a practical art that uses science, was inconsistent with the second part of the definition, technology as applied science. But that confusion did indeed reflect usage; *engineering*, for example, was understood both as a practical art and as an applied science. Furthermore, the first part of the definition did not reduce *technology* to the application of science. Instead, this definition was fully compatible with an interactive model of the science-technology relationship, a model that was accepted tacitly by most engineers and scientists, and explicitly by the engineer Bovey, the scientist Lawrence, the sociologist Merton, and the historian Sarton. Yet this interactive model would not thrive in the post–World War II world.

TWELVE

Suppression and Revival: *Technology* in World War II and the Cold War

By the end of the 1930s, the concept of technology seemed on the verge of becoming a keyword in academic discourse. But conceptual change is rarely linear, just like technological change. In a sense, technology was in competition with other concepts for the mantle of modernity, most important science. Science received a huge boost in prestige during World War II and the early Cold War. As science became even more identified with technological modernity, it overshadowed the concept of technology. But even as modern technology was relying more heavily on science, natural scientists and many humanist intellectuals doubled down on the ideology of pure science. This ideology undermined the arguments of postwar scholars who tried to make sense of the increasingly intimate connections between science and technology.

But countervailing trends during the early Cold War encouraged wider use of the concept of technology. First was a new economic discourse on technological innovation. The second trend was an emerging critique of technology that drew from Continental critiques of technique. Third, the rise of the history of technology as an academic field also contributed to the discourse of technology. Historians of technology in particular helped resurrect a cultural view of technology, but with almost no awareness of the history of the concept.

Learning the Wrong Lesson from World War II: Technology Subordinated

At the end of the 1930s, just when discourse about technology appeared to be reaching a critical mass in a variety of academic fields, the concept was pushed into the background by the rising prestige of science. Even at the eve of the intense technological preparations for World War II, scholars wrote little about technology and war.[1] World War II witnessed the greatest harnessing of industrial technology for the military ever, yet this horrifying success story did not enhance the cultural status of technology. Instead, in the world war and the Cold War that followed, what benefitted most was the prestige of science.

Two factors help explain this shift. First was the mobilization of academic scientists for war, and the successful efforts of elite scientists to capitalize on this role. Second was the intellectual reaction against totalitarianism during World War II and especially the Cold War. This reaction was especially strong among historians of science, most of whom abandoned the interactive model of the science-technology relationship in favor of a pure-science ideal.

"Physicist's war" captures this shift of cultural capital to science. The phrase was coined by James B. Conant in November 1941, just before the bombing of Pearl Harbor. Conant, an organic chemist by training, had been president of Harvard University since 1933 and was a strong advocate for America's entry into the war. In 1941, he had just taken over as chair of the National Defense Research Committee, which had been demoted to an advisory board under the Office of Scientific Research and Development.[2] Conant coined "physicist's war" to distinguish the Second World War from the First. World War I had been termed the "chemist's war" because of chemists' role in developing poison gases, research that Conant himself had worked on. But as the historian David Kaiser explains, when Conant coined the phrase, he was not primarily referring to physicists' contributions to weapons development. Instead, he was seeking draft exemptions for physics instructors who were needed to teach elementary physics to hundreds of thousands of new military recruits.[3]

In terms of symbolic capital, teaching elementary physics hardly compared with developing war-changing weapons. It's not surprising, therefore, that the physicist's war became identified with new weapons in the popular press (and also among academics).[4] Yet as Kaiser argues, military secrecy prevented a realistic description of weapons programs, making it very difficult to assess the actual contribution of physics to the war effort.

Secrecy was especially stringent for the Manhattan Project, which designed and produced the world's first atomic bombs. Well before the first bombs were ready, the project head, General Leslie Groves, ordered the physicist Henry Smyth to compile a technical report on the project that was suitable for public release. Smyth was allowed to include only information either "widely known to working scientists and engineers" or of limited relevance to bomb production. Excluded were details of the massive apparatus that produced the fissionable materials. Also omitted was information about the complex interplay of physicists, chemists, metallurgists, engineers, and skilled machinists who worked together to design and manufacture the bombs. The result was a report that stressed basic nuclear physics, leading most readers to conclude "that physicists had built the bomb."[5]

The Smyth report was made public only after atomic bombs had been dropped on Hiroshima and Nagasaki in August 1945. Subsequent US government reports drew from the report to portray "nuclear weapons as the latest in a series of development in theoretical physics."[6] Other military research and development efforts were also described as scientific triumphs, radar in particular, but also developments in aviation and rocketry.[7] Scientists had indeed made unprecedented contributions to the war effort, but the idea of a "physicist's war" seriously distorted the role of academic scientists in military technology. David Edgerton's remarks about Britain apply equally to the United States. The science-centered account of the war, he says, "painted an exceptionally partial and misleading picture of the complex world of wartime invention, research and development."[8]

When leaders in this "physicist's war" bothered to discuss technology, they subordinated the concept to that of science. In 1944, Frank B. Jewett gave a talk to the Institute of Postwar Reconstruction at New York University that was titled "The Promise of Technology." Jewett spanned the worlds of engineer and scientist. He earned his PhD in physics from the University of Chicago under Albert A. Michelson, and then spent decades promoting industrial research as a manager of research for AT&T. He was elected to the National Academy of Sciences in 1918, appointed chairman of the industrial research division of the National Research Council in 1923, and made president of the National Academy of Sciences in 1939. He was also president of the American Institute of Electrical Engineers from 1922 to 1923, and he generally accepted the title of engineer.[9]

Jewett was clearly well positioned to embrace a nuanced view of the science-technology relationship, yet his talk made it clear that *technology* was to be understood purely as the application of science. "All that we call 'technology' is nothing but the application of fundamental science discoveries and the employment of scientific methods for useful or desirable purposes," he said. Jewett argued for using *technology* to refer to applications of not just physical sciences but also biological sciences. "Such things as medicine, public health and agriculture are really technology in the sense that they are utilitarian applications of fundamental biological science." He insisted that there was, in normal times, a sharp divide between fundamental science and technology. Fundamental scientists sought to explore unknowns and expand the horizons of knowledge, "without any particular thought as to their possible ultimate utility." These scientists were mostly housed in universities, where they also trained new scientists both to continue the work of fundamental research and to staff industrial research establishments. In contrast, technology was concerned not with creating new knowledge, but rather with "new or better applications of fundamental science." This distinction remained sharp, even though the students received "very similar" training for work in either fundamental science or technology.[10]

Jewett's description of the science-technology relationship is quite surprising, considering that wartime research had largely erased the already fuzzy distinction between pure and applied research. As historians of technology have repeatedly shown, technological innovation generates new knowledge as well as new artifacts and processes. Well before the war, Jewett's own Bell Laboratories showed that applied research could produce fundamental knowledge. Even the Manhattan Project was far more than mere application of fundamental research to a practical problem.[11] Jewett, however, defined wartime research as an anomaly, likely to disappear once the emergency had passed. But sensing the possibility of a future cold war, he acknowledged that wartime conditions might continue "if peace is but an armistice."[12]

Jewett's rhetoric became conventional wisdom in the postwar world. Yet his applied-science definition of *technology* was new at the time and, like the rhetoric of pure science, deeply ideological. Jewett's reductive definition of *technology* was a far cry from E. O. Lawrence's 1937 description of the science-technology relationship as "the great partnership of modern times."[13] Of course, the rhetoric of pure science, now transferred to fundamental science, was not new, having served as a standard justification for the professional autonomy of science since the mid-nineteenth century. But in 1944, Jewett's ideal of pure science could not have been further from reality, with the veneer of pure science having been torn away by a massive infusion of federal research funds. Throughout the 1950s, that funding would continue at levels two orders of magnitude greater than in the mid-1930s, a possibility that Jewett clearly had foreseen.[14]

Jewett's reassertion of the pure-science ideal, rebranded as fundamental science, reflected his role not as a manager of industrial research but rather as a representative of the American community of scientists. This community was dominated by academic scientists and a few academic engineers, who usually allied with the scientists. One of these engineers was Vannevar Bush, former dean of engineering at MIT and director of the wartime Office of Scientific Research and Development. As head of the OSRD, Bush helped mobilize academic science for the war effort. Near the end of the war, he released a report on postwar science written with input from a wide range of scientific leaders, mainly in academia. This report, *Science, the Endless Frontier*, remains among the best-known science-policy documents of the twentieth century.[15]

Bush's report was a plea for increased government support for basic research after the war. He did not define *basic research*. This term was, like Jewett's *fundamental science*, essentially a synonym for *pure science*: science pursued without concern for practical utility.[16] Bush insisted that basic research was "the pacemaker of technological progress," the source of "scientific capital," creator of "the fund from which the practical applications of knowledge must be drawn." He argued that "new products and processes" depended on new knowledge "painstakingly developed by research in the purest realms of science."[17]

Bush was reasserting an ideal of pure science just before the secret of one of the greatest achievements of applied science, the atomic bomb, was revealed to the public. More than any other person, he was responsible for the decision to build the bomb.[18] Yet *Science, the Endless Frontier* contained almost nothing about problems raised by such massive R&D projects, most of which did not require much pure science. Instead, as David Edgerton argues, Bush was making an argument for continued support of academic science, which he feared would be swamped by the massive increase in applied science funding for industry and government.[19]

This fear was widespread among leaders of American science at the time. Viewed cynically, academic scientists were simply demanding increased government funding without much oversight. However interpreted, Bush and Jewett were engaged in the same double-boundary work performed by Francis Bacon and John Tyndall, insisting on the utility of science while defending its autonomy. Bush and Jewett failed even to pose the science-technology relationship as a problem worthy of study, treating claims about the utility of basic science as self-evident truths, even though both men intimately knew the vast amount of work required to derive practical applications from scientific principles. But their goal was to support university science, and thus they had to elide the complexity of the science-technology relationship. Jewett and many other scientists avoided this complexity by defining *technology* as the application of science. In the report, Bush barely mentioned *technology* (or even *engineering*), except in the phrase "science and technology," which was another way of skirting the crucial problem of application.[20]

Bush and Jewett both insisted that "freedom of inquiry" be maintained despite increased government support for basic research. This autonomy was primarily meant not to give freedom to individual scientists, but rather to give autonomy to universities and nonprofit research institutes that were the "centers of basic research." In addition, Bush was not concerned with freedom of inquiry for researchers in government agencies or industrial laboratories, despite the massive expansion of jobs in applied research. Such researchers faced real pressures to tailor their results to the wishes of their employers, in violation of professional norms.[21] Jewett at least understood that industrial researchers required some autonomy for their creativity to flourish. Bush, in contrast, demanded that university researchers be "free to pursue the truth wherever it may lead." As long as basic researchers had this freedom, "there will be a flow of new scientific knowledge to those who can apply it to practical problems in Government, in industry, or elsewhere." Bush insisted that this research had to come from the United States. "A nation which depends upon others for its new basic scientific knowledge will be slow in its industrial progress and weak in its competitive position in world trade, regardless of its mechanical skill."[22]

But the taxpayers had to accept this claim as a matter of faith, since neither Jewett nor Bush nor any other spokesman for American science could provide evidence that support for basic research was essential to national progress. In fact, innovation studies since the 1960s have consistently found no relationship between economic growth and total national R&D spending, much less the far smaller spending for basic research.[23] Furthermore, if the results of basic research were indeed open to all, as Bush demanded, then nothing could stop other nations from exploiting the practical possibilities of this knowledge. The success of the American war effort did not depend on support for basic research, but rather relied on the nation's industrial strength and its expertise in engineering and applied science.[24] After all, two German chemists (and an Austrian-Jewish physicist) discovered nuclear fission, but it was Americans, with substantial British help, who succeeded in converting this discovery into a nuclear weapon.[25]

Technology in the Cold War: Subordination Continued

In the fifteen years that followed World War II, the United States experienced an unparalleled period of technological enthusiasm. The popular press depicted a future full of domed, climate-controlled cities, commercial space flight, airplanes in every garage, energy too cheap to meter, automated factories, and vast increases in leisure time.[26] This enthusiasm was already on the rise even before the war ended, as revealed in a 1944 *Saturday Evening Post* cartoon depicting the troubles of a middle-class family in the "Utopia" of 1955. The wife had to deal with balky electronic door sensors, a misbehaving automatic kitchen, overcrowded flight lanes for her private airplane, and a grumpy husband exhausted from his four hours at the office.[27]

Technological optimism really kicked into high gear after August 6, 1945. That date marks the atomic bombing of Hiroshima, when the US Army revealed a new technology of fearful power but also great promise. Now, it seemed, total human mastery of nature was within reach, no longer a utopia but a future to be realized within a lifetime. According to one popular newsweekly, "even the most conservative scientists and industrialists" were predicting a future that would "make the comic-strip prophecies of Buck Rogers look obsolete."[28] This hubris extended far beyond nuclear technology. Leading American entomologists, for example, confidently predicted the total extermination of insect pests.[29]

Yet this enthusiasm was expressed almost entirely in terms of science, not technology. Advertisements of the era, whether for televisions or airliners, do not contain the term *technology*.[30] The tremendous cultural authority of science continued to push technology into the background, especially in the ongoing debates concerning atomic energy. These debates were almost always framed in terms of science, not technology. Paul Boyer has described in detail the widespread technological utopianism of the early atomic age, yet his numerous sources almost never used the term *technology* when discussing the atomic bomb or other applications of atomic energy.[31] The same was true of academic authors. Even William Ogburn barely mentioned technology in his 1946 article "Sociology and the Atom."[32] When technology was linked to atomic energy, it was typically in the phrase "science and technology," which, as I noted, generally meant "science and its applications."[33] The emergence of the Cold War around 1948 only strengthened this subordination of technology to science.

It's no surprise that elite scientists did little to clarify the relationship between basic research and technology. Almost everyone who mattered accepted the claim that basic research was essential for technological progress, so empirical research on the subject could only undermine this belief. When the topic was examined empirically, as in the Project Hindsight study conducted by the US Department of Defense in the mid-1960s, the benefits of basic research proved difficult to establish.[34] In essence, the scientific establishment had an interest in cultivating ignorance about the relationship between basic research and technological change.

At least one group of scholars should have known better: historians of science and technology. Before the Cold War, most historians of science accepted as uncontroversial the idea that economic and technological problems had a significant effect on pure science, at least in some areas. But this idea carried a Marxist taint, largely as a result of the 1931 Cross Roads conference. With the emergence of the Cold War, any ideas associated with Marxism became suspect, especially in the United States.

Even during World War II, the defense of scientific freedom was already being linked to the fight against totalitarianism. In Britain, the Society for Freedom in Science, founded in 1941, sought to keep government planning away from scientific research. One founding member was Michael Polanyi, a Hungarian-born physical chemist. According to the historian William McGucken, Polanyi subscribed to the "'liberal' view which saw science as munificiently [*sic*] showering its gifts on mankind when allowed freely to pursue its own spiritual aims but collapsing into barren torpor if required to serve the needs of society." Polanyi's chief antagonist was the Marxist scientist J. D. Bernal, who believed that scientific research should be directed toward meeting fundamental social needs.[35]

Such views spread well beyond the community of natural scientists. Members of the Society for Freedom in Science lectured frequently to the influential Cambridge University program in the history of science.[36] In the United States, Robert Merton turned away from his research on the social foundations of science. Instead, he developed his famous norms of science in a classic 1942 paper. This paper, which stressed the self-regulating character of scientific institutions, was written in part as an explicit reaction against the intellectual threat posed by National Socialism. Merton even adopted an applied-science concept of technology, arguing that "every new technology bears witness to the integrity of the scientist"—that is, technology served to validate scientific expertise.[37]

Among historians of science during the Cold War, there was an equally sharp shift away from examining the science-technology relationship, as they largely abandoned the idea of science and technology as autonomous but interacting fields of knowledge and practice. The Cold War university did not welcome ideas tinged with Marxism. In particular, the Zilsel thesis about the artisanal origins of the scientific revolution was in effect declared heretical. I recall bringing up Paolo Rossi's formulation of the Zilsel thesis in a history of science seminar taught by Elie Zahar at the London School of Economics in 1977. Zahar, the sole remaining student of Karl Popper at LSE, reacted with shock, dismissing the argument as Marxist drivel, even though Rossi was no Marxist.[38]

This anti-Marxist animus encouraged historians to neglect the social context of science. After 1945, Anglo-American history of science came under the spell of Alexandre Koyré, who treated science as a product of the world of ideas.[39] A. R. Hall, who made important contributions in both the history of science and technology, took up Koyré's approach. In 1957, Hall argued that the scientific revolution was a revolution in theory, and had very little to do with technology. He continued to argue for a fundamental separation between science and technology for the rest of his fruitful career.[40] Another influential historian of science, Charles C. Gillespie, also stressed the strict separation between science and technology in his history of the Leblanc process. This process was actually the product of a complex, two-way interaction of science and technology. However, Gillespie argued for a sharp distinction between science and technology based on the intentions of the historical actors.[41] The net result was that among historians of science, science and technology were treated as fundamentally different worlds. Between 1945 and 1960, few articles in *Isis*, the leading American history of science journal, explored the science-technology relationship, though it was undoubtedly one of the key markers of late modernity.[42]

Technology Resurrected: Countervailing Trends in the 1950s

The legacy of the 1930s was not completely forgotten, however. Other intellectual, political, and professional trends helped prepare the way for the concept of technology to emerge as a powerful keyword in the 1960s. The Cold War itself involved technological as well as military and political competition. In 1955, for example, the head of the National Science Foundation conjured up the threat of Soviet technology to argue for training additional American scientists and engineers. But before the 1957 launch of Sputnik, the first artificial satellite, few Americans feared technological competition from the Soviet Union. Even in the media frenzy that followed Sputnik, most of the debate was framed in terms of science.[43]

But three other trends were more significant for technology than competition with the Soviets. First was continued work by economists and business historians related to technological innovation. Second was the translation of work by Continental scholars, mainly in philosophy, on *Technik* or *la technique*. And third was the rise of a new cohort of humanities scholars who taught in engineering schools. These scholars embraced the concept of technology, and specifically the history of technology, as an intellectual calling that also served their institutional interests.

Benoît Godin provides the best background to the history of the concept of innovation, showing that the present-day use of the term arose mainly after World War II. In the early Cold War, a number of American scholars, mainly in economics, began to link innovation to technological change. They also developed multistage models of innovation that critics later termed the linear model. Scholars of innovation based their ideas on the prewar work of Joseph Schumpeter, who distinguished between the concepts of innovation and invention. Schumpeter did not, however, connect these concepts to technology.[44]

One of the most influential models of innovation came from the MIT economist W. Rupert Maclaurin. Maclaurin, whose father was the former MIT president Richard C. Maclaurin, became interested in the topic of technology while serving on a committee advising Vannevar Bush for *Science, The Endless Frontier*. With advice from Schumpeter at Harvard University, Maclaurin began a program of historical research into the economics of technological change. His own research focused on the radio industry.[45] In 1951, he generalized this research into a model of technological advance, which consisted of five "pro-

pensities": "to develop pure science," "to invent, "to innovate," "to finance," and finally "to accept" technological change.[46]

Maclaurin's model was significantly more sophisticated than later
caricatures of the linear model suggest. He argued that "pure science"
was a misnomer, because "practical objectives" had surely influenced
major contributors to modern science. He also pointed out that pure
science did not lead directly to technological change. However, he did
argue that pure science was playing an increasing role in innovation,
becoming "the father, if not actually the mother, of invention." Overall, his model fit with the tenor of the times, placing research in pure
science at the start of the process of technological change.[47]

Maclaurin helped make *technological change* and *innovation* closely
related terms.[48] A few other economists and business historians also
took up the study of technological innovation. One was Yale Brozen,
who in the early 1950s published a series of articles based on his dissertation in economics at the University of Chicago, "Some Economic
Aspects of Technological Change." These articles helped spread the
concept of technological change, but did little to make the study of
technology an integral part of economics. Brozen returned to Chicago
in 1957 to take a faculty position in the School of Business, where he
became a strong proponent of the Chicago school, opposing regulation
and defending economic concentration.[49]

Other key studies of technology emerged from what later became
known as business history. Significant work came out of the Center
for Research in Entrepreneurial History at Harvard Business School,
an organization that Schumpeter helped found. One important work
sponsored by the center was Harold Passer's 1953 study of electrical
manufacturers.[50] Another major contribution was the 1949 monograph
Steamboats in the Western Rivers by the economic historian Louis C.
Hunter. Hunter's book remained a model for historians of technology
into the 1980s, and he later became active in the Society for the History
of Technology.[51] In all these works, technology was a central concept.
Overall, these writings were rich in historical detail, and their authors
displayed a sophisticated understanding of technology in relation to
other fields of human activity. However, research on technology by
business historians remained peripheral to the neighboring disciplines
of economics and history.

Meanwhile, another trend helped bring technology to the attention
of academics: a largely European discourse critical of the perceived dehumanizing effects of modern technology. This literature emerged in

part from the horrors of World War II, though its roots were in the Weimar-era debates over Technik. Much of it was written by philosophers and theologians who drew from the French and German concepts of la technique and die Technik. In English, these works were often discussed as critiques of *technique*, reflecting the continuing dilemma of translation.[52] But as the 1950s progressed, translators increasingly framed this literature in terms of *technology*. Whatever term was used, these critiques overwhelmingly conceptualized technology in instrumental terms as the principal source of disenchantment in the modern world.[53]

Two prime examples of this genre are *The Decline of Wisdom* (1955) by Gabriel Marcel and *The Failure of Technology* (1949) by Friedrich Jünger. These works were translated from French and German respectively, and both were deeply pessimistic about modern technological society. Marcel, a Catholic philosopher, questioned the dominance of technique and technical modes of thought. He did not deny the benefits of technical progress, which helped liberate humans from the forces of nature. But, he insisted, "this liberation is in danger of itself being turned into slavery." Technique had its proper place, but it created an environment that was "artificial and inhuman in the strongest sense." It led to a "singularly barren," dehumanized, and abstract world that was inimical to the use of wisdom. Marcel's solution was to reaffirm the spiritual nature of man, and thus support the struggle against the "dissociation of life from spirit" produced by a "bloodless rationalism." Throughout the English translation, however, *technology* was absent. Instead, the original French *technique* was translated most often as "technique" and sometimes as "technics."[54]

There was little new about this Continental fear of technical rationality as dehumanizing. The fear had been captured decades earlier in Weber's idea of the disenchantment of the world. But because Marcel's analysis was translated in terms of *technique* rather than *technology*, it had little bearing on the concept of technology in English.

In contrast, Friedrich Jünger's small book, *The Failure of Technology*, consistently rendered *Technik* as "technology." Jünger was a reactionary humanist, a conservative literary intellectual, and the lesser-known brother of the writer Ernst Jünger. The German title of the work was *Die Perfektion der Technik*, but the irony would have been lost in English. Jünger, like Marcel, viewed technology as a powerful, dehumanizing force rooted in mechanical thinking, far more than just methods and machines. This force strove for perfection in a never-ending process of rationalization that took over all life, destroying life in the process. In Jünger's account, technology was the historical agent, and its minions, technicians, merely carried out technology's imperatives.[55]

Jünger's vision of technology was completely instrumental. "Within the realm of technology there exist solely technical purposes." As a result, "the technician rejects everything that does not correspond to his ideas of efficiency and purpose." The state had to support technological progress in order to survive, but consequently, "technology infiltrates and usurps . . . the state." Technology took over sports and permeated popular culture. Ultimately, "technology's striving for power is unbroken, . . . its spearheads are driving farther. . . . The state itself is now conceived by technology as a organization which must be brought to perfection, . . . an all-embracing machine which nothing escapes."[56] Ultimately, the state itself became a technocracy.

This extreme form of criticism was new in English, if not in German. But even in German, most critiques of Technik were more nuanced, often accepting Technik as a valid component of culture. Jünger completely ignored the German engineering philosophers who sought to elevate the cultural standing of Technik. Instead, technology was for him the antithesis of culture. In a sense, he is heir to Weber's view of Technik as pure means-ends rationality. Jünger's critique is also a mirror image of the technological determinism of Beard and Ogburn. Like Jünger, they viewed technology as uniquely powerful, but unlike Jünger, they also regarded it as ultimately beneficial.

Jünger's critique reflected a general unease with modernity that was widespread in the 1950s. Even Pope Pius XII had something to say about technology. In his Christmas message for 1953, he discussed the conflict between technology and spirituality in language quite similar to Marcel's. The pope condemned the "excessive . . . esteem for what is called 'progress in technology.'" This "omnipotent myth" came to be seen as "the final end of man and of life, substituting itself . . . for every kind of religious and spiritual ideal." There was nothing wrong with technology itself, which ultimately "comes from God and . . . ought to lead to God." But "technological progress" became a threat "when it is accepted in the thinking of men as something autonomous and an end in itself." This kind of thinking resulted in a "technological spirit" that posed "a grave spiritual danger," making men unreceptive to spiritual truths.[57]

The pope's Christmas message is a fascinating statement of unease about technology, portraying it not as an instrument of human progress but as a substitute for the human spirit. The translation of his remarks, presumably made by the Vatican, rendered the Italian term *tecnica* as "technology."[58] The pope's condemnation of technological determinism, the idea of technology as "autonomous and an end in itself," was particularly perspicacious. In contrast to Jünger, Pope Pius did not impute this determinism to technology itself. Instead, the problem with technology was a false faith in the ideology of techno-logical progress.

Technology Institutionalized: The Society for the History of Technology

The pope's nuanced but critical attitude toward technology found echoes in the late 1950s in a surprising place: the new Society for the History of Technology (SHOT). The history of SHOT's founding has been ably told by Bruce Seely, who demonstrates that the push for this new organization came from humanities professors teaching in Amer-ica's booming engineering schools.[59] Yet these professors were not nar-rowly focused on engineering, as indicated by their decision to call the new field history of technology rather than history of engineering.

In the 1950s, the engineering profession as a whole did not seem eager to embrace the term *technology*, even though many engineering schools were located in institutes of technology. After all, engineers al-ready had engineering as a core concept for their diverse and divided field. When Congress proposed creating a cabinet-level Department of Science and Technology in the late 1950s, engineering organizations opposed the move. Engineers did not object in principle to such a de-partment but rather to its name. In congressional testimony, one engi-neering spokesman, Enoch R. Needles, complained that most people did not understand the term *technology*, quoting a dictionary definition of *technology* as "industrial science" and insisting that the term was not an acceptable alternative for *engineering*. Speaking as president of the Engineers Joint Council, which represented twenty major engineering societies, Needles stated that "we deplore the trend to use 'technology' as a substitute for engineering." He also complained about the "undue emphasis on 'science'" in the bill, reflecting engineers' postwar status anxieties.[60]

It was in this environment that SHOT was born. Of course, not all the society's founders were connected to engineering. But the man who was the driving force behind the organization, Melvin Kranz-berg, certainly was. Kranzberg was a historian of modern France who received his PhD from Harvard University in 1942. In 1952, he moved to the Case Institute of Technology in Cleveland, where he was hired to teach the Western Civilization course sequence. Even before moving to Case, Kranzberg had been working on a study about teaching nontechnical courses to engineers that was sponsored by the American Society for Engineering Education. He continued this work at Case.[61]

Kranzberg's work for this study connected him with a number of scholars interested in the history of technology who would later play a key role in creating SHOT. This early network included Carl Condit, John Rae, and Thomas P. Hughes, all of whom would go on to make major contributions to the history of technology. In 1956, Kranzberg contacted the History of Science Society (HSS) to propose collaboration between HSS and the Humanistic-Social Division of the American Society for Engineering Education. In a letter to Dr. Marie Boas, secretary of HSS, Kranzberg broached, "off the top of my head," the idea of a "History of Technology Society."[62] In her reply, Boas ignored Kranzberg's comment about forming a new society. Instead, she offered to help him promote "interest in the history of technology," adding that "we have always included this [history of technology] within the Society's interest in any case."[63]

Yet as I discussed above, the history of science as an academic field had actually become less receptive to the history of technology during the Cold War. Kranzberg decided to forge ahead regardless of the response from historians of science. He invited several members of his scholarly network to form an advisory committee to promote the history of technology, including the possibility of a scholarly society and journal. The invitees were an odd collection, consisting mainly of engineers and historians, but also including Lewis Mumford and William Ogburn. Kranzberg invited only one economic historian, Warren Scoville, who had written a book on the American glassmaking industry under the sponsorship of Maclaurin's Committee on Technological Change at MIT. But Scoville apparently did not respond, leaving economic and business historians at the margins of SHOT.[64]

At this point in his career, Kranzberg actually had limited knowledge of the history of technology; his expertise was in nineteenth-century French history. He admitted as much in a 1957 letter to Henry Guerlac, editor of the HSS journal *Isis*. Regarding the history of technology, he wrote, "I am virtually ignorant in the subject, for it is only in the past two years that I have begun to delve in it." As Seely notes, Kranzberg was being somewhat disingenuous, playing down the threat of a new academic organization that would compete with HSS or its journal.[65] But he was not feigning ignorance; he had done no serious research in the history of technology, and in fact never would. Instead, he contributed his organizational brilliance to the field, forging a new academic discipline despite the lack of a natural institutional home.

SHOT was founded a year later, in 1958. One of the society's first tasks was coming up with a name for its journal. Kranzberg favored *Technology and Culture*, but some SHOT members objected, fearing that the term *culture* would alienate engineers, and Kranzberg took this objection seriously. From the beginning, he sought to maintain close ties between SHOT and the engineering profession, since engineering schools were an obvious source of employment for new PhDs in the field. To build a consensus, he circulated a somewhat slanted survey, asking for advice on a list of possible titles. Mumford replied with a new suggestion, *Technics and Culture*, and scribbled an explanatory note on the form: "I prefer 'technics' to the narrower technology. The low-brow objection to culture should be disregarded. But the anthropologist's possible objection, that technics itself is part of culture should perhaps be heeded."[66] Mumford's note to Kranzberg, along with those of other scholars, shows how much the concept of technology was still in flux.

In the end, Kranzberg got his way and the journal was named *Technology and Culture*. But debates over how to organize the new field demonstrate just how little Kranzberg and his colleagues knew about broader issues in the history of technology, as well as about the concept of technology more generally. Mumford had apparently forgotten the German debates about Technik that he had read some twenty-five years earlier for *Technics and Civilization*. He had lived through and even contributed to major shifts in the meaning of *technology*, yet he seemed oblivious to them. Of the other scholars involved, only the medievalist Lynn White had a thorough grounding in the European literature of the history of technology. But White was president of Mills College until 1958 and was not a central participant in the early discussions about SHOT and its journal.[67]

SHOT's strong connection to academic engineering had two main effects. Without really knowing what they were doing, Kranzberg and his compatriots reproduced the attitudes of Weimar-era German engineering toward Technik, even recreating the name of one of the key journals of that era, *Technik und Kultur*, the monthly magazine of the elite German engineering society Verein Deutscher Diplom Ingenieure.[68] At the same time, SHOT reflected, in a very real way, the status anxieties of American engineers in the Cold War. Kranzberg, in his opening essay for *Technology and Culture*, lamented that appreciation for "technology did not result in greater esteem for the engineer, the man responsible for this progress."[69] Like engineers, historians of technology too felt underappreciated in the Cold War cultural climate.[70]

This cultural climate, which turned scientists into heroes, was reflected in how historians of technology understood the science-technology relationship. As I noted above, the Cold War helped shift the historiography of science toward pure science, divorcing the history of science from technology just when science had become more relevant to technology than ever. But this separation of science from technology also served the needs of the new Society for the History of Technology, whose members sought legitimacy for an organization separate from the history of science. In a sense, SHOT members justified their new organizational boundary by arguing for the autonomy of technology from science.[71]

As might be expected, the first few years of *Technology and Culture* devoted significant attention to defining *technology*. One frequently invoked definition appeared in the inaugural volume of the first comprehensive encyclopedia of the history of technology in English, *A History of Technology*, which appeared in 1954. The editors of the initial volume, Charles Singer, E. J. Holmyard, and A. R. Hall, eschewed a detailed study of the term. They did briefly acknowledge that the term had a history, noting *technology*'s original definition as "systematic discourse about the (useful) arts," and claiming that beginning in the nineteenth century, *technology* gradually came to be seen as "almost identical with 'applied science.'" However, for their purposes, *technology* was defined "as covering the field of how things are commonly done or made," and by extension "what things are done or made."[72] This definition was so broad as to be practically meaningless, and quite incompatible with the definition of *technology* as applied science.

Kranzberg, in his role as the first editor of *Technology and Culture*, addressed Singer, Holmyard, and Hall's definition of *technology* in the opening article of the new journal. He quoted the definition critically, noting that it led to a treatment of technology as if it were "insulated from the rest of society." He did not offer an alternative definition, but rather stressed the need to understand technology in terms of its relations to society.[73] In the same opening issue, the management expert Peter Drucker suggested revising the Singer, Holmyard, and Hall definition from passive to active voice, so that *technology* would cover not "how things are done or made" but "how man does or makes." Such a change would shift the focus of technology from material things to human work. As an outsider to the field, Drucker expressed hope that historians of technology would provide "a better understanding of technology," but despaired of the prospect given the limitation of the historians' existing concept of technology.[74]

Both Kranzberg and Drucker were articulating, to different degrees, the tension over the role of human agency implicit in the modern concept of technology. Both scholars expressed dissatisfaction with the dominant, innovation-centered concept of technology present in the Singer, Holmyard, and Hall volumes. Yet this deep tension over the concept failed to provoke much debate in the pages of *Technology and Culture* before the mid-1970s, except on the question of the boundaries between technology and science.[75] When meaningful discussions did arise, they remained in the realm of assertions, devoid of historical context.[76]

These early discussions about the concept of technology are illuminating in two ways. First, they reveal profound historical amnesia. Historians of technology appeared to know nothing about the history of the concept of technology itself, or even to be aware that the concept had a history. In particular, they did not recognize that the term was relatively novel. This lack of awareness is clear both in private correspondence and in published works.

Second, the discussions in *Technology and Culture* reveal a wide range of opinions about what *technology* did (or should) mean. Some authors restricted the definition of *technology* primarily to material artifacts, especially tools and machinery, while others argued for including immaterial "tools" such as language. Many definitions emphasized technology as process, with varying degrees of human agency, sometimes including skills and craft. Other definitions described *technology* as a field of study, a type of scientific knowledge, or the application of scientific knowledge. Some authors understood technology as roughly coterminous with engineering, while others included agriculture and even medicine. A few viewed technology as covering forms of organization, or even as the dominant cultural system of our age.[77] Such disunity was rather remarkable for a newly created academic field, which seemed to lack a basis for shared discourse.

Perhaps these historians should be excused for their lack of conceptual clarity, as they were more concerned with building up a respectable body of historical research than with definitions. Yet without any understanding of their own historical context, historians of technology had little basis for claiming authority over the core concept of their field.

Into the 1960s: Stable Confusion

At the start of the 1960s, technology was still overshadowed by the tremendous cultural authority of science in the English-speaking world,
especially in the United States. Survey data demonstrate the public's
esteem for science. In a survey from the late 1950s undertaken by the
prestigious Survey Research Center at the University of Michigan, only
2 percent of Americans agreed that science made the world "worse off."
A total of 83 percent answered that science made the world "better off,"
while only 8 percent choose "both" better and worse, which in an era
of nuclear weapons was clearly the right answer. In this particular survey, subjects were given a definition of *science* that also encompassed
practical activities, such as "curing a disease, or the invention of a
new auto engine, or making a new fertilizer." *Technology* was not mentioned. The very framing of the survey demonstrates the cultural hegemony of science in this era. And this particular survey was sponsored
by the National Association of Science Writers, suggesting that science
journalists themselves rarely distinguished *science* from *technology*.[78]
 Despite being overshadowed by science, the concept of technology
became increasingly widespread by the early 1960s. Yet as a term it remained ambiguous, with three primary but incompatible meanings
in circulation, primarily among academics. First was the definition of
technology as the application of science. In the popular mind, science
already included a wide range of technological activities, making it
easy to accept this definition of *technology*. A second definition also remained common, *technology* as industrial arts, that is, an autonomous
body of knowledge, practices, and even artifacts. This meaning found
an academic home in the new Society for the History of Technology.[79]
And bubbling up through the froth of academic discourse was a third
meaning, *technology* as technique, or instrumental reason.
 These three distinct meanings were rarely contested, because technology was only weakly implicated in scholarly boundaries. Scientists
were happy to include technology within the scope of science when
they desired funding, and to exclude technology when they sought autonomy. Engineers likewise had little use for the concept of technology, as the public already saw technology as subsidiary to science. And
humanities scholars had few reasons to clarify *technology*, as the term's
ambiguity made it easier to connect it with Continental critiques of
technique. By eliding the distinction between various meanings of

technology, humanist critics could make their arguments seem broader and more profound.

Only one academic organization had a real stake in the definition of *technology*: the Society for the History of Technology. Yet for SHOT, it made sense to construe *technology* as broadly as possible. A broad definition helped recruit more members for the organization, as well as contributors to its journal. For example, *Technology and Culture*, under Kranzberg's editorship, helped introduce American audiences to the work of Jacques Ellul, whose influential book *Technological Society* (in French *La technique*) was a wholesale condemnation of modern technology as instrumental reason, much like Jünger's *The Failure of Technology*. Even though Kranzberg had little sympathy for this definition of *technology*, in 1962 he devoted a special issue of *Technology and Culture* to a conference largely about Ellul's work, thus helping familiarize English-speaking academics with Continental critiques of technology.[80]

All these factors help explain why the concept of technology never coalesced around a clear set of meanings. A variety of factors combined to inhibit clarity: problems in translation, historical amnesia, the dominant role of science, and disciplinary boundaries. Even in the 1960s, when critiques of technology transformed the concept into a true keyword, blurred discourses continued. Just like in Weimar Germany, many humanist critics embraced an instrumental understanding of technology as technique, while engineers and their allies, including many members of SHOT, defended a cultural view of technology as a creative expression of human values. Both these views were on display in the counterculture of the late 1960s and 1970s, which pitted critics of technological civilization against defenders of alternative technologies.[81]

But in terms of scholarly discourse, the core meanings of *technology* had become stable by the early 1960s—or to be more precise, the confusion had stabilized. Since then, many forests have been consumed by the publication of writings on technology. But as long as academics lack a historical understanding of the concept of technology, these trees will have died in vain.

THIRTEEN

Conclusion: *Technology* as Keyword in the 1960s and Beyond

At the end of the 1950s, *technology* was on the verge of becoming a keyword for understanding the modern world. In the decades that followed, the term spilled out of its academic niche, spreading into other scholarly fields and diffusing into popular culture. *Technology* became politically powerful, a symbol of hopes and fears, and an ideological tool in debates about modernity. Yet despite all this sound and fury, it remained unclear just what *technology* signified. Was it the knowledge and practices of the industrial arts? A form of applied science? An oppressive system of instrumental rationality? A powerful external force shaping society? Or a fundamental expression of cultural values and political choices?

All these contradictory meanings predated the emergence of *technology* as a keyword. As a result, when the term erupted into public discourse in the 1960s, it seemed fully grown, a mature concept requiring little examination. Few people who used the term were even aware of its conflicting meanings. Debates about the concept of technology repeatedly foundered in incomprehension and misunderstanding. This confusion was a direct product of the concept's history, reflecting the deep but unrecognized tension between instrumental and cultural approaches. In particular, the forgotten rupture between technology and art continued to undermine the cultural understanding of technology.

To tell this story in any detail would require another book.[1] Here I sketch out this story to show how the history of concepts helps explain the continued confusion over the meanings of *technology*. *Technology* as a keyword spread from three distinct, but not isolated, fields of discourse. The first was the rise of innovation studies after World War II. Second was the emergence of the field of "technology and society" and other broadly humanistic approaches to technology, such as the history of technology. And third, critiques of technology also rose from academic obscurity to command broad public attention. Each of these three fields was shaped in different ways by tensions between the instrumental and the cultural approaches to technology.

Technology as Innovation

In the previous chapter, I discussed how studies of technological innovation arose after World War II, beginning with the work of the MIT economic historian Rupert Maclaurin. These studies had a strong historical focus and were often connected with schools of business. Researchers in this field analyzed the innovation process as a sequence of stages, describing the entire sequence as both *technological innovation* and *technological change*. But there was no consensus among scholars on what these stages were.[2]

In *Narratives of Innovation*, his pioneering study of technological innovation as a concept, Benoît Godin describes how these stage models of innovation became reified in statistical categories beginning in the 1950s. One of the most common sequences consisted of three steps: basic research, applied research, and development. After World War II, the US government vastly increased its role in promoting new technologies in areas such as digital computers and civilian nuclear power. The funding of such technologies was generally subsumed under the rubric research and development. Classifying innovation into stages helped national governments keep track of this spending.

But these categories, and the linear process they embodied, did not accurately reflect the actual process of technological change. Beginning in the 1960s, studies repeatedly pointed out the flaws in this sequence, most famously the 1967 Project Hindsight sponsored by the Department of Defense. This study showed that major new military technologies arose not from undirected scientific exploration but rather from a confluence of related innovations, most of which were directed toward specific military technologies. In summarizing these findings, two leaders of Project Hindsight noted that the relationship between science and technology "is clearly not the simple, direct sequence taught by the folklore of science."[3] Similar studies in the 1960s repeatedly showed that public and private needs, not basic research, spurred new technologies.[4]

Despite these criticisms, most models of innovation remained implicitly linear, in part because of the reified categories used in official statistics. These categories implied that new technology arose from basic research, usually understood as unconstrained scientific exploration. Such models of innovation thus supported a definition of *technology* as the application of science. These models also gave short shrift to the social needs and cultural values that shape innovation, thus encouraging, at least implicitly, an instrumental view of technology.[5]

What models of technological innovation did not encourage was sustained reflection on the concept of technology itself. Innovation was the focus; technology was a mere modifier. Some authors did provide explicit definitions of *technology*, but these definitions tended to be ad hoc constructions suited to specific projects. For example, an interim report from Project Hindsight defined *technology* with a confusing list of meanings. The report then noted that this definition was "substantially synonymous with the current DOD category of Exploratory Development," also known as 6.2 funding.[6] In other words, the Project Hindsight researchers tailored their definition of *technology* to fit an existing budget category.

Economists, who came rather late to studies of technological innovation, supplied their own definitions. At first, they continued to use *technique* and *technology* more or less interchangeably after World War II. In the early 1960s, however, the economist Vernon Ruttan proposed an important distinction between these two terms. Ruttan defined *technology* as "the body or stock of techniques, procedures, or ways of conducting economic activity." He described the "level of technology" in terms of the surface of a "meta-production function," that is, the economic output possible given various inputs to the production process. Changes in technique were possible, in fact common, even absent changes in technology. For example, if the cost of labor increased relative to the cost of capital, it could become economical for a manufacturer to buy industrial robots to automate production. The robots represented a change in *technique* for the manufacturer, but not a change in *technology*. In economic terms, a change in technique represented movement to a different point on the surface of the production function. But true technological change, such as the introduction of a new technique, represented a change in the production function itself.[7]

Ruttan's distinction between *technique* and *technology* had real advantages. Using this distinction, economists could treat the choice of techniques as an economic problem, an "endogenous" variable to be calculated using economic models. Technological change, in contrast, remained outside the economic system, an "exogenous" variable that shaped the economy without itself being shaped by economic choices.[8]

But this distinction between technological and technical change was untenable. There is no sharp division between changes in technological knowledge and changes in techniques of production. The very act of adopting or using new techniques involves a learning process, thus changing technological knowledge. Economists have acknowledged this effect since the 1950s with the concept of the learning curve, a concept incompatible with the idea that technological change is exogenous.[9]

In the 1960s, the discourse of technological innovation spread beyond academia into politics and business. In politics, competition with the Soviet Union was discussed increasingly in terms of technology and technological innovation, a shift that started soon after the Soviet Union launched Sputnik in October 1957. Initially, Sputnik was viewed primarily as a failure of American science, not American technology.[10] Nevertheless, the top-secret Gaither Report written the following month framed the threat in terms of Soviet technology, not Soviet science. Details of this report were soon leaked to the press and discussed extensively, though often without mentioning *technology*.[11]

Only in the 1960s did Cold War competition come to be framed more consistently as a technological problem. For example, *technology* was a key element in the famous memo that James Webb and Robert McNamara wrote to vice president Lyndon Johnson in 1961. In it, Webb and McNamara argued for a crash program to put a man on the moon, claiming that such projects "symbolize the technological power and organizing capacity of a nation."[12] Then in 1964, leading Republicans began attacking the Johnson administration for permitting a "technology gap" between the United States and the Soviet Union, echoing the "missile gap" that the Democrats had used against the Eisenhower administration during the 1960 election.[13]

In the mid-1960s, various branches of the US government commissioned reports about technological innovation. Some were connected with military technology, such as Project Hindsight. Others were more focused on economic growth. In 1967, the Commerce Department issued a report titled *Technological Innovation*, which argued for giving "much more attention to the social and business climate which creates the possibility" of technological change. This report stressed the role of small businesses in fostering innovation.[14] Reports such as this received significant notice in the popular press, thus helping spread the language of technological change and technological innovation.[15]

A number of academics also helped bring innovation studies to the business press. One of these was James R. Bright, a business professor at Harvard University who wrote a number of articles about technological change for the *Harvard Business Review*. In a 1963 article, Bright embraced deterministic language to argue that "technological change . . . is the most powerful factor in business today, and its power seems to be growing." In language reminiscent of William F. Ogburn, he made questionable claims about the economic effects of guided missiles and television. He argued that "the missile" had decimated employment in aircraft manufacturing, and that television had "almost destroyed the traditional form of the movie industry."[16] None of this was true.

More sober advice came from the philosopher Donald Schön, also writing for the *Harvard Business Review* in 1963. Schön, then a consultant with Arthur D. Little, reported the results of a study funded by the "military services" on resistance to radical inventions. He considered the matter not just from the viewpoint of the inventor, but also from the perspective of organizations that resisted "new technologies." He argued that resistance to radical technological change was usually justified, because "changes in technology tend to carry with them major changes in social organization." In other words, Schön was pointing out that innovators needed to be sensitive to social context and organizational structures, not just technical needs.[17] He expanded on these ideas a few years later in his popular book *Technology and Change*, which put even more emphasis on technological change and technological innovation.[18] While linear models of innovation like Bright's drew from an instrumental view of technology, contextual models like Schön's were much closer to the cultural approach.

Technology and Social Change

During the 1960s another topic, technology and society, rose from academic obscurity into public consciousness. While the discourse of technological innovation and technological change focused on the creation of new technologies, that of technology and society dealt with the social effects of these technologies.

This field was not new. Its roots lay in the 1920s with the work of William F. Ogburn on the social effects of mechanical inventions. Although Ogburn acknowledged that social factors could influence technology, his work remained doggedly deterministic, treating technological change as an external factor shaping society. He continued to write about the social impacts of new technologies into the 1950s, but few scholars followed his lead.[19]

The early 1960s saw renewed scholarly and popular interest in the social effects of technological change. This interest was spurred by a growing public awareness of technological risks across a range of issues, concerns reflected in congressional hearings on nuclear fallout, environmental pollution, auto safety, and automation. In addition, industry responded to growing criticism of its products. The chemical industry, for example, responded with considerable hostility toward Rachel Carson's 1962 critique of pesticides. One industry spokesman even implied that Carson was a Communist dupe.[20] Yet other corporate leaders were more measured in their reactions. In a statement to a congressional committee, Thomas J. Watson, president of IBM, admitted that technological change was not all positive. Although he insisted that technological change was "essential to our Nation's progress," he acknowledged that "the problems created by technology touch all levels and aspects of our national life." He called for further study of automation and technological change in order to "find new economic and social solutions to the problems they create."[21]

In support of Watson's vision, in 1964 IBM awarded Harvard University a generous ten-year grant of five million dollars to create the Harvard University Program on Technology and Society. To head the program, Harvard hired a little-known philosopher, Emmanuel G. Mesthene, who at the time was an analyst for the RAND Corporation. Between 1964 and 1972, the Harvard program supported over one hundred scholars who produced some fifteen books as well as 165 articles published in academic and popular periodicals.[22]

Despite its generous funding, the Program on Technology and Society did little to develop a general understanding of technology. Under Mesthene's uninspiring leadership, the program could not disentangle the multiple meanings of *technology* circulating in the literature. Its research failed to clarify the science-technology relationship, and it never developed a framework for understanding the relationship between technology and society. Mesthene did acknowledge the possibility of two-way interaction between technological change and social factors. Nevertheless, most of the work supported by the program focused on "how technological change generates social change." In the final report of the Harvard program in 1972, Mesthene explicitly embraced a "soft determinism," which held that "technology determines the general direction of change," but not "the specific forms that the change will take." Such thin conclusions were typical of the program. As Ruth Cowan notes, most reviewers of Mesthene's writings "found them vacant."[23] IBM was not too happy with the results either, and canceled the grant two years early.

Also in the program's final report, Mesthene did express hope for a different future, one less subjected to the undirected pressures of technological change. He claimed that with better knowledge of how "society is shaped by its technology," it would be possible to "strengthen the ability of society to control and direct technological change."[24] Here we can see the unrecognized tension between instrumental and cultural approaches to technology. The Harvard program's research posited technology as an external force, the product of an instrumental logic operating without conscious human choice. Yet exercising control over technology required a different approach altogether, one that acknowledged technology as a manifestation of human culture. Under these circumstances, it is no surprise that Mesthene did little to advance this goal.

In acknowledging the social control of technology as a worthy goal, Mesthene was reacting to widespread criticism of technology in the 1960s—criticism that had probably inspired IBM to fund the Harvard program in the first place. These negative reactions intensified in the late 1960s as activists began linking technology to the Vietnam War and the destruction of the environment. In 1969, Mesthene himself became a target of such criticism in a *New York Review of Books* article with the provocative title "Technology: The Opiate of the Intellectuals." Its author, an obscure New Left writer named John McDermott, argued that Mesthene and his ilk represented the interests of a technocratic elite, though McDermott never used the term *technocracy*. Given IBM's role in funding the Harvard program, Mesthene was an easy target. But McDermott went further, insisting that Mesthene advocated a policy of "laissez innover" that precisely paralleled the nineteenth-century theory of "laissez-faire." For all the limitations of the Harvard Program on Technology and Society, this accusation was quite unfair. As McDermott's own quotations show, Mesthene was no technological enthusiast, but rather quite ambivalent about technological change. He acknowledged that the social effects of technological change were both good and bad, and that the bad and good were intertwined.[25]

In sum, McDermott's essay shared one of the main flaws of Mesthene's work: an almost complete lack of historical awareness about prior scholarship on technology. McDermott seemed completely ignorant of anything apart from recent writings. He never mentioned Veblen's proposal for a "Soviet of technicians," despite its obvious relevance to his arguments. But even more striking was his failure to engage with, or even mention, the philosophical critics of technology in the 1960s.

Critiques of Technology in the 1960s

By the late 1960s, critiques of technology had gone increasingly mainstream. Some of them focused on specific technologies, such as Ralph Nader's investigation of auto safety and Rachel Carson's indictment of pesticides. But a more philosophical critique aimed directly at the core of modern technology: defining *technology* as an oppressive system of technical rationality. As I pointed out in chapter 12, such critiques were already present in the 1950s, but only at the margins of popular discourse. But in the 1960s, they found a much broader audience, thus helping popularize the concept of technology.

These critiques of technology appeared most starkly in the works of a number of public intellectuals, among them Jacques Ellul, Herbert Marcuse, and our old friend Lewis Mumford. These men shared little in common aside from their attitudes toward technology. Marcuse was a heterodox Marxist philosopher, a member of the so-called Frankfurt school. He fled Nazi Germany in the 1930s and eventually settled in Southern California, becoming a philosophy professor at the University of California, San Diego. Ellul was a deeply religious provincial French academic originally trained in law who wrote voluminously on philosophy, theology, and sociology. Mumford was an American intellectual who wrote mostly about technology, art, and urban planning. Moreover, these three did not even agree on the term for the phenomenon they were attacking: Ellul preferred *technique*, Marcuse *technology*, and Mumford *technics*.[26] But all three were united in their vision of technology as a system of instrumental reason that subordinated ends to means.

There was very little novel in any of these critiques, which harkened back to Thomas Carlyle's 1829 essay on the "mechanical age." Like later critics, Carlyle acknowledged the great achievements of that age. Yet, he lamented, "men are grown mechanical in head and in heart, as well as in hand."[27] Ellul, Marcuse, and Mumford echoed his critique of mechanical thinking, but transformed it into a critique of technology.

This shared vision of technology as technique clearly reflected the tension between instrumental and cultural approaches to technology. Ellul, Mumford, and Marcuse were elitist humanist intellectuals, deeply hostile toward popular culture and profoundly troubled by the direction of modern life. Their critiques were in essence lamentations for a lost world of authenticity, community, and freedom that had been destroyed by the relentless march of all-encompassing technological systems. None of these men acknowledged the human values, however distorted by capitalism and militarism, expressed in the creative acts that produced modern technology. And they all argued more through assertions than evidence. Marcuse and Ellul seemed especially estranged from the real world of technology, drawing their knowledge from a handful of secondary works. Mumford, in contrast, had a deep well of knowledge about the history of technology, but his grasp of recent technology remained limited.

Ellul's critique was the first to gain wide public recognition, at least among English-speaking academics. In March 1962, his work was the focus of a major conference jointly sponsored by the editorial board of the *Encyclopaedia Britannica* and the Center for the Study of Democratic Institutions (CSDI). The CSDI was a liberal think tank headed by Robert M. Hutchins, former president of the University of Chicago and a leading advocate for the humanities in higher education. Melvin Kranzberg also played a significant role in organizing the conference, and he published its proceedings in a special issue of *Technology and Culture* later that year. Many prominent intellectuals participated, including Hutchins, but only one engineer was present.[28]

Ellul attracted the conference organizers' interest because of his book *La technique, ou l'enjeu du siècle* (*Technique, or the stake of the century*) (1954). The book had garnered little attention outside France, but in 1961 the CSDI decided to sponsor an English translation, which was published in 1964. In the preface to the English edition, Ellul cautioned his readers that *technique* was not the same as *technology*. But that warning proved disingenuous, as Ellul freely confounded the industrial and instrumental meanings of *la technique* in French. In any case, the English title, *The Technological Society*, made it clear that the book was about technology.

Ellul's argument was clear as well. He noted that *technique* refers to the system of technological rationality, a system that starts with machines but expands in the modern world to encompass all forms of human thought and action. Ends lose their meaning, and means become ends. As summarized by Robert K. Merton in the preface, *technique* represents "the never-ending search for 'the one best way' to achieve any designated objective." As a result, technique rejects all moral values, removing human judgment from technological change. As Ellul insisted, "Technique has become autonomous; it has fashioned an omnivorous world which obeys its own laws and which has renounced all tradition." Furthermore, technique is an expansive force, invading nontechnical domains, converting "every nontechnical activity . . . into technical activity." And even though Elull denied that he was a pessimist, he offered no way out of this state of affairs. "Enclosed within his artificial creation, man finds that there is 'no exit'; that he cannot pierce the shell of technology to find again the ancient milieu to which he was adapted for hundreds of thousands of years."[29]

Ellul offered little evidence to support these extreme claims, but simply asserted them as self-evident. "How can anyone fail to see?" he asked rhetorically at one key point. His admiring translator, John Wilkinson, dealt with the weakness of Ellul's argument by rather preposterously comparing the book to Hegel's *Phenomenology of Mind*. Wilkinson suggested that like Hegel's prose, Ellul's "must drive the literal-minded reader mad." *The Technological Society* made no attempt to delve into the details of technical choice, and provided no comparisons of technical and nontechnical human activities. Ellul insisted that technical rationality "excludes spontaneity and personal creativity," and "destroys, eliminates, or subordinates the natural world." But he declined to discuss these aspects of technique in any detail "because of their obviousness."[30]

Why, then, did such a deeply flawed intellectual work garner so much attention in a foreign country, and a decade after its publication? The book seems to have resonated with American intellectuals uncomfortable with the rabid technological enthusiasm of the postwar era, with its images of commercial space travel and domed, climate-controlled cities.[31] This enthusiasm was peaking in the early 1960s, despite the beginnings of a powerful critique emerging from the environmental movement. But *The Technological Society* had nothing substantive to add to such critiques. Instead, the book's success reflected the weakness of American thinking about technology in this era, a weakness directly linked to the confused meanings of the concept.

Ellul was no leftist, though he is often described as a Christian anarchist. However, the critique of technology as instrumental reason knew no political boundaries. Herbert Marcuse developed a similar analysis of technology that became closely associated with the New Left. Traditionally, Marxists have tended to be technological optimists. After all, it was only through the development of the productive forces that socialism and then communism would be able to replace the capitalist order. Orthodox Marxists tended to forget Marx's extensive analysis of how capitalist technology harmed workers. Instead, most Marxists celebrated technological change as a force that would help bring into being the communist utopia.[32]

Marcuse challenged this uncritical approach toward technology. Born in Berlin in 1892, he received a PhD in philosophy from the University of Freiburg in 1922, returning in 1928 for postgraduate work with Martin Heidegger. In his studies under Heidegger, Marcuse combined Marxism with other trends of European philosophy, such as existentialism, and he rejected the rigidity and dogmatism of Stalinist Marxism. But he did not reject Marxism itself. Instead, he argued for restoring it by bringing it back in touch with people's real experiences.[33]

In 1933, Marcuse left Freiburg to join the Institute for Social Research in Frankfurt. This research center had been founded in 1923 by a group of unorthodox Marxists. It later became known as the Frankfurt school and included some of the most creative philosophers in the twentieth century, such as Theodor Adorno, Max Horkheimer, and Walter Benjamin. Like Marcuse, these philosophers also developed critiques of modern technology. But when the National Socialists seized power in 1933, Germany became decidedly hostile toward Marxist Jewish philosophers. Consequently, in 1934 Marcuse fled Germany, never to return, and immigrated to the United States.[34]

In 1964, Marcuse published his best-known book, *One-Dimensional Man*, a fully articulated philosophical critique of technology. Unlike orthodox Marxists, Marcuse rejected the belief that human freedom would be achieved through a proletarian revolution. He argued that workers had lost their revolutionary role because they had become enslaved to artificial needs created and fulfilled by modern technology. According to Marcuse, technology had become a form of totalitarianism, a system for controlling society, preventing change and suppressing dissent. Technology controlled people by manipulating their needs. These false needs kept individuals enslaved to the very machines that had the potential to liberate them.[35]

Marcuse's understanding of technology was explicitly instrumental.

Like Ellul, he argued that technology was not just a material process but rather a narrow form of rationality that focused on means instead of ends. In the one-dimensional society, technology defined its own ends. It expanded instrumental rationality to every corner of society; all life became subject to calculation. Hence in a great historical irony, "the liberating force of technology—the instrumentalization of things—turns into a fetter of liberation; the instrumentalization of man." The net result was a total loss of freedom. "Technological rationality thus protects rather than cancels the legitimacy of domination, and the instrumentalist horizon of reason opens on a rationally totalitarian society."[36]

One-Dimensional Man was pessimistic, but not completely so. Marcuse claimed that technology made it almost impossible to imagine alternatives. Yet in his conclusion, he held out some hope for transcending the domination of technology. That would require "a break with the prevailing technological rationality" while maintaining the "technical base." By increasing human productivity, this base "remains the very base of all forms of human freedom." Overcoming the domination of technology would require shifting the technical base toward different ends, guided by different values. Eventually, such values could produce a sort of techno-utopia, "the rational enterprise of man as man, of mankind."[37]

But who would lead such a transformation? In a gesture toward the civil rights movement, Marcuse pinned his hopes on "outcasts and outsiders, the exploited and persecuted of other races and other colors." By "refusing to play the game," the resistance of such individuals suggested the possibility of change to him. Yet Marcuse admitted that his theory offered little hope to these people, providing "no concepts which could bridge the gap between the present and its future."[38] Although he tempered this pessimism in later writings, *One Dimensional Man* offered little prospect of fundamental change.

Neither Ellul nor Mumford had significant influence on the New Left, but Marcuse embraced the student protesters, speaking at rallies and writing essays in support of the movement. The New Left had already begun to develop a critique of technology before the publication of *One Dimensional Man*. In 1964, this critique was captured in the political activist Mario Savio's famous address at a rally supporting the Free Speech Movement at Berkeley. In this speech, he denounced "the machine" as "odious," and encouraged students to (metaphorically) throw their bodies into the gears to make it stop. Berkeley students also incorporated defaced computer punch cards into their protests. To

CHAPTER THIRTEEN

them, the punch card symbolized corporate rigidity and the alienation produced by modern technology.[39] In effect, Marcuse provided such students with a philosophical language to express what many of them already felt. Through his speeches and articles in the underground press, Marcuse's critique of technology was widely disseminated to the New Left during the late 1960s.[40]

Lewis Mumford developed a similar instrumental critique of technology during the 1960s. This critique marked a significant shift from his understanding of technology in *Technics and Civilization* some three decades earlier. Early in his career, Mumford embraced a cultural and, on balance, optimistic view of technics and technology. Although he criticized technology's past ("paleotechnics"), he expressed an almost utopian faith in technology's future ("neotechnics").[41]

But Mumford's writings about technology became decidedly more pessimistic after the atomic bombing of Hiroshima. He believed that the scientific community, by helping create atomic weapons, had betrayed the promise of neotechnics.[42] In the 1960s, he began describing modern technology as a megamachine, an all-encompassing system of tightly controlled human and material components, dedicated solely to its own perpetuation and aggrandizement.[43] His writings became increasingly bleak, culminating in a two-volume work, *The Myth of the Machine*. The first volume, *Technics and Human Development*, located the origins of the megamachine in the construction of the ancient Egyptian pyramids.[44] The second volume appeared in 1970, titled *The Pentagon of Power*. In this book, Mumford stridently denounced modern technology as the ultimate megamachine. Just like Ellul and Marcuse, his bleak vision relied on a concept of technology as instrumental reason run amok, not a means to an end but an end in itself. In the form of the megamachine, technology was no longer an integral aspect of culture but rather its enemy.[45] But this argument only made sense when technology was conceptualized as instrumental rationality, a view that contrasted sharply with Mumford's earlier understanding of technics as a sensuous material practice animated by the human spirit.

Technology Contested: Instrumentalism versus Culture

In the late 1960s and early 1970s, the sense of technology as threat pervaded stories about technology in the popular press. In October 1970, extensive excerpts from Mumford's *The Pentagon of Power* filled four consecutive issues of the *New Yorker*.[46] Other authors grappled with the dehumanizing potential and unintended harms of modern technology. Theodore Roszak's well-known book from 1969, *The Making of a Counter Culture*, drew heavily from Marcuse and Ellul to assert that modern America had become a technocracy. He argued that critiques of technology were central to the antiestablishment youth of the 1960s. Even a few members of Congress adopted this language. In 1969, Rep. Cornelius Gallagher of New Jersey, a champion of the right to privacy, delivered a speech on technology and society to a meeting of managers. In language reminiscent of Ellul and Marcuse, Gallagher likened modern technology to a "heathen idol." To appease this "God of technology," Americans were sacrificing clean air, clean water, and ultimately "our individuality, our dignity, and our privacy."[47] Prince Philip of Great Britain contributed to the debate as well, warning of the threat to the environment from "this insatiable technological monster."[48]

Leading engineers and scientists were taken aback by this pessimistic rhetoric. Engineers, claimed *Fortune* magazine in 1971, were "no longer heroes. They are distressed and angered by the sudden strong wave of opposition to the great engineering creations—power plants, highways, dams—that Americans universally used to regard as monuments to progress."[49] Many engineers and scientists reacted defensively to the pessimism, reaffirming their faith in the inevitable benefits of technological change.[50] One such engineer was Simon Ramo, a founder of the defense contractor TRW, who resolutely defended technology against its critics. A prolific and often poorly informed writer, he embraced a version of Ogburn's cultural lag thesis, arguing that concerns about modern technology arose because social change failed to keep pace with technological change. Such deterministic arguments served as a defense of the status quo, and a refusal to even consider reorienting the direction of technological change.[51]

Many reform-minded scientists and engineers were more open to critiques of technology, accepting the need to deal with the negative consequences of technological change, and adopting the critics' theme of "technology out of control." But these reformers also rejected the instrumental pessimism of Mumford and Marcuse. Some believed that the solution lay in the "democratic control" of technology.[52]

This interest in controlling technology had a significant effect on public policy. From 1966 to 1972, Congress held multiple hearings related to mitigating the dangers of technological change, eventually leading to the creation of the Office of Technology Assessment (OTA) in 1972. As Sylvia Fries has explained, the 130 witnesses who testified at these hearings did not have a clear or unified definition of *technol-*

ogy. Some witnesses embraced an instrumental definition of *technology* as problem-solving methods, while others portrayed technology as an uncontrolled, dangerous force in need of human guidance. Nevertheless, most witnesses agreed that the government had a legitimate role in limiting the problems created by new technologies.[53]

Whatever the merits of the OTA, the debate over its creation helped popularize the idea that technology could be controlled through democratic means. The idea of democratic control expressed a cultural vision of technology, one that made technological change a matter of political choice, subject to interests and values expressed through democratic politics. In contrast, from an instrumental viewpoint, society could at best accommodate itself to the negative consequences of technological change. In a sense, the OTA put into practice what Mesthene and the Harvard program had advocated: a systematic structure for mitigating the negative consequences of technological change. Yet the political vision that produced the OTA failed to address the dominance of corporations and the military over technological innovation. For this reason, the OTA never actualized its potential for democratic control.[54]

Another set of critics emerged from the consumer and environmental movements and proved more eager to challenge corporate and bureaucratic interests. In contrast to the philosophers who in effect rejected technological reason itself, these critics valued scientific research and technical expertise. The risks of modern technology could be tamed, they argued, not by rejecting technical expertise, but by redirecting it toward alternative human values. In essence, these critics advocated the reform of modern technology. One reformist critic was Ralph Nader. In his book *Unsafe at Any Speed* (1965), Nader condemned the automobile industry for failing to use modern science and technology to improve the safety of automobiles. He was quite explicit in laying the blame on the industry, not technology itself. "A great problem of contemporary life is how to control the power of economic interests which ignore the harmful effects of their applied science and technology." Environmentalists such as Rachel Carson and Barry Commoner made similar arguments, marshaling scientific research to demonstrate the risks posed by industrial technologies.[55]

Nader, Carson, and Commoner embraced a cultural view of technology, at least implicitly. They all accepted that cultural values shape technological change, and argued for the creation of alternative technological paths in the service of progressive human values. Nader insisted that aviation safety research could be applied to automobiles, if the automobile industry was willing to place human lives above short-term profit. Carson did not reject pesticides completely, but argued instead for an ecologically minded approach to pest control that respected the balance of nature. By endorsing such alternative paths, Carson and Nader were rejecting both the pessimism of philosophers and the enthusiasm of technology's boosters.

A cultural view of technology could provide support for radical change as well as for piecemeal reforms. In the 1960s, the environmental anarchist Murray Bookchin used an explicitly human-centered understanding of technology to sketch out a technological utopia. Drawing from the humanist anarchism of Peter Kropotkin, Bookchin argued that modern technology could form the material basis for a decentralized, environmentally sound society. He explicitly attacked the pessimists with their "blanket rejection of technology," their belief that "technology is . . . imbued with a sinister life of its own." Instead, he insisted that technology offered the possibility of "a new dimension of human freedom" rooted in environmental values.[56] Yet Bookchin remained little read outside anarchist circles until the 1990s. Much better known were the similar though less radical arguments of the British economist E. F. Schumacher. In the early 1970s, Schumacher's book *Small Is Beautiful* provided a manifesto for the appropriate technology movement, which advocated using alternative values to create alternative technologies.[57]

Even within the counterculture itself, antitechnology rhetoric was never dominant. Most countercultural critics of technology in the 1960s were hostile not toward technology per se but rather toward specific types of technology, such as nuclear power or anything made of plastic. They happily embraced alternative technologies, such as bicycles or solar water heaters. Radical activists like Abbie Hoffman proclaimed their technological enthusiasm in best-selling books such as his pseudomemoir *Revolution for the Hell of It* (1968), in which Hoffman insisted that "the only pure revolution in the end is technology."[58] Similar sentiments can be found among countercultural entrepreneurs like Stewart Brand, whose *Whole Earth Catalog* became a fixture among young environmentalists in the early 1970s.[59]

A few engineers also grappled seriously with the works of Ellul and Marcuse. Matthew Wisnioski describes one of them, James Horgan, who in 1973 argued for a reciprocal relationship between technology and culture ("human values") in *Mechanical Engineering*, the monthly magazine of the American Society of Mechanical Engineers.[60] Another leading advocate of the cultural view was the civil engineer Samuel

Florman. In his book *The Existential Pleasures of Engineering* (1976), Florman patiently dismantled the "antitechnology dogma" of scholars such as Ellul, Marcuse, and Mumford. Technology, he maintained, was inextricably human, a type of human activity. At some level, humans always made a choice to engage in a particular technological activity. According to Florman, technology was no monolithic entity that forced humans to do its bidding.[61]

Historians of technology also maintained faith in a cultural approach. Melvin Kranzberg never wavered in his human-centered concept of technology. He insisted, against Ellul, that technology was "profoundly human."[62] Yet Kranzberg's journal, and historians of technology more generally, did little to develop such ideas until the 1970s. Only then did they firmly embrace what John Staudenmaier has called the "contextual" approach, which views technology as fundamentally shaped by its social context.[63]

After the 1970s, deterministic and instrumentalist voices reasserted themselves in the popular discourse of technology. Ronald Reagan's election in 1980 represented in part a repudiation of attempts to steer technological change in socially beneficial directions. Reagan hailed the first launch of the Space Shuttle *Columbia* as "a victory for the American spirit," while gutting government programs for renewable energy and environmental protection.[64] Meanwhile, the rise of the personal computer helped sustain technological enthusiasm, in part by combining the personal computer with values of the counterculture.[65] Some political conservatives, for example Rep. Newt Gingrich, displayed as much technological enthusiasm as countercultural activists.[66]

The hegemony of technological determinism became especially clear in the discourse around digital computers. Since the end of World War II, computers had served as vessels for hopes and fears about technology. Although enthusiasts dominated this discourse, there were always critics who saw computers as the ultimate threat to human freedom. But for the most part, these critics expressed little more than a mirror image of the enthusiasts' vision, reframing the coming digital utopia as a dystopia.[67]

It was in this era that computers became firmly linked to the concept of technology, in part through the phrase "information technology." Ronald Kline shows how these two powerful keywords of the postwar era were combined in the 1960s. *Information technology* was at first used to refer to methods for managing organizations. But by 1970, this meaning had largely disappeared, replaced by a definition of *informa-*

tion technology as the industrial arts of digital computers. By the 1980s, says Kline, *"information technology* was widely viewed as unstoppable."[68]

In the 1990s, enthusiasm for information technology surged to a fever pitch after the US government turned the Internet over to private control. This enthusiasm was often explicitly deterministic. In 1998, the merger of the Exxon and Mobil oil behemoths provided *Business Week* with the opportunity to editorialize about old versus new industries. The magazine's editors contrasted the old petroleum industry with a new online shopping company, Amazon. The editorial began with a resounding statement of technological determinism: "Technology is a relentless force that creates and destroys with little pity." The editors discussed the contrast between the Exxon-Mobil merger and the rise of Internet businesses such as Amazon. "Technology," they insisted, was "changing the very rules of the economic game." It had made possible the phenomenal market capitalization of Amazon despite its lack of profits. It would eventually transform Amazon too, because the "only certainties about technological change are that it is constant, painful and, in the end, positive for economic growth." *Business Week* was so proud of this editorial that it was reprinted as an advertisement on the Op-Ed page of the *New York Times*.[69]

Such rhetoric is strikingly reminiscent of Charles Beard's 1927 essay, "Time, Technology, and the Creative Spirit" (see chapter 9). In that essay, Beard praised technology as a key agent in human history. "Technology marches in seven-league boots from one ruthless, revolutionary conquest to another, tearing down old factories and industries, [and] flinging up new processes with terrifying rapidity." But Beard's rhetoric was decidedly precocious, using a concept of technology that had little currency even within academia. By the time *Business Week* published its editorial, however, this language had become thoroughly conventional.

History for the Present

This book is about more than just reconstructing the past; it is also an intervention in the present. At a minimum, my conceptual history of technology can help clarify the present-day meanings of the term. If scholars would merely think critically about what they mean when they use the term, so that they understood which definition of *technology* they were using, it would change how we write and talk about technology.

For example, the type of determinism expressed by *Business Week* has annoyed historians of technology for decades, yet the belief seems immune to criticism. Thus, the *New York Times* columnist Thomas Friedman can write, apparently without embarrassment, nonsense like this: "Just as the world was getting flattened by globalization, technology went on a rampage—destroying more low-end jobs and creating more high-end jobs faster than ever." And the former *Wired* editor Kevin Kelly devoted a whole book to "what technology wants." For Kelly and Friedman, collective decisions about regulation, research, trade, and macroeconomic policy all matter little compared with the onslaught of technology. The reality that humans choose which technologies to pursue, that people express diverse and conflicting human values in the technologies they create—this reality fades into the background, disappearing from the discourse of modernity.[70]

Academics have railed helplessly against this deterministic discourse for years. Yet in many ways, their helplessness is a direct result of their own lack of clarity about the concept of technology. Determinism, whether expressed by enthusiasts or by pessimists, is tightly linked to an instrumental conception of technology that divorces it from culture. To overcome this determinism, scholars need to consciously recreate and also popularize a cultural view of technology. Such a view would reject the divorce of technology from art, and restore the idea of technology as a creative expression of human values and strivings, in all their contradictory complexity.

Conceptual history also helps clarify the continued confusion between technology as industrial arts and technology as technique. Technology as industrial arts refers to means and methods for transforming the material world, while technology as technique is far broader, encompassing all skills and procedures for achieving a specific end. The history of concepts shows how both meanings entered English-language discourse in part through the mistranslation of the Continental concept of technique, which refers both to the industrial arts and to instrumental action in general. Thorstein Veblen first added the industrial meanings of *Technik* to the English-language concept of technology, followed decades later by Talcott Parsons, who translated the instrumental meaning of *Technik* as "technology."[71] Today, academics still blithely confound both of these meanings, equating bodily practices with "technologies of the self," and bureaucratic methods with "technologies of power." This usage stems directly from Foucault, who in his later works, in both English and French, treated *technique* and *technology* interchangeably.[72] Inspired by Foucault's usage, scholars have produced an entire genre of academic discourse that has little in common with work on technology as industrial arts. Yet these scholars never acknowledge that they are using the concept of technology in a fundamentally different way.[73] There is, in fact, no reason to use *technology* rather than *technique* in this context, aside from academic pretense.[74]

Conceptual history also sheds light on the dogged persistence of the definition of *technology* as the application of science. Historians have shown that the concept of applied science is deeply ambiguous; it sometimes implies the subordination of technology to science, while at other times it supports claims for the autonomy of technology. Historians of technology typically reject the definition of *technology* as applied science, without acknowledging this ambiguity. Nevertheless, the rhetoric of subordination continues into the twenty-first century, for example in persistent, largely evidence-free claims about the essential role of basic science for technological progress.[75] These claims have declined somewhat since the end of the Cold War, but they endure, in part because of increasing demands that research deliver practical results. For academic scientists in particular, defense of basic research still evokes the ancient tensions between scholars and technicians that frame so much of my story.[76] And this belief in technology as the application of basic science also remains central to popular culture, and not just in the English-speaking world. For example, in Cixin Liu's Hugo Award–winning *The Three-Body Problem*, meddlesome aliens seek to halt human technological progress by sabotaging basic research in particle physics.[77]

Despite all this confusion over its meaning, *technology* remains a consequential word. In its instrumental and deterministic senses, it continues to be invoked to legitimate the choices of the powerful. Humanity faces grave threats as a result of the choices embedded in our technologies, from the ever-present danger of nuclear war to the now-certain calamity of global warming. The least scholars can do is to create a coherent historical narrative that challenges the hegemony of these meanings.

But this book is not just about the meaning of words. A historical critique of the concept of technology also helps change our understanding of technology. And changing our understanding of technology is a step toward transforming actual technologies.

I've used this history to argue for a humanistic concept of technology as inextricably cultural. Technology is an expression of human values, both good and bad, of the desire to create and of the urge to destroy, of love and hate, of peace and war. Yet the dominant instrumental approach to technology denies the connection between technological and moral choices. By rejecting instrumentalism, we affirm the need to think ethically about technology.

Changes in thought are necessary but not enough for changing the material world. Nevertheless, we need to imagine alternative technological futures before we can create them. In these alternative futures, technology furthers human creativity, community, sustainability, and self-expression. To achieve such futures, we must reimagine technology as a conscious expression of the best of the human spirit.

Rehabilitating Technology: A Manifesto

This book is not a neutral work of scholarship but rather an intervention in the present, a first step in rehabilitating technology as a concept for history and social theory, with an eventual goal of shaping technologies toward more humane ends. With that in mind, I've included a brief manifesto that sets out my hopes for what a reformed concept of technology might look like.

We should seek to

1. Liberate technology from scholars who reduce it to instrumental reason, to the process of finding the best means to achieve a specific end. By rejecting instrumentalism, we also reject the belief that technology lacks its own moral compass.

2. Rescue technology from determinists, people who view technology as driven by its own ends, as a self-directed system isolated from conscious human control.

3. Reassert conceptual links between technology and art. We should view technology as a type of art in the older sense of the term, before the reduction of art to fine art. Understanding technology as art has the potential to resolve much of the conceptual confusion about technology.

4. Rethink the nature of application to develop a new understanding of the relationship between science and technology. The problem of application in technology is similar to that in ethics. Both fields deal with the application of universal principles in endlessly varying contexts.

5. Reclaim craft as an essential element of technology. Craft cannot be reduced to manual skill; it always involves cognitive judgments, judgments that rest in part on ethical principles. Science can never eliminate craft—not from technology, and not even from science.

6. Correct the unbalanced and often biased understanding of technology among scholars, their tendency to elevate theory over practice, discourse over materiality, principles over applications.

Acknowledgments

I've worked on this book for far too long, and I've accumulated more debts than I can possibly acknowledge. Inspiration for this project came from Ruth Oldenziel, whom I met just as she was finishing her dissertation on the history of gender, technology, and engineering in the United States. She let me read a key chapter that presented a cultural history of the concept of technology in America. Oldenziel was the first historian to take such an approach, which bore fruit a few years later in her monograph *Making Technology Masculine*. Over a decade later, my colleague at the University of Wisconsin–Madison, Ron Numbers, asked me to contribute an overview of the idea of technology for a Festschrift in honor of the historian of science David Lindberg. I planned to simply elaborate on Oldenziel's pioneering work, but soon found myself lost in a maze of primary sources. My chapter for Lindberg's Festschrift never appeared, but my research for it started me down the long path to this book.

Many people supported me early in this project, starting with Oldenziel. Leo Marx read some of my first work on the subject, and provided both inspiration and concrete suggestions. Rosalind Williams was very helpful from the start, and later lent me archival photocopies from her research on Lewis Mumford. I promise to return these as soon as I find them. Mikael Hård gave me crucial feedback on a draft of my article for *Technology and Culture* in 2006, "*Technik* Comes to America." Although I was unable to incorporate much of Mikael's advice in the article, I did follow many of his suggestions for the book.

Ronald Kline's article on the concept of applied science set a model for me to follow, and Ron provided helpful comments for my 2006 article. Charles Camic, now at Northwestern University, gave me detailed advice, especially about Thorstein Veblen. Howard Segal also gave me helpful feedback on my 2006 article. Guido Frison's articles were crucial to the early framing of my project, and I've had many stimulating exchanges with him since 2006. Karen Rosneck at UW–Madison's Memorial Library went far beyond her job description to help me make sense of cognate Russian terms.

I benefited tremendously from research assistants at the early stages of the book, even when their research didn't end up in it. History of science graduate students Amrys Williams, Joshua Kundert, and Erika Milam scoured the UW–Madison libraries for useful references, and summarized much of what they found. History of science major Christopher Hallquist did excellent research on John Dewey and Bertrand Russell for a summer internship. Ann Myers helped me with translations of Sombart and Schmoller, while Maria Giulia Carone translated one of Guido Frison's Italian articles for me.

When I began turning my article for *Technology and Culture* into a book, I leaned heavily on the work of a number of scholars who also encouraged me in the project, among them Peter Dear, Serafina Cuomo, and especially Pamela Long. Pam's masterly monograph *Openness, Secrecy, and Authorship* was essential for my book, and she gave me detailed feedback on my early chapters. Thomas Misa invited me to give a talk in 2011 at the University of Minnesota, where I first presented my ideas about instrumental versus cultural visions of technology. Since then, Tom has never stopped urging me to finish the book. David Edgerton has been a good friend and a strong advocate. He gave me crucial advice on my introduction, even though I didn't follow all of it. Skuli Sigurdsson also gave me helpful feedback on my draft introduction. Ellie Truitt read my medieval chapter and caught some howlers. Benoît Godin's work on the conceptual history of innovation was especially useful for the last two chapters of my book; Ben also provided me with helpful comments on the draft book manuscript. Several colleagues at UW–Madison helped me repeatedly, doing their best to answer my obscure questions; among them are Tom Broman, Lynn Nyhart, Richard Staley, and especially Mike Shank.

I benefited as well from many conferences and graduate seminars where my work was discussed. There are far too many to list, but I would specifically like to thank Guido Frison and Ludovic Coupaye at University College London for inviting me to their 2013 workshop,

"Epistemologies of Technology and Techniques"; Liliane Pérez and Guillaume Carnino for inviting me to a 2015 workshop, "La Technologie entre l'Europe et les États-Unis aux XIXe et XXe siècles," at L'École des hautes études en sciences sociales; and David Edgerton, who invited me to give a talk in the 2017 seminar series of the Centre for the History of Science, Technology and Medicine at King's College London. Thanks also to the many graduate students in my seminars who commented on various aspects of this project.

Karen Darling at the University of Chicago Press has been a strong supporter since I first described this project to her in a long conversation at the History of Science Society annual meeting in 2011. I especially thank Karen for her patience. Thanks also to the extraordinarily helpful anonymous readers for the press, who commented on multiple versions of the manuscript.

I am grateful as well for the financial support I received for research on this book. I completed the final chapter of the manuscript in the spring of 2017 while a visiting scholar at the Max Planck Institute for the History of Science in Berlin. During my time as a faculty member in the Department of the History of Science at the University of Wisconsin–Madison, my work on this project benefited from two sabbaticals, a Vilas Associates fellowship, and a resident fellowship at the Institute for Research on the Humanities. Early work was supported by National Science Foundation grant #0646788.

A special thanks to my parents, Tobalee Schatzberg and Paul Schatzberg. They never stopped believing in me, and my father was always ready to help with German translations. My children, Simon and Madeline, were much younger when I started working on this project. They were much more patient than a parent has any reason to expect. Madeline also helped me catalog the hundreds of books and articles that I accumulated.

Finally, a special thanks to Hallie Lieberman, who makes me happy. Without her unflagging support, I would not have been able to finish this book.

Notes

CHAPTER ONE

Unless otherwise indicated, all translations are my own.

1. In a typical example, the Technology web page of the *New York Times* describes itself as covering "the Internet, telecommunications, wireless applications, electronics, science, computers, e-mail and the Web." Metadata from "Technology News—The New York Times," http://www.nytimes.com/pages/technology/index.html, accessed 4/28/2016 (discontinued).
2. Friedman, *The World Is Flat*.
3. Forman, "The Primacy of Science," 10.
4. Castells, *The Rise of the Network Society*, 29–30; Kline, "Construing 'Technology' as 'Applied Science,'" 194–95.
5. Ellul, *The Technological Society*; Davis, *Means without End*; Mumford, *Pentagon of Power*.
6. Foucault, *The Foucault Reader*, 268; Foucault, *L'herméneutique du sujet*, 46–48. For a nuanced discussion of Foucault's "technology of the self," see Behrent, "Foucault and Technology," 90–92.
7. Winner, "Do Artifacts Have Politics?," 121–36.
8. On the idea of technology as practice or action, see Mitcham, *Thinking Through Technology*, chap. 9.
9. Edgerton, *The Shock of the Old*, xvii; Roberts and Schaffer, preface to *The Mindful Hand*, xv–xvi.
10. Schick and Toth, *Making Silent Stones Speak*.
11. Brey, "Theorizing Modernity and Technology," 54–55; Feenberg, "Modernity Theory and Technology Studies." Some scholars have attempted to extract concepts of technol-

ogy from these theorists, thus demonstrating the absence of an explicit theory. Shields, "Reinventing Technology in Social Theory"; Gerrie, "Was Foucault a Philosopher of Technology?"; Burkitt, "Technologies of the Self"; Behrent, "Foucault and Technology."

12. For example, Scott Lash's undertheorized use of *technology* in Lash, "Technological Forms of Life," esp. 107–8.

13. I draw this argument from Cuomo, *Technology and Culture in Greek and Roman Antiquity.*

14. Mumford, *Technics and Civilization*, 6; Engelmeyer, "Allgemeine Fragen der Technik," 311:103.

15. At least, this is the argument of Landes, *Revolution in Time.*

16. Parsons, "Some Reflections on 'The Nature and Significance of Economics,'" 525. The oil-pan example is from M. Weber, *Economy and Society*, 66.

17. Aristotle, *Aristotle's Nicomachean Ethics*, 121 (1140b25).

18. On these contradictions in the history of art, see Shiner, *The Invention of Art.*

19. For a critique of the instrumental (or, more broadly, essentialist) approach in the philosophy of technology, see Feenberg, *Questioning Technology*, viii, 14–17. An exemplar of the reduction of technology to technique is Ellul, *The Technological Society.*

20. Kevin Kelly typifies the instrumental enthusiast. E.g., Kelly, *What Technology Wants.*

21. E.g., Lienhard, *The Engines of Our Ingenuity.*

22. I draw this idea of technology as belonging to all humans from Thorstein Veblen. See chapter 8.

23. There are far fewer English-language works in the Library of Congress catalog under the subject heading "technology—philosophy" (275) than under similar headings for other concepts, such as "art—philosophy" (847), "science—philosophy" (2,660), and "political science—philosophy" (1,284). Library of Congress Online Catalog, expert mode search with the string "6500 'art philosophy'| Language: English," accessed 1/15/2017.

24. Winner, "Technologies as Forms of Life," 3–10.

25. P. Edwards, "Infrastructure and Modernity," 187.

26. Edgerton, *The Shock of the Old*; Godin, "In the Shadow of Schumpeter."

27. For a survey of theories about language, see Formigari, *A History of Language Philosophies.*

28. Dewey, *The Quest for Certainty*, 4–5; also Feenberg, *Questioning Technology*, 1; and more generally Lobkowicz, *Theory and Practice.*

29. The focus on science as knowledge is present even in works by philosophers sympathetic to problems of practice, e.g., Hacking, *Representing and Intervening.*

30. The classic work is Bijker, Hughes, and Pinch, *The Social Construction of Technological Systems.*

31. An exception to this reductive use of *technoscience* is Pickstone, *Ways of Knowing.*

32. For overviews, see Durbin, "Philosophy of Technology"; Meijers, *Philosophy of Technology and Engineering Sciences*.
33. For the 2017 meeting of the American Historical Association, see https://www.historians.org/annual-meeting/2017-program, accessed 2/11/18. For the 2017 meeting of the Eastern Division of the American Philosophical Association, see http://www.apaonline.org/page/2017E_Program, accessed 2/11/18.
34. E.g., Castells, *The Rise of the Network Society*.
35. Watkins, "Rodney Dangerfield, Comic Seeking Respect, Dies at 82."
36. The most prominent exception is Mitcham, *Thinking Through Technology*. Mitcham's work has had a major influence on my understanding of *technology*.
37. On philosophical approaches to definition, see Mitcham and Schatzberg, "Defining Technology and the Engineering Sciences," 28–32.
38. Hickman, *Philosophical Tools for Technological Culture*, 11–14; quotation is from p. 11.
39. Ibid.; quotations are from p. 14.
40. See chapter 7.
41. Heidegger, *The Question concerning Technology*, 4; Heidegger, *Die Technik und die Kehre*, 5. Heidegger's essay is definitely overrated as a contribution to the philosophy of technology.
42. Some Continental languages, for example Danish and Spanish, are shifting to cognate forms of *technology* for this field, no doubt in response to English usage.
43. For a discussion of translating *technique-technology* into English, see Redondi, "History and Technology," 1; Salomon, "What Is Technology?," 113–56. In the 1960s, Guillerme and Sebestik commented on "the Anglo-Saxon meaning of *technology*, which can be identified with *la technique* in general and with statements about applied science." Guillerme and Sebestik, "Les commencements de la technologie," 69.
44. "Technologie" (1998), 960, referring to the influence of "angelsächsischen Sprachgebrauchs" on German usage. For a list of (mainly French) definitions of *technology* and the term's Continental cognates, see Beaune, *La technologie introuvable*, 253–63.
45. "Technik," *Wörterbuch Duden* online, http://www.duden.de/rechtschreibung/Technik, accessed 2/11/18; "technique," *Dictionnaire de Français Larousse*, http://www.larousse.fr/dictionnaires/francais/technique/76950, accessed 2/11/18.
46. Whitney and Smith, *The Century Dictionary*, revised and enlarged ed., s.v. "technology"; Hammond, "[Review of *Century Dictionary*]," 112.
47. P. Long, *Openness, Secrecy, Authorship*, 18–19.
48. In contrast, *machines* and *machinery* were much discussed by nineteenth-century economic writers. See chapter 10.
49. L. Marx, "*Technology*: The Emergence of a Hazardous Concept" (2010).

50. For examples, see van der Pot, *Die Bewertung des technischen Fortschritts*, 1:200–209, 1:15.

51. Many scholars have produced similar divisions of the meaning of *technology*. Mitcham has divided these meanings into objects, activities, knowledge, and volition. Mitcham, *Thinking Through Technology*. See also Laudan, "Natural Alliance or Forced Marriage?," S18.

52. "Technology, n."

53. Koselleck edited an eight-volume German encyclopedia of key concepts in political thought, the *Geschichtliche Grundbegriffe* (*GG*). The *GG* did not have entries on *science* or *technology*. Although the *GG* was focused on concepts used in German, it was not a monolingual enterprise, as all its concepts moved repeatedly across linguistic borders. Typical entries also took a long view, examining ancient and medieval discourse about earlier forms of modern concepts. Furthermore, Koselleck stressed that basic concepts have ambiguous, multiple meanings, requiring constant reinterpretation in varying social contexts. Bevir, "Review: Begriffsgeschichte"; Richter, "Begriffsgeschichte and the History of Ideas."

54. Foucault, "Nietzsche, Genealogy, History," 81, 82. Anglo-American philosophers and intellectual historians made similar arguments in the 1960s and 1970s, though using quite different terminology. E.g., Skinner, "Meaning and Understanding in the History of Ideas," 3–53.

55. Nietzsche, for example, exposed the ancient aristocratic norms that shaped the concept of the good. And Foucault himself devoted the second and third volumes of *The History of Sexuality* to the classical world, searching not for similarities with current concepts but rather for differences that expose present-day sexual hypocrisy. Nietzsche, *On the Genealogy of Morality*, 12; Foucault, *The History of Sexuality*.

56. Translators from ancient Greek continue to render *techne* as "art," but they admit that this translation is not just imperfect but misleading. E.g., Plato and Aristotle, *Plato, Gorgias, and Aristotle, Rhetoric*, 30n1.

57. I borrow this ideas from Camic, "Reputation and Predecessor Selection."

58. For an example that borders on self-parody, see Law, *Aircraft Stories*. Many STS scholars have turned away from the linguistic turn, rediscovering materiality. Historians of technology find this move puzzling, having never forgotten about the material world. For a critique of the recent discourse of materiality, see Ingold, "Materials against Materiality."

59. MacIntyre, *A Short History of Ethics*, 2–3; quoted in Ball, *Transforming Political Discourse*, 17.

CHAPTER TWO

1. On the teaching of Greek and Latin in the United States, see Winterer, *The Culture of Classicism*.

2. Or in German, "Die Frage nach der Technik." Heidegger, *The Question concerning Technology*.

3. Cuomo, *Technology and Culture in Greek and Roman Antiquity*, 166–67.

4. Mitcham, *Thinking Through Technology*, 117–18; Roochnik, *Of Art and Wisdom*, 18–26. James Scott, drawing from Detienne and Vernant, pits *metis*, cunning intelligence, against *techne*. I doubt that such an opposition existed among the ancient Greeks. J. Scott, *Seeing Like a State*; Detienne and Vernant, *Cunning Intelligence in Greek Culture and Society*.

5. Roochnik, *Of Art and Wisdom*, 43–44; Cuomo, *Technology and Culture in Greek and Roman Antiquity*, 12–18. See also Burford, *Craftsmen in Greek and Roman Society*.

6. Cuomo, *Technology and Culture in Greek and Roman Antiquity*, 12–14; quotation is from p. 12.

7. Plato and Aristotle, *Plato, Gorgias, and Aristotle, Rhetoric*, 79 (491A); Roochnik, *Of Art and Wisdom*, 1, 200.

8. Roochnik, *Of Art and Wisdom*, 45–46.

9. Ibid., 51; Cuomo, *Technology and Culture in Greek and Roman Antiquity*, 15,19–21.

10. See Plato and Aristotle, *Plato, Gorgias, and Aristotle, Rhetoric*.

11. Roochnik, "Is Rhetoric an Art?," 127–28.

12. Plato and Aristotle, *Plato, Gorgias, and Aristotle, Rhetoric*, 33–34 (450B–C); Mitcham, *Thinking Through Technology*, 118.

13. Plato and Aristotle, *Plato, Gorgias, and Aristotle, Rhetoric*, 49 (465A).

14. Roochnik, *Of Art and Wisdom*, 90; Mitcham, *Thinking Through Technology*, 119; M. Nussbaum, *The Fragility of Goodness*, 94. See also Atwill, *Rhetoric Reclaimed*, esp. chap. 5, "Plato and the Boundaries of Art."

15. For a summary, see Martin Ostwald, "Translator's Introduction," in Aristotle, *Nicomachean Ethics*, xiv–xvi. In other works, Aristotle does not maintain this sharp division between *techne* and *episteme*. M. Nussbaum, *The Fragility of Goodness*, 444n11.

16. Aristotle, *Aristotle's Nicomachean Ethics*, 118 (1139b20).

17. Ibid., 119 (1140a8).

18. Ibid., 120 (1140b6), see also 117 (39b4).

19. Ibid., 120 (1140a19), see also 119 (40a1).

20. For a more detailed discussion of Aristotle's views on these three types of knowledge, see Atwill, *Rhetoric Reclaimed*, 164–74.

21. Cuomo, *Technology and Culture in Greek and Roman Antiquity*, 36–28, 166–67.

22. Aristotle, *Aristotle's Nicomachean Ethics*, 1 (1094a5). Note that Aristotle links the nature of ends to techne and praxis only later in the work.

23. Ibid., 121 (1140b25); Aristotle, *Nicomachean Ethics*, 154n20.

24. Aristotle asserted a slightly different hierarchy of knowledge in *Metaphysics*, but with episteme always on top. Aristotle, *The Metaphysics*, 981b1–30, consulted through Perseus.

25. Ibid., 982b, consulted through Perseus; Atwill, *Rhetoric Reclaimed*, 77–79; Lobkowicz, *Theory and Practice*, 6–8.
26. MacIntyre, *After Virtue*, 158; MacIntyre, *A Short History of Ethics*, 98. See also Long's discussion of Xenophon, who elevated praxis over productive techne. P. Long, *Openness, Secrecy, Authorship*, 21–23.
27. Lobkowicz, *Theory and Practice*, 35–46; quotation is from p. 42.
28. Gooday, "'Vague and Artificial'"; Bud, "'Applied Science.'"
29. R. Bernstein, "Heidegger's Silence?," 120–24.
30. P. Long, *Openness, Secrecy, Authorship*, 18–19.
31. Roochnik, *Of Art and Wisdom*, 57–58.
32. Ibid., 58–60.
33. In a sense, I am restating Melvin Kranzberg's first law: "Technology is neither good nor bad; nor is it neutral." Kranzberg, "Technology and History," 545.
34. MacIntyre, *After Virtue*, 159.
35. Burford, *Craftsmen in Greek and Roman Society*, 185.
36. Cuomo, *Technology and Culture in Greek and Roman Antiquity*, 36–28, 166–67.
37. Aristotle, *Politics*, 1337b8–18, as quoted and translated by Cuomo, *Technology and Culture in Greek and Roman Antiquity*, 9.
38. Burford, *Craftsmen in Greek and Roman Society*, 198–207; P. Long, *Openness, Secrecy, Authorship*, 77–78.
39. E. Whitney, "Paradise Restored," 27n17; Burford, *Craftsmen in Greek and Roman Society*, 25–26.
40. Burford, *Craftsmen in Greek and Roman Society*, 153–57.
41. Ibid., 130.
42. Cuomo, *Technology and Culture in Greek and Roman Antiquity*, 25–26; Plato, *Republic*, 28, 92, 109 (34B–C, 95D, 414D–415C).
43. E. Whitney, "Paradise Restored," 28–30; Cuomo, *Technology and Culture in Greek and Roman Antiquity*, 9–11.
44. Cicero, *On Obligations*, 150.
45. P. Long, *Openness, Secrecy, Authorship*, 36. For an overview of classical divisions of the arts, see Tatarkiewicz, "Classification of Arts in Antiquity."
46. Cuomo, *Technology and Culture in Greek and Roman Antiquity*, 33–34.
47. This argument is condensed from chapter 3 of ibid., 77–102. See also Joshel, *Work, Identity, and Legal Status at Rome*.
48. Galen, "Exhortation to the Study of the Arts," 521. Special thanks to Lindsey Morse at University of Puget Sound for providing me with a more literal translation of this passage.
49. Quintilian, *The Institutio Oratoria of Quintilian*, 2:346–47, consulted through Perseus.
50. E. Whitney, "Paradise Restored," 38–40.
51. On the origins of the liberal arts tradition, see Stahl, Johnson, and Burge, *Martianus Capella and the Seven Liberal Arts*.

52. Tatarkiewicz, "Classification of Arts in Antiquity," 234–35; see also Kristeller, "The Modern System of the Arts," 505–6.
53. Shiner, *The Invention of Art*, 19–27; E. Whitney, "Paradise Restored," 51.

CHAPTER THREE

1. E.g., Eco, *The Aesthetics of Thomas Aquinas*; Maritain, *Art and Scholasticism*. This focus is present even in exemplary works such as Shiner's *The Invention of Art*, which argues against projecting the category of fine arts onto pre-Enlightenment discussions of art. Shiner, *The Invention of Art*; see also Kristeller, "The Modern System of the Arts" (part 1).
2. E.g., Hicks, "Martianus Capella and the Liberal Arts"; van den Hoven, *Work in Ancient and Medieval Thought*; E. Whitney, "Paradise Restored."
3. Lynn White, "Technology and Invention in the Middle Ages."
4. P. Long, *Openness, Secrecy, Authorship*, 36.
5. On social and economic conditions in the early Middle Ages, see Wickham, *The Inheritance of Rome*, esp. chap. 4. On the historiography of the period, see R. Collins, "Making Sense of the Early Middle Ages."
6. Lynn White, "Cultural Climates and Technological Advance in the Middle Ages"; Mumford, *Technics and Civilization*, 111–12.
7. Ovitt, "The Cultural Context of Western Technology," 490.
8. Ovitt, *The Restoration of Perfection*, 58–69.
9. Ovitt, "The Cultural Context of Western Technology," 494–97.
10. Van den Hoven, *Work in Ancient and Medieval Thought*, e.g, 198–99; Allard, "Les arts mécaniques aux yeux de l'idéologie médiévale."
11. Ovitt, "The Cultural Context of Western Technology," 493–94.
12. Ovitt, *The Restoration of Perfection*, 64–65, 69–70.
13. For example, Whitney versus van den Hoven. E. Whitney, "Paradise Restored," 71; van den Hoven, *Work in Ancient and Medieval Thought*, 254–55.
14. Although van den Hoven has made a strong argument for continuity, her detailed examination of early Christian monastic texts does point to a significant elevation of the status of manual labor, and by implication crafts associated with such labor. Van den Hoven, *Work in Ancient and Medieval Thought*, 128–58.
15. John the Scot as quoted in E. Whitney, "Paradise Restored," 70–71.
16. Ibid., 71.
17. Lynn White, *Medieval Religion and Technology*, 248. For a critique of this comment, see van den Hoven, *Work in Ancient and Medieval Thought*, 162.
18. E. Whitney, "Paradise Restored," 83.
19. Compare van den Hoven, *Work in Ancient and Medieval Thought*, 176.
20. Hugh of Saint-Victor, *The Didascalicon*, 54.
21. Van den Hoven, *Work in Ancient and Medieval Thought*, 160.
22. J. Taylor, introduction to Hugh of Saint-Victor, *The Didascalicon*, 4–5.
23. Hugh of Saint-Victor, *The Didascalicon*, 51, 54.

24. E. Whitney, "Paradise Restored," 88; van den Hoven, *Work in Ancient and Medieval Thought*, 162–71.
25. Hugh of Saint-Victor, *The Didascalicon*, 74–75.
26. He actually referred to them as "sciences" (*scientiae*), that is, knowledge rather than practice of specific arts. Van den Hoven, *Work in Ancient and Medieval Thought*, 164n18.
27. Hugh of Saint-Victor, *The Didascalicon*, 75–79.
28. Ibid., 88.
29. Halliwell, *Aristotle's Poetics*, 84.
30. Hugh of Saint-Victor, *The Didascalicon*, 58–59.
31. Ibid., 55.
32. Van den Hoven, *Work in Ancient and Medieval Thought*, 166–67; Hugh of Saint-Victor, The Didascalicon, 56.
33. Van den Hoven, *Work in Ancient and Medieval Thought*, 165; E. Whitney, "Paradise Restored," 84–85.
34. Hugh of Saint-Victor, *The Didascalicon*, 52, 55.
35. Ibid., 56.
36. Allard, "Les arts mécaniques aux yeux de l'idéologie médiévale," 21; van den Hoven, *Work in Ancient and Medieval Thought*, 173; Hugh of Saint-Victor, *The Didascalicon*, 140–42.
37. On Hugh's subsequent influence in medieval thought, see E. Whitney, "Paradise Restored," 99–123.
38. Allard, "Les arts mécaniques aux yeux de l'idéologie médiévale," 21.
39. In another work, Hugh did discuss the connections between theoretical and applied geometry, but he described applied geometry as *practical*, not *productive* or *mechanical*, thus obscuring any connection to the category of mechanical arts. Lobkowicz, *Theory and Practice*, 83–84.
40. E. Whitney, "Paradise Restored," 101. In contrast, Lobkowicz argued that the category of mechanical arts was less influential in the twelfth century. Lobkowicz, *Theory and Practice*, 84.
41. E. Whitney, "Paradise Restored," 108–9.
42. Ibid., 112–14; Ovitt, *The Restoration of Perfection*, 123.
43. Spade, "Medieval Philosophy," sec. 5.1.
44. E. Whitney, "Paradise Restored," 139–40.
45. Ovitt, *The Restoration of Perfection*, 121–22.
46. Thanks to Elly Truitt for this observation.
47. Glick, "Technology."
48. E. Whitney, "Paradise Restored," 131.
49. Ibid., 122–24; van den Hoven, *Work in Ancient and Medieval Thought*, 178–85.
50. For overviews, see Eco, "In Praise of Thomas Aquinas"; D. Turner, *Thomas Aquinas*.
51. Thomas Aquinas, *Summa Theologica*, 367–82 (ST I–II, Q55–66). There is a significant but somewhat misleading literature about art in Thomas, most

of which focuses on aesthetic questions rather remote from the medieval concept of ars. E.g., Maritain, *Art and Scholasticism*; Eco, *The Aesthetics of Thomas Aquinas*. Other authors deal briefly with the role of the mechanical arts in Thomas. See Ovitt, *The Restoration of Perfection*, 131–33; E. Whitney, "Paradise Restored," 139–41.

52. Thomas Aquinas, *Summa Theologica*, 369–70 (ST I–II, Q56 A3).
53. Thomas Aquinas, *Commentary on the Metaphysics of Aristotle*, vol. 1, bk.1, lesson1, ¶28, ¶31, http://dhspriory.org/thomas/Metaphysics1.htm, accessed 2/11/18.
54. Thomas Aquinas, *Summa Theologica*, 377–78 (ST I–II, Q57 A3).
55. Ibid., 376–77 (ST I–II, Q57 A2).
56. Ibid., 377–78 (ST I–II, Q57 A3).
57. Similarly, Thomas distinguished between prudence and art on the same basis. He argued that "the good of an art is thought of as existing not in the craftsman himself, but rather in the artifact itself. . . . By contrast, the good of prudence exists in the very agent whose perfection is the acting itself." Ibid., 380 (ST I–II, Q57 A5).
58. Maritain, *Art and Scholasticism*, 9.
59. Ibid., 5–7, 11–15.
60. Thomas Aquinas, *Summa Theologica*, 385, 87 (ST I–II, Q58 A2, A4).
61. Thomas's *prudence* is a translation of Aristotle's *phronesis*, which some translators render as "practical knowledge." On the role of phronesis in philosophical ethics, see R. Bernstein, *Beyond Objectivism and Relativism*.
62. Thomas Aquinas, *Summa Theologica*, 388 (ST I–II, Q58 A5).

CHAPTER FOUR

1. Bennett, "The Mechanical Arts"; Dear, "Mixed Mathematics." An alternative neo-Pythagorean tradition from antiquity through the Middle Ages valued mathematics as the heart of natural philosophy, but this tradition was largely obscured by the dominant Aristotelian tradition. See Albertson, *Mathematical Theologies*. Thanks to Mike Shank for this reference.
2. E.g., Gimpel, *The Medieval Machine*.
3. For a well-researched popular survey of medieval technology, see Gies and Gies, *Cathedral, Forge, and Waterwheel*.
4. S. A. Epstein, "Urban Society"; McNeill, *The Pursuit of Power*.
5. The best-known text on medieval crafts is the twelfth-century manuscript *On Diverse Arts* by the pseudonymous Theophilus, probably the Benedictine monk and skilled metalworker Roger of Helmarshausen. Lynn White suggested that *On Diverse Arts* was a response to critics of the luxurious artifacts that adorned Benedictine churches. Against such criticism, Theophilus insisted on the virtuousness of the crafts that produced these artifacts, crafts that he portrayed as gifts of God. P. Long, *Openness, Secrecy, Authorship*, 85–87; Lynn White, *Medieval Religion and Technology*, 97–100.

NOTES TO CHAPTER FOUR

6. McNeill, *The Pursuit of Power*, chap. 6.
7. P. Long, *Openness, Secrecy, Authorship*, 96–101.
8. Ibid., 141.
9. Bacon, *The Advancement of Learning*, 42–43.
10. P. Long, *Openness, Secrecy, Authorship*, chaps. 4, 6, 7.
11. Ibid., 104.
12. On Vesalius, see Rossi, *Philosophy, Technology, and the Arts*, 7–8.
13. P. Long, *Openness, Secrecy, Authorship*, 184.
14. Ibid., 178–81; Smith and Gnudi, introduction to *The Pirotechnia of Vannoccio Biringuccio*.
15. P. Long, *Artisan/Practitioners*, 101–5.
16. Rossi, *Philosophy, Technology, and the Arts*, 22; Shiner, *The Invention of Art*, 35–56.
17. Rossi, *Philosophy, Technology, and the Arts*, 10–11.
18. Ibid., 6; Casini, "Juan Luis Vives [Joannes Ludovicus Vives]."
19. Rossi, *Francis Bacon*, 8.
20. Ibid., 607; Gilbert, *Queene Elizabethes Achademy*.
21. P. Long, *Openness, Secrecy, Authorship*, 119–20.
22. For a survey of these activities, see Galluzzi, *Renaissance Engineers*.
23. P. Long, *Openness, Secrecy, Authorship*, 129–32; quotation (from Filarete) is from p. 132.
24. Shapin, *A Social History of Truth*, 57–58.
25. Dear, *Revolutionizing the Sciences*, 50.
26. The metaphor of the "fruits" of basic research was widespread after World War II. See, e.g., National Academy of Sciences (U.S.) Committee on Science and Public Policy, *Basic Research and National Goals*.
27. This awareness was expressed in the idea of the arts as a unity of theory and practice, for example in architecture. See P. Long, *Openness, Secrecy, Authorship*.
28. On Bruno as an influence on Bacon, see Rossi, *Philosophy, Technology, and the Arts*, 77–80.
29. Agassi, *The Very Idea of Modern Science*, 3–13.
30. Ash, *Power, Knowledge, and Expertise*, 192–93; quotation is from p. 193.
31. Bacon, *The New Organon*, bk. 1, aphorism 73, as quoted in Ash, *Power, Knowledge, and Expertise*, 189.
32. Nanni, "Technical Knowledge and the Advancement of Learning," 53; Bacon, *The New Organon*, 61, bk. 1, aphorism 74.
33. Bacon, *The New Organon*, 100, bk. 1, aphorism 29.
34. Bacon, *The Advancement of Learning*, 88–89 (quotation); Bacon, *The New Organon*, 21, 227–28; Rossi, *Philosophy, Technology, and the Arts*, 117–21.
35. Rossi, *Philosophy, Technology, and the Arts*, 86.
36. Although Long does not make this point explicitly, I draw this conclusion from her detailed discussion of the relevant sources. P. Long, *Openness, Secrecy, Authorship*, chaps. 4, 6.

37. Bacon, *The New Organon*, 66, bk. 1, aphorism 80, 69–70, bk. 1, aphorism 85; Nanni, "Technical Knowledge and the Advancement of Learning," 54–55.
38. Gieryn, *Cultural Boundaries of Science*, 37–64.
39. The classic work in this genre is Bush, *Science, the Endless Frontier.*
40. Ash, *Power, Knowledge, and Expertise*, 199–203.
41. Bacon, *The New Organon*, 70, bk. 1, aphorism 85, 61, bk. 1, aphorism 74; Weeks, "Francis Bacon and the Art–Nature Distinction," 122–23.
42. Ash, *Power, Knowledge, and Expertise*, 207–11.
43. E.g., Shapin and Schaffer, *Leviathan and the Air-Pump*; Galison, *How Experiments End*; Edgerton, *The Shock of the Old*; H. M. Collins, *Tacit and Explicit Knowledge.*
44. Bacon, *The New Organon*, 85–90, bk. 1, aphorism 109; Nanni, "Technical Knowledge and the Advancement of Learning," 59–61; Weeks, "The Role of Mechanics in Francis Bacon's *Great Instauration*," 174.
45. Veblen, "The Place of Science in Modern Civilization," 595.
46. Zilsel, "The Sociological Roots of Science," 554; P. Long, *Artisan/Practitioners*, 10–22; Cohen, *The Scientific Revolution*, 336–42.
47. P. Long, *Artisan/Practitioners*, chap. 1. For an example of deeply misleading historical defense of pure science, see Forman, "The Primacy of Science."
48. P. Long, *Openness, Secrecy, Authorship*, esp. chaps. 4, 6–7.
49. Bennett, "The Mechanical Arts."
50. P. Long, *Artisan/Practitioners*. In addition to Long, see Ash, *Power, Knowledge, and Expertise*; Klein and Spary, *Materials and Expertise*; McNeill, *The Pursuit of Power*; Roberts, Schaffer, and Dear, *The Mindful Hand*, esp. sections 1 and 2.
51. P. Smith, *The Body of the Artisan*; Harkness, *The Jewel House.*
52. Roberts and Schaffer, preface to *The Mindful Hand*, xiii.
53. Shapin, *A Social History of Truth*, 38–40, 58.
54. Ibid., esp. chap. 8.
55. Ibid., 367. British science became more open to contributions from artisans in the eighteenth century. Sorrenson, "George Graham, Visible Technician." But, I would argue, this openness did not produce a fundamental change in the hierarchy of knowledge.
56. Shapin, *A Social History of Truth*, 395–97.
57. Dear, *Revolutionizing the Sciences*, 50.
58. Harrison, "'Science' and 'Religion'"; Shapin, "The Virtue of Scientific Thinking."

CHAPTER FIVE

An earlier version of this chapter was published as Schatzberg, "From Art to Applied Science."

1. Mokyr, *The Gifts of Athena*, 34–41. For critiques of Mokyr, see Berg, "The Genesis of 'Useful Knowledge,'" 127–30; Hilaire-Perez, "Technology as a Public Culture in the Eighteenth Century"; Ashworth, "The Ghost of Rostow."

2. Klein, "Artisanal-Scientific Experts in Eighteenth-Century France and Germany"; Klein and Spary, *Materials and Expertise*, pt. 3.

3. Alder, *Engineering the Revolution*, 60.

4. Johnson, *A Dictionary of the English Language*, s.v. "art," "science."

5. Horn, "The Privilege of Liberty."

6. A few dissenters, especially engineers, tried to develop a more nuanced understanding of the relationship between scientific theory and industrial practice, but their voices were drowned out by the rhetoric of subordination. E.g., Rankine, *A Manual of Applied Mechanics*, 1–3; Channell, "The Harmony of Theory and Practice."

7. On encyclopedias in the era, see Yeo, *Encyclopaedic Visions*. On the eighteenth century as an "age of improvement," see Friedel, *A Culture of Improvement*.

8. Yeo, *Encyclopaedic Visions*, chap. 5.

9. Chambers, *Cyclopædia*, 1; quotations are from p. viii. For the broader context of Chambers's discussion of art and science, see Yeo, *Encyclopaedic Visions*, 147–52.

10. Chambers, *Cyclopædia*, 1:viii.

11. Ibid., vol. 1; quotations are from pp. viii, ix.

12. D'Alembert, *Preliminary Discourse*; Diderot and d'Alembert, *Encyclopédie*. For an annotated English version of the article on art, see Diderot, "Art (Applied Natural History)." For a more detailed analysis of art and industry in the *Encyclopédie*, see Poni, "The Worlds of Work"; also Darnton, "Philosophers Trim the Tree of Knowledge."

13. Diderot and d'Alembert, *Encyclopédie*, 1:714, 1:717; d'Alembert, *Preliminary Discourse*, 40–43. For similar analyses, see Roberts, introduction to section 3 of *The Mindful Hand*, 189–90; Yeo, *Encyclopaedic Visions*, 152–54.

14. Shiner, *The Invention of Art*, esp. 80; quotation is from p. 111.

15. Ibid., 81–83.

16. D'Alembert, *Preliminary Discourse*, 43 (quotation), 55, 68–70; Shiner, *The Invention of Art*, 84.

17. Bacon, *The Advancement of Learning*, 85, 101 (quotation), 104 (quotation).

18. D'Alembert, *Preliminary Discourse*, 144–45, 156–57; Shiner, *The Invention of Art*, 83–85; Diderot and d'Alembert, *Encyclopédie*, 1:xvii–xviii.

19. D'Alembert, *Preliminary Discourse*; quotation is from p. 123; J. Mason, *The Value of Creativity*, 117–20. Kant further endorsed the Enlightenment connection of *fine art* with the inexplicable nature of creative genius. Kant, *Kant's Critique of Judgement*, §46, 188–190; Shiner, *The Invention of Art*, 147–48.

20. Chisholm, *The Encyclopædia Britannica*, 2:657–60. Note that the process by which *art*, when used without a modifier, came to signify *fine art* appears

to be more gradual and more complicated that previous scholars have implied. E.g., Kristeller, "The Modern System of the Arts . . . (II)," 23–24, 30–31, 43–46.

21. Mill, *A System of Logic*, 2:520–31. In the early American Republic, a discourse on the relationship between mind and hand mirrored the discourse of science and art, but with a distinctive American cast related to the democratic values of the early Republic. Rice, *Minding the Machine*.

22. For an example of this discourse of art divorced from utility, see Jarves, *The Art Idea*.

23. Shiner, *The Invention of Art*, 99–120.

24. A relevant work is R. Williams, *The Triumph of Human Empire*, although her book's scope is much broader than the Arts and Crafts movement.

25. See Schabas, *The Natural Origins of Economics*.

26. For a thorough analysis of this issue, see MacLeod, *Heroes of Invention*.

27. The classic work is Landes, *Unbound Prometheus*.

28. See Sabel and Zeitlin, *World of Possibilities*; Berg, *The Machinery Question*, 154, 250; Berg, "The Genesis of 'Useful Knowledge,'" 127–30. There is a large and complex literature on skill. See, e.g., S. R. Epstein, "Craft Guilds in the Pre-Modern Economy," 166.

29. S. Edwards, "Factory and Fantasy in Andrew Ure"; Farrar, "Andrew Ure."

30. Interestingly, Ure also distinguished "art or manufacture" from "mechanical engineering," which concerns the provision of mechanical power, and from "handicraft," which modifies manufactures "into objects of special or local demand." Ure, preface to *A Dictionary of Arts, Manufactures, and Mines*, 1:iii.

31. Ure, *The Philosophy of Manufactures*, 20, 23.

32. Berg, *The Machinery Question*, 184, 199 (quotation); Babbage, *On the Economy of Machinery and Manufactures*.

33. Ure, *The Philosophy of Manufactures*; quotations are from pp. 20, 21, 368.

34. K. Marx, *Capital* (1976), 1:470n23, 544–45, 563–64.

35. Ure, *The Philosophy of Manufactures*; quotations are from pp. viii, 23–24, 32, 25.

36. Gooday, "'Vague and Artificial,'" 548.

37. Robert Bud notes that Ure's understanding of *applied science* was picked up by the French founders of the Conservatoire nationale des arts et métiers. Bud, "'Applied Science,'" 542.

38. Ure, *The Philosophy of Manufactures*, 2 (quotation), 16 (quotation), 32–33, 37–38.

39. Babbage, *On the Economy of Machinery and Manufactures*, 206.

40. MacLeod, *Heroes of Invention*, esp. 45–52, 59–90.

41. Bud, "'Applied Science' in Nineteenth-century Britain," 18.

42. Layton, *Revolt of the Engineers*; Kline, "Construing 'Technology' as 'Applied Science'"; Kline, "Science and Technology"; Gooday, "'Vague and Artificial'"; Lucier, "The Origins of Pure and Applied Science in Gilded Age America."

43. J. Turner, "Le concept de science dans l'Amérique du XIXe siècle"; quotations are from pp. 771–72. Denise Phillips makes similar points in her analysis of nineteenth-century English translations of *Wissenschaft*. Phillips, "Francis Bacon and the Germans."
44. See the special issue of *History of Science* 45, no. 2 (June 2007).
45. Paul Forman repeatedly commits this error. Forman, "The Primacy of Science."
46. Bigelow, *Elements of Technology*, 1–2.
47. "Science, Art, Discovery," 10.
48. E.g., Partington, *The British Cyclopaedia of the Arts and Sciences*. On eighteenth- and nineteenth-century scientific encyclopedias more generally, see Yeo, *Encyclopaedic Visions*.
49. Edward Everett, *Orations and Speeches on Various Occasions* (Boston: Little, Brown, 1860), 1:275; quoted in Staiti, *Samuel F. B. Morse*, 224.
50. Forman, "The Primacy of Science," 39, with variants of this phrase repeated throughout the article.
51. H. Williams, *A History of Science*; quotations are from 1:3, 1:5, 6:1–2.
52. LaFollette, *Making Science Our Own*, 51. LaFollette's chart contains a good number of inventors and engineers, often misidentified in the press as physicists (such as Charles Steinmetz and Vannevar Bush).
53. Gieryn, *Cultural Boundaries of Science*, 37–64.
54. Ibid., 43, 46–49.
55. On the interaction of science and technology in the invention of telegraphy, see Hindle, *Emulation and Invention*.
56. As quoted in Gieryn, *Cultural Boundaries of Science*, 54.
57. Ibid., 63.
58. Kline, "Construing 'Technology' as 'Applied Science.'" For an updated version of this article, see Kline, "Science and Technology."
59. Rowland, "A Plea for Pure Science," 242, 246.
60. Tyndall helped create the myth of Faraday's selfless research in his 1868 book, *Faraday as Discoverer*. Gieryn, *Cultural Boundaries of Science*, 59.
61. Rowland, "A Plea for Pure Science," 242, 245; Kline, "Construing 'Technology' as 'Applied Science,'" 198–200; Dennis, "Accounting for Research"; Hounshell, "Edison and the Pure Science Ideal."
62. For a more temperate version of Rowland's arguments, see the address of Rowland's colleague at Hopkins: Remsen, "The Age of Science." For further examples, see Kline, "Construing 'Technology' as 'Applied Science.'"
63. Forman, "The Primacy of Science," 28–34.
64. Kline, "Construing 'Technology' as 'Applied Science'"; Channell, "The Harmony of Theory and Practice"; Rankine, *A Manual of Applied Mechanics*, 1–11.
65. Kline, "Construing 'Technology' as 'Applied Science,'" 198.
66. Historians of technology have produced a vast body of work demonstrating this opportunistic use of science in technical change during the nine-

teenth and early twentieth centuries. Kline, "Forman's Lament," 164–65. For a broad survey, see Hughes, *American Genesis*, chap. 4.

67. Rankine, *A Manual of Applied Mechanics*, 6–7; Spencer, "The Genesis of Science," 158–59.
68. E.g., E. Knight, *Knight's New Mechanical Dictionary*.
69. Search on Google Books Ngram Viewer, 9/24/2015, using terms *applied science* and *mechanical arts*,
70. F. Taylor, *Scientific Management*. Of the extensive literature on Taylor, I recommend these two works: D. Nelson, *Frederick W. Taylor and the Rise of Scientific Management*; Aitken, *Scientific Management in Action*.
71. L. Marx, "*Technology*: The Emergence of a Hazardous Concept" (2010).
72. H. Adams, "The Dynamo and the Virgin," 381.
73. Kasson, *Civilizing the Machine*.
74. Colvin, "Art," 660.

CHAPTER SIX

1. W. Whitney, *The Century Dictionary* (1897), 8:6209.
2. Mitcham and Schatzberg, "Defining Technology and the Engineering Sciences," 36; Blount, *Glossographia*, 309; Oldenziel, "Gender and the Meanings of Technology."
3. E.g., Flower, "On the Relative Ages of the Stone Implement Periods in England," 275.
4. Wilson, *What Is Technology?*; Wilson, "On the Relations of Technology to Agriculture"; Wilson, "On the Physical Sciences Which Form the Basis of Technology"; Bigelow, *An Address on the Limits of Education*; Vincent, "On Some Recent Processes for the Manufacture of Soda."
5. Scott and Liddell, *An Intermediate Greek-English Lexicon*, 476–77, 804.
6. The following analysis is based primarily on Mitcham, *Thinking Through Technology*, 128–30.
7. "τεχνολογέω," Scott and Liddell, *An Intermediate Greek-English Lexicon*, 804; Sachs, "Glossary," in Plato and Aristotle, *Plato, Gorgias, and Aristotle, Rhetoric*, 289.
8. Lippmann, *Beiträge zur Geschichte der Naturwissenschaften und der Technik*, 199–201.
9. Mitcham, *Thinking Through Technology*, 129.
10. Cicero, *Letters to Atticus* 4.16.3.
11. "Soweit wir den Wörterbüchern vertrauen können, hat das spätgriechische Wort τεχνολογια samt seinen nächsten Verwandten in der Wortfamilie (τεχνολογος, τεχνολογικός) keinen unmittelbaren Einfluß auf die lateinische Gelehrtensprache des Mittelalters und der angehenden Neuzeit gehabt." Seibicke, *Technik*, 99. Ong, however, suggests a link between the Greek tradition and Ramus in the sixteenth century. Ong, *Ramus*, 353n4.
12. Sellberg, "Petrus Ramus."

13. Ong, *Ramus*, 197–204.
14. P. Miller, *The New England Mind*, 123–80.
15. Ibid., 162.
16. Perrin, "Possible Sources of Technologia at Early Harvard."
17. Ames, *Technometry*, 118–19; P. Miller, *The New England Mind*, 175–76.
18. P. Miller, *The New England Mind*, 143, 80.
19. Mitcham and Schatzberg, "Defining Technology and the Engineering Sciences," 35.
20. "Scientia artium et operum artis, aut, si mavis, scientia eorum, quae organorum corporis, manuum potissimum, opera ab hominibus perficiuntur." Wolff, *Philosophia rationalis sive logica*, 33.
21. Mitcham and Schatzberg, "Defining Technology and the Engineering Sciences," 36.
22. On the nature of cameralism, especially as distinguished from mercantilism, see Tribe, "Cameralism and the Science of Government." For an excellent overview of cameralism and its role in shaping Beckmann's *Technologie*, see Frison, "Linnaeus, Beckmann, Marx," 141–48.
23. Frison, "Linnaeus, Beckmann, Marx," 141–54, 161–67.
24. Banse and Müller, *Johann Beckmann und die Folgen*; W. Weber, "Grosse Technologen"; Ropohl, "Prolegomena zu einem neuen Entwurf der allgemeine Technologie"; Müller and Troitzsch, *Technologie zwischen Fortschritt und Tradition*.
25. Beckert, *Johann Beckmann*, 79.
26. Ropohl, "Prolegomena zu einem neuen Entwurf der allgemeine Technologie," 153.
27. Frison, "Linnaeus, Beckmann, Marx," 141–54, 161–67.
28. "Technologie ist die Wissenschaft, welche die Verarbeitung der Naturalien, oder die Kentniß der Handwerke, lehrt." Beckmann, *Anleitung zur Technologie*, 2nd ed., 17.
29. Phillips, *Acolytes of Nature*, 3–10.
30. Lindenfeld, *The Practical Imagination*, 31.
31. Ibid.; Beckert, *Johann Beckmann*, 81–82; Beckmann, *Anleitung zur Technologie* (1777), xxix–xxxiv.
32. W. Weber, "Grosse Technologen," 235.
33. Matschoss, *Geschichte der Dampfmaschine*, 89.
34. Grignon, *Mémoires de physique sur l'art de fabriquer le fer*, 199–200. Also quoted in Matschoss, *Geschichte der Dampfmaschine*, 86.
35. Justi et al., *Schauplatz der Künste und Handwerke*. On the origins of the *Déscriptions*, see Sheridan, "Recording Technology in France."
36. Halle, *Werkstäte der heutigen Künste*, vol. 1, "Vorrede."
37. Ibid.
38. Ibid.
39. Beckmann, *Anleitung zur Technologie*, "Vorrede."; Frison, "Linnaeus, Beckmann, Marx," 142.

40. Halle, *Werkstäte der heutigen Künste*, 1:281–360.
41. Bigelow, for example, discussed only the technical aspects of painting, noting that the "inventive part" of painting "demands original genius," which "is attained by few" and "not taught by any rules of art." Bigelow, *Elements of Technology*, 89.
42. W. Weber, "Grosse Technologen," 236–38.
43. Poppe, *Handbuch der Technologie*; Karmarsch, *Geschichte der Technologie*.
44. W. Weber, "Grosse Technologen," 235.
45. As can be seen in a search on the Google Books Ngram Viewer in the German corpus for *Anleitung zur Technologie*, accessed 2/11/18.
46. W. Weber, "Grosse Technologen," 241; Lindenfeld, *The Practical Imagination*, 31–32.
47. Frison, "Some German and Austrian Ideas," 107, 113–14. On similar French concepts of technologie, see Mertens, "Technology as the Science of the Industrial Arts."
48. *Allgemeine deutsche Real-Encyklopädie*, 7th ed., 8:683–84; Knapp, *Lehrbuch der chemischen Technologie*, vol. 1; Karmarsch, *Die polytechnische Schule zu Hannover*, 3–4, 60–67; "Technologie" (1885), 780–83.
49. *Allgemeine deutsche Real-Encyklopädie*, 7th ed., 8:683–84.
50. Karmarsch, *Die polytechnische Schule zu Hannover*, 3–4, 60–67.
51. Beckmann, *A History of Inventions and Discoveries*.
52. Bud and Roberts, *Science versus Practice*, 108.
53. Knapp et al., *Chemical Technology*, 1st American ed., 13. A similar work in chemische Technologie translated into English is Wagner and Crookes, *A Handbook of Chemical Technology*.
54. Lucier, *Scientists and Swindlers*, chap. 5.
55. E.g., Buchanan, *A Technological Dictionary*.
56. Treadwell, "[Review of] Elements of Technology . . . by Jacob Bigelow," 188–89.
57. Bigelow, *Elements of Technology*, iv.
58. It is not entirely clear that this sentence refers to *technology* as opposed to the entire title of the book. The full sentence is, "Under this title it is attempted to include such an account as the limits of the volume permit, of the principles, processes, and nomenclatures of the more conspicuous arts, particularly those which involve applications of science, and which may be considered useful, by promoting the benefit of society together with the emolument of those who pursue them." In the context of the preceding sentence, "this title" most likely refers to *technology* rather than to the book itself, but the implication remains that as a category, technology subsumes the topics of Bigelow's book.
59. Crabb, *Universal Technological Dictionary*, s.v. "technology."
60. Good, "On Medical Technology."
61. Poppe, *Handbuch der Technologie*; Beckmann, *Anleitung zur Technologie*.

62. "Half Yearly Retrospect of German Literature," 628, 637; Griscom, "Foreign Literature and Science, Extracted and Translated," 398. Another possible source for Bigelow is Jeremy Bentham's book on reform of the schools, *Chrestomathia*. In this book, Bentham designated technology ("or the arts and manufactures in general") as a field of study for students in the "fifth stage," and he discussed the details in some length. Originally published between 1815 and 1817, *Chrestomathia* was little known until the posthumous 1843 edition, and it remained obscure in the United States even after that. Bigelow is unlikely to have encountered this book directly. Bentham, *Chrestomathia*.

63. Meier, "The Technological Concept in American Social History, 1750–1860," 22–24. The MIT mathematician and historian of science Dirk Struik preceded Meier in briefly mentioning the Bigelow myth. Struik, *Yankee Science in the Making*, 169–70, 338.

64. Meier, "Technology and Democracy, 1800–1860," 618.

65. E.g., G. Allen, *Master Mechanics and Wicked Wizards*, 13; Schweber, "The 'Science' of Legal Science," 432.

66. P. Miller, *The Life of the Mind in America*, 289.

67. Segal, *Technological Utopianism*, 198n7; Ferguson, *Bibliography of the History of Technology*, 62; Oldenziel, *Making Technology Masculine*, 23.

68. J. [Thomas P. Jones], "Bigelow's Elements of Technology," 216.

69. Emerson, "[Review of Bigelow, *Elements of Technology*]," 337–38.

70. Treadwell, "[Review of] Elements of Technology . . . by Jacob Bigelow," 187.

71. James and Weiss, "An Assessment of Google Books' Metadata"; Google Books Ngram Viewer, accessed 2/11/18.

72. The classic work is Jeremy, *Transatlantic Industrial Revolution*. For a broader overview, see Hindle and Lubar, *Engines of Change*.

73. L. Stewart, *The Rise of Public Science*.

74. L. Marx, "*Technology*: The Emergence of a Hazardous Concept," 574. Middle class is, of course, a problematic concept that differs by national context and era.

75. Oldenziel, *Making Technology Masculine*, 195n8.

76. Stratton and Mannix, *Mind and Hand*; Angulo, *William Barton Rogers*.

77. For an overview and critique of this historiography, see Reynolds, "The Education of Engineers in America before the Morrill Act of 1862."

78. Stratton and Mannix, *Mind and Hand*, 36–38, 42–45; Calhoun, *The American Civil Engineer*.

79. Stratton and Mannix, *Mind and Hand*, 81.

80. W. Rogers, "A Plan for a Polytechnic School."

81. Ibid., 1:420–21.

82. Ibid., 1:422, 423.

83. Ibid., 1:421.

84. See, e.g., Olson, "Science, Technology, and the Industrial Revolution."

85. Rice, *Minding the Machine*.

86. W. Rogers, "A Plan for a Polytechnic School," 1:422.
87. E. Rogers, *Life and Letters*, 1:333–34.
88. Ibid., 2:416, 419–20; Angulo, *William Barton Rogers*, 90–93; Stratton and Mannix, *Mind and Hand*, 172–80.
89. Committee of Associated Institutions of Science and Arts, *Objects and Plan of an Institute of Technology*, 13–14.
90. Ibid., 22, 24.
91. In 1855, Benjamin Franklin Greene, president of Rensselaer, developed a far more detailed proposal for a school of higher technical education. Greene, *The Rensselaer Polytechnic Institute*.
92. Stratton and Mannix, *Mind and Hand*, 378, 394, 407–14.
93. E. Rogers, *Life and Letters*, 2:41.
94. Ibid., 2:404, 412; my emphases.
95. This is an inference drawn from the lack of discussion about *technology* in the detailed account of the legislative debates in Stratton and Mannix, *Mind and Hand*, 200–220.
96. Marx was referring to a somewhat later period in the nineteenth century, but the point still holds. L. Marx, "*Technology*: The Emergence of a Hazardous Concept" (2010), 572.
97. Stratton and Mannix, *Mind and Hand*, 190.
98. Ibid., 239, 242.
99. Bigelow, *An Address on the Limits of Education*, 3–4. In his pathbreaking cultural history of technology in this era, Segal misinterprets this passage, arguing that Bigelow distinguished *technology* from *science*, based on Bigelow's statement that technology "has done more than any science to enlarge the boundaries of profitable knowledge" (3). In context, this statement does not exclude technology as a science. Segal, *Technological Utopianism*, 81.
100. Bigelow, *An Address on the Limits of Education*, 6, 12.
101. Mueller, "How Brockhaus' *Conversations-Lexicon* Became the *Encyclopaedia Americana*," 211–13.
102. E. Rogers, *Life and Letters*, 2:220–21.
103. Grant, *The Story of the University of Edinburgh*, 1:354–56; Anderson, "'What Is Technology.'"
104. See also Wilson, "On the Relations of Technology to Agriculture," 255–56. The name of the chair may well have come from Lyon Playfair, who had studied chemistry in Germany and had recently published a report on technical education on the Continent. Anderson, "'What Is Technology,'" 175–77; Playfair, *Industrial Instruction on the Continent*.
105. Wilson, *What Is Technology?*, 3–4.
106. Ibid., 4, 16, 22.
107. Grant, *The Story of the University of Edinburgh*, 1:356–59; Anderson, "'What Is Technology,'" 182.
108. Only much later, after the middle of the twentieth century, did universities create degree programs in new fields like information technology.

Kline, "Cybernetics, Management Science, and Technology Policy,"
520–29.

109. E. Rogers, *Life and Letters*," 1:361, 1:80–82.
110. Wilson, "On the Physical Sciences Which Form the Basis of Technology."
111. E. Rogers, *Life and Letters*, 2:166.
112. Exceptions are the Georgia School of Technology (1888), the Armour Institute of Technology (1892), and the Clarkson Memorial School of Technology (1896). A few universities added departments of technology in the same era. Thomas, *Where to Educate, 1898–1899*. In England, the Society of Arts organized a set of Technology Examinations in 1872 to encourage better training of skilled craftsmen. "Proceedings of the Society: Conference on Technological Education." (See chapter 11.)
113. On this tension in general, see Rice, *Minding the Machine*.
114. Onions and Murray, *A New English Dictionary*, 9:137.
115. Burton, "The Lake Regions of Central Equatorial Africa."
116. Burton, *Mission to Gelele*, 2:134–35.
117. Adas, *Machines as the Measure of Men*, 154, on Burton.
118. Based on a JSTOR search of the journal.
119. Murray, Wallace, and Bradford, *The Encyclopædia of Geography*, 1:267.
120. Winchester, *The Meaning of Everything*, 226–29.
121. E.g., Vincent, "On Some Recent Processes for the Manufacture of Soda."
122. Lucier, *Scientists and Swindlers*, 143.
123. Powell, "Human Evolution," 181–82.
124. Powell and Boas, "Museums of Ethnology and Their Classification," 614; Powell, "The Evolution of Religion," 185; Powell, "Technology, or the Science of Industries," 319.
125. Onions and Murray, *A New English Dictionary*, 9:137.

CHAPTER SEVEN

1. There are important exceptions to this neglect of the engineering discourse of Technik. E.g., Hård, "German Regulation"; Mitcham, *Thinking Through Technology*; Rohkrämer, *Eine andere Moderne?*; Dietz, Fessner, and Maier, *Technische Intelligenz und "Kulturfaktor Technik"*; Voskuhl, "Engineering Philosophy."
2. Heidegger's essay completely ignored the existing literature on Technik. For a careful critique, see R. Bernstein, "Heidegger's Silence?"
3. For a thoughtful recent analysis of Marx's thinking about technology, see Wendling, *Karl Marx on Technology and Alienation*.
4. K. Marx, *Capital*, 3rd ed., 1:47; Marx and Engels, *Werke*, 23:54.
5. Wendling notes the need to "speculatively construct" aspects of Marx's account of technology. Wendling, *Karl Marx on Technology and Alienation*, 168. See also Donald MacKenzie's classic article, in which he has argued

persuasively that Marx was not a technological determinist. MacKenzie, "Marx and the Machine."

6. Frison, "Some German and Austrian Ideas," 106, 114. See chapter 10 for a more detailed discussion.

7. Jevons, *The Theory of Political Economy*, vii; Sperber, *Karl Marx*, 460–61.

8. My analysis here owes much to the pioneering research of Guido Frison, who was the first scholar to understand the significance of Marx's use of the terms *Technik* and *Technologie*, although my interpretation of this significance differs in some aspects from Frison's. Frison, "Technical and Technological Innovation in Marx"; Frison, "Smith, Marx and Beckmann," 17–36; Frison, "Linnaeus, Beckmann, Marx."

9. Wendling, *Karl Marx on Technology and Alienation*, 178–82; K. Marx, *Capital*, 3rd ed., vol. 1, chap. 15, "Machinery and Modern Industry." Marx's extensive notes on his technological sources have been published as K. Marx, *Die technologisch-historischen Exzerpte*. On the importance of these sources to the machinery chapter in *Kapital*, see Karl Marx to Friedrich Engels, 28 January 1863, in Marx and Engels, *Werke*, 30:319–22.

10. K. Marx, *Capital*, 3rd ed., 1:47, 456–57, 584; Marx and Engels, *Werke*, 23:55, 510, 652. The standard German text of *Das Kapital* is the fourth German edition of 1890, edited by Engels, online at http://www.mlwerke.de/me/me23/me23_000.htm, accessed 2/11/18.

11. See K. Marx, *Die technologisch-historischen Exzerpte*, 197–205.

12. K. Marx, *Capital*, 3rd ed., 1:352n2; K. Marx, *Capital* (1976), 1:493; Marx and Engels, *Werke*, 23:392n89. Note that for the last quotation I used the Fowkes translation, which more accurately captures the sense of the original German.

13. K. Marx, *Capital*, 3rd ed., 1:352n2; Marx and Engels, *Werke*, 23:392n89.

14. K. Marx, *Capital: A Critique of Political Economy*, 1:51–52, also 191, 220–21, 462, 472–73.

15. See chapter 5.

16. For a careful analysis of this issue, see Wendling, *Karl Marx on Technology and Alienation*, 183–91.

17. Ibid., 189.

18. Voskuhl, *Androids in the Enlightenment*.

19. Frison, "Some German and Austrian Ideas," 107, 108–9, 115.

20. K. Marx, *Capital*, 3rd ed., 1:567, 475; Marx and Engels, *Werke*, 23:631–32, 530.

21. Moore and Aveling's translation of *Technik* as "technology" in the second quotation could very well be in error; Fowkes translates the term as "techniques," and the German is ambiguous. K. Marx, *Capital* (1976), 1:638.

22. Basic political and social concepts are typically represented by nouns, as scholars tend to reify actions to portray them as objects in the world, a practice discussed in Billig, *Learn to Write Badly*.

23. E.g., K. Marx, *Das Kapital: Kritik der politischen Oekonomie*, 1st ed., 1:168, 178, 182, 288, 289; K. Marx, *Das Kapital*, 2nd ed., 190, 200, 204, 315, 316.

24. *Technologisch* continued to be used in the mid-nineteenth century, but mainly in the context of technical education or terminology and not as Marx did, as a synonym for *technisch*. In any case, *technologisch* remained relatively rare in nineteenth-century German; in the 1860s, it was roughly sixty times less common than *technisch*. Based on a search on Google Books Ngram Viewer for "technologisch_INF, technisch_INF," with smoothing of 5, viewed 10/28/2015.

25. Seibicke, "Technica aut Technologia," 167–73, 186–89.

26. Seibicke, *Technik*, 181.

27. Ibid., 181–211; quotation is from p. 211.

28. Ibid., 211, 276. These meanings remained very much current in the interwar period. *Der große Brockhaus*, 18:509.

29. Frison, "Some German and Austrian Ideas," 119. Frison's argument is supported by German encyclopedias, even though these are lagging indicators of actual usage. *Allgemeine deutsche Real-Encyklopädie*, 10th ed., 14:719–20; *Conversations-Lexikon*, 14: 406–7; *Brockhaus' Konversations-Lexikon*, 15: 651–55; *Der große Brockhaus*, 18: 509–12; *Brockhaus Enzyklopädie*, 18: 517–23, 526.

30. Seibicke, *Technik*, 212–16, 226–27, 276. This dual meaning was maintained through the nineteenth century and into the present. Engelmeyer, "Allgemeine Fragen der Technik," 312:97; *Brockhaus Enzyklopädie*, 18:517, s.v. "Technik."

31. Gispen, *New Profession, Old Order*, 44–48; Hortleder, *Das Gesellschaftsbild des Ingenieurs*, 20; "Statut des Vereins deutscher Ingenieure," 4.

32. See, e.g., Messinger, *Langenscheidt's New College German Dictionary*, 526, s.v. "Technik"; Breul, *Cassell's New German and English Dictionary*, 591, s.v. "Technik."

33. McClelland, *The German Experience of Professionalization*, 115, 124; Manegold, *Universität, Technische Hochschule und Industrie*, 249–305.

34. Mitcham, *Thinking Through Technology*, 20–27; Gispen, *New Profession, Old Order*.

35. On *Bildung*, see Gadamer, *Truth and Method*, 8–16; Ringer, *The Decline of the German Mandarins*, 86–87.

36. Gispen, *New Profession, Old Order*, 78–85; Kundert, "German Engineers and Bildung during the Nineteenth Century."

37. Hård, "German Regulation," 34. On boundary work, see Gieryn, *Cultural Boundaries of Science*.

38. Ringer, *The Decline of the German Mandarins*, 89–90; Herf, *Reactionary Modernism*, 1 (quotation).

39. Mitcham, *Thinking Through Technology*, 25–26. For a discussion of Engelmeyer's philosophical writings in Russian, see Nikiforova, "The Concept of Technology and the Russian Cultural Research Tradition," 192–95. For

a biography of a similar pre-revolutionary Russian engineer, see Graham, *The Ghost of the Executed Engineer.*

40. Engelmeyer, "Allgemeine Fragen der Technik," 311:21–22, 69–71, 101–3, 133–34, 149–51; 312:1–2, 65–67, 97–99, 129–30, 145–47; 313:17–19, 65–67; Braun, "Allgemeine Fragen der Technik."

41. Braun, "Allgemeine Fragen der Technik," 308.

42. Engelmeyer, "Allgemeine Fragen der Technik," 311:102–3; 312:66–67. These themes are central to the "social shaping" approach to technology studies. MacKenzie and Wajcman, *The Social Shaping of Technology.*

43. Engelmeyer, "Allgemeine Frogen der Technik," 311:21.

44. Ibid., 311:21, 101, 102, 134.

45. Ibid., 311:97, 98, 101, 103, 49; 312:18. Reuleaux's original title was spelled "Cultur und Technik." Reuleaux, "Cultur und Technik." In 1890, a version of this article appeared in English under the title "Technology and Civilization." Reuleaux, "Technology and Civilization."

46. Dessauer, *Streit um die Technik,* 34; Braun, "Technik als 'Kulturhebel' und 'Kulturfactor,'" 42–43. Hård has provided the best summary of this literature in English. Hård, "German Regulation."

47. Hård, "German Regulation," 36–44; Dessauer, *Streit um die Technik,* 32–33.

48. E.g., Götz [Niekisch], "Menschenfresser Technik (1931)." See Herf, *Reactionary Modernism,* 38–40.

49. For a discussion of German ambivalence toward technology in popular culture, see Rieger, *Technology and the Culture of Modernity.*

50. Sieferle, *Fortschrittsfeinde?,* 158.

51. Herf committed a basic error in his pathbreaking book by classifying all scholars who took a cultural view of technology as reactionary modernists. Thus, he lumped liberals and leftists with reactionary modernists, such as the social democratic engineer Viktor Engelhardt and the liberal aristocrat Richard Coudenhove-Kalergi, both of whose books were burned by the Nazis. Coudenhove-Kalergi in particular was a pioneer of the idea of a united Europe, an outspoken critic of anti-Semitism, and a principled anti-fascist. Herf, *Reactionary Modernism,* 40, 178, 210n; Engelhardt, "Technik und soziale Ethik"; Levenson, "The German Peace Movement and the Jews," 294–99; Wyrwa, "Richard Nikolaus Graf Coudenhove-Kalergi." For a critique of Herf, see Rohkrämer, "Antimodernism, Reactionary Modernism and National Socialism."

52. An unsystematic search of JSTOR shows that major German works on Technik received almost no attention in English-language scholarly journals between the world wars, aside from occasional brief reviews.

53. Frison, "Some German and Austrian Ideas," 107, 119.

54. Whimster, *Understanding Weber,* 32.

55. Engels, "Engels an W. Borgius."

56. E.g., Kautsky, "Was will und kann die materialistische Geschichtsauffassung leisten?," 231–32; Bukharin, *Historical Materialism,* 134–42.

57. For a thoughtful discussion of Sombart's Nazi affiliation, see Stehr and Grundmann, "Introduction: Werner Sombart," xxxii–xliii. Note that Sombart did shift to a more instrumental view of Technik in his last book in 1938. Stehr and Grundmann, "Introduction: Werner Sombart," xxxi.

58. Sombart, "Technik und Kultur."

59. E.g., Ringer, *The Decline of the German Mandarins*, 261–63; Herf, *Reactionary Modernism*, 133–34; T. Meyer, "Zwischen Ideologie und Wissenschaft," 82–86.

60. For widespread antimodernism among German intellectuals, see Ringer, *The Decline of the German Mandarins*.

61. Hård, "German Regulation," 56–60; see also Joerges, "Soziologie und Maschinerie," 53–55. The most thorough examination of Sombart's 1911 article is T. Meyer, "Zwischen Ideologie und Wissenschaft," which discusses both the article's methodological contributions and its conservative cultural critique.

62. Sombart, "Technik und Kultur," 305–6. For Sombart's original 1910 paper, including audience comments, see Sombart, "Technik und Kultur"; for a somewhat problematic English translation of the 1910 paper, see Sombart, "Technology and Culture," *Verhandlung*; and for a translation of Weber's comments to the 1910 talk, see M. Weber, "Remarks on Technology and Culture." On the *Archiv*, see Ghosh, "Max Weber, Werner Sombart and the Archiv für Sozialwissenschaft."

63. Sombart, "Technik und Kultur," 307–8.

64. Ibid.

65. Ibid., 309.

66. For a more detailed analysis, see T. Meyer, "Zwischen Ideologie und Wissenschaft," 74–76.

67. Sombart, "Technik und Kultur," 310–11.

68. Ibid., 314.

69. Meyer concluded that "in the end, . . . Sombart did not understand Marx." T. Meyer, "Zwischen Ideologie und Wissenschaft," 76–77n66.

70. Not all contemporary Marxists shared this determinist view. For a less determinist reading of Marx's materialist conception of history, see Eduard Bernstein's influential 1899 work of Marxist revisionism, where he discussed the same passages of Marx that Sombart used for his determinist interpretation. Bernstein, *The Preconditions of Socialism*, 12–22.

71. Sombart, "Technik und Kultur," 315–16. My analysis here draws especially from Hård, "German Regulation," 58–60. On the critique of technological determinism in the 1970s, see, e.g., Hughes, "Emerging Themes in the History of Technology."

72. Sombart, "Technik und Kultur," 320.

73. Ibid., 321–22.

74. Ibid., 324, 325, 336, 340, 343–44.

75. Ibid., 342–46. On the circuit of culture, see Du Gay, *Doing Cultural Studies*.

76. Sombart's remarks refute Paul Forman's claim that Sombart "reascribes to science an unqualified foundational primacy to and for technology." Forman, "The Primacy of Science," 22.
77. Sombart, "Technik und Kultur," 337, 341.
78. Whimster, *Understanding Weber*, 30.
79. Grundmann and Stehr, "Why Is Werner Sombart Not Part of the Core of Classical Sociology?"
80. Weber's most detailed discussion of Technik occurred in extemporaneous remarks to Sombart's article "Technik und Kultur," which were not particularly insightful, as Sombart himself noted. M. Weber, "Remarks on Technology and Culture." Weber included a bit of the history of technology in his *General Economic History*, but he devoted far less attention to the topic than Sombart. M. Weber, *General Economic History*. The literature on Weber's understanding of Technik is surprisingly limited. See Frison, "Some German and Austrian Ideas," 120–22; Swedberg, *Max Weber and the Idea of Economic Sociology*, 148–50. For an argument that Weber was deeply interested in technology, see Sprondel and Seyfarth, *Max Weber und die Rationalisierung sozialen Handelns*, 170–74, which nevertheless shows that Weber subordinated industrial Technik to Technik in the sense of formal rationality.
81. Camic, Gorski, and Trubek, *Max Weber's Economy and Society*.
82. For a detailed discussion, see Angus, "Disenchantment and Modernity."
83. M. Weber, *Economy and Society*, 65–67.
84. Ibid., 65.
85. Ibid., 66.
86. M. Weber, *Wirtschaft und Gesellschaft*, 3:3, 32–33; M. Weber, *The Theory of Social and Economic Organization*, 160–62 (quotation); M. Weber, *Economy and Society*, 65–67.
87. M. Weber, "'Objectivity' in Social Science and Social Policy," 90; emphasis removed.
88. Ibid., 111; emphasis removed.
89. Recall that Weber commented at length on Sombart's original "Technik und Kultur" paper. M. Weber, "Remarks on Technology and Culture." Also, Weber's *The Protestant Ethic and the Spirit of Capitalism* was in many ways a response to Sombart's *Der moderne Kapitalismus*. Whimster, *Understanding Weber*, 33–40.
90. M. Weber, "Remarks on Technology and Culture," 26.
91. Sombart, "Technik und Kultur," 314.
92. Carlyle, "Signs of the Times."
93. For surveys of literature from this era, see Schneider, "Uber Technik, technisches Denken und technische Wirkungen"; and Dessauer, *Streit um die Technik*.
94. Rieger, *Technology and the Culture of Modernity*, chap. 2.
95. McClelland, *The German Experience of Professionalization*.

96. Exceptions include Hobson, *The Evolution of Modern Capitalism*; Marshall, *Industry and Trade*; Mumford, *Technics and Civilization*.
97. This gap between German and American scholarship concerns a narrow set of concepts. It does not support the questionable historiography that portrays American social sciences as a theoretical wasteland before Parsons introduced its scholars to European theory. See Camic, "Alexander's Antisociology," 173.

CHAPTER EIGHT

An earlier version of this chapter was published as Schatzberg, "*Technik* Comes to America."

1. For an overview of the British Industrial Revolution, see Berg, *The Age of Manufactures, 1700–1820*. On lighting, see Schivelbusch, *Disenchanted Night*.
2. See Hughes, *American Genesis*. For a similar assessment of the economic impact of the Second Industrial Revolution, see Gordon, *The Rise and Fall of American Growth*.
3. Siemens, "Science in Relation to the Arts," 49–50; see also Rankine, *A Manual of Applied Mechanics*, 1, 10–11.
4. Spencer, "The Genesis of Science," 157–58.
5. Powell, "Human Evolution," 181–82, 188–91; see also Tylor, *Primitive Culture*, 1:56–62.
6. For example, Alfred Marshall referred repeatedly to "arts of production" in Marshall, *Principles of Economics*. See also Jevons, *The Theory of Political Economy*, 239.
7. E.g., Kames, *Elements of Criticism*, 16.
8. U.S. Const., art. I, sec. 8.
9. *OED Online*, 3rd ed. (September 2015), s.v. "industry," "industrial," accessed 2/11/18, http://www.oed.com/.
10. Carlyle, *Past and Present*, 242, 335.
11. Chaussard, *Ode philosophique sur les arts industriels*.
12. Wilson, *What Is Technology?*, 6–7, 9–10, 16 (quotation).
13. Butterworth, *The Growth of Industrial Art*.
14. E.g., Veblen, *The Vested Interests and the State of the Industrial Arts*, 57.
15. Veblen did so at least once, but this translation appears to be the exception. See below.
16. On counterfactuals, see Evans, *Altered Pasts*.
17. E.g., Maskell, *The Industrial Arts*.
18. For an early example, see Rhode Island General Assembly, *The Industrial Arts in the Public Schools*. See also Foster, "Industrial Arts/Technology Education as a Social Study"; Foster, "The Founders of Industrial Arts in the US." Thanks to Magnus Hultén for providing me with copies of the Foster articles and other references on this topic.

19. See chapter 5.
20. Kevles, *The Physicists*, 279–83; Herbst, *The German Historical School in American Scholarship*, esp. 130–31; Ross, *The Origins of American Social Science*, 58, 109; Dorfman, "The Role of the German Historical School in American Economic Thought."
21. This failure of American reviewers to deal with Technik in depth is particularly clear in reviews of the later editions of Sombart's *Der moderne Kapitalismus*. A major revision of the first two volumes was published in 1916–17 and of the third volume in 1927. See, e.g., Commons and Perlman, "Review of *Der moderne Kapitalismus*, by Werner Sombart"; Usher, "The Genesis of Modern Capitalism."
22. E.g., Fellowes, "What Technique Does for a Picture"; Edebohls, "The Technique of Vaginal Hysterectomy."
23. Seligman, "The Economic Interpretation of History. I."
24. On Seligman and his economic interpretation of history, see Hofstadter, *The Progressive Historians*, 197–200; Ross, *The Origins of American Social Science*, 186–89.
25. Seligman, "The Economic Interpretation of History. I"; quotations are from pp. 613, 623.
26. Seligman, "The Economic Interpretation of History. II"; quotations are from pp. 71, 72 (emphasis added). For the original German of the Engels letter quoted by Seligman, see Engels, "Engels an W. Borgius."
27. This conflation continued. See, e.g., Mitchell, "Sombart's Hochkapitalismus"; Usher, "The Genesis of Modern Capitalism," 529. See also my discussion of Talcott Parsons in chapter 10.
28. Bücher, *Industrial Evolution*, 53, 57, 60, 325; Bücher, *Die Entstehung der Volkswirtschaft*, 62–63, 68, 71, 379.
29. Gille, *Histoire des techniques*; Gille, *History of Techniques*.
30. F. Nussbaum, *A History of the Economic Institutions of Modern Europe*. *Technique* was also used in this sense by the economist Alfred Marshall, for example in the phrases "industrial technique" and "mechanical technique." Marshall, *Industry and Trade*.
31. B. Russell, *Sceptical Essays*, 36, 234; B. Russell, *Unpopular Essays*, e.g., 43, 130, 143. Thanks to Chris Hallquist for alerting me to Russell's usage of *technique*.
32. L. Marx, "*Technology:* The Emergence of a Hazardous Concept," 976–77; Oldenziel, *Making Technology Masculine*, 42–46; Veblen, *The Engineers and the Price System*, chap. 6.
33. Aside from Veblen's translation of Gustav Cohn, discussed below, he rarely used *technology* before 1900, and *technological* only on occasion, roughly as a synonym for *industrial*. For Veblen's early use of *technological*, see Veblen, "Review of *Einführung in der Socialismus* by Richard Calwer," 271; Veblen, "Why Is Economics Not an Evolutionary Science?," 380, 397; Veblen, "The Preconceptions of Economic Science, III," 249.

34. Edgell and Tilman, "The Intellectual Antecedents of Thorstein Veblen." Edgell and Tilman have separately produced two excellent assessments of Veblen's life and work. Edgell, *Veblen in Perspective*; Tilman, *The Intellectual Legacy of Thorstein Veblen*. On the difficulty of tracing Veblen's sources, see also Camic, "Veblen's Apprenticeship," 691.

35. Edgell, *Veblen in Perspective*, 8–10, 76–99; Veblen, *The Instinct of Workmanship and the State of Industrial Arts*, 1–4, 145; Tilman, *The Intellectual Legacy of Thorstein Veblen*, 225; Veblen, "Industrial and Pecuniary Employments."

36. Veblen, "The Socialist Economics of Karl Marx and His Followers, I." A substantial literature compares Marx and Veblen: e.g., Edgell, *Veblen in Perspective*, 135; Diggins, *The Bard of Savagery*. On forces versus relations of production, see K. Marx, preface to *A Contribution to the Critique of Political Economy*.

37. For example, in his discussion of the "industrial efficiency of the group" as a driver of social evolution. Veblen, *The Theory of the Leisure Class*, 4–5; Veblen, "The Instinct of Workmanship and the Irksomeness of Labor"; quotation is from p. 198.

38. Veblen, "The Instinct of Workmanship and the Irksomeness of Labor."

39. On the German historical school, see Grimmer-Solem, *The Rise of Historical Economics and Social Reform in Germany, 1864–1894*.

40. Camic, "Veblen's Apprenticeship." As Camic notes, however, Cohn was more theoretically inclined than many members of the historical school, particularly Schmoller, and took a more balanced position in the *Methodenstreit* between Schmoller and Carl Menger.

41. Cohn, *System der Finanzwissenschaft*, 42, 45, 59–60; Cohn, *The Science of Finance*, 53, 57, 73–75.

42. Veblen, "Gustav Schmoller's Economics," 69–93, esp. 80–81.

43. Ibid., 89, 90, 91; Schmoller, *Grundriss der allgemeinen Volkswirtschaftslehre*, 1:187–228; quotation is from p. 211. On Schmoller, see Balabkins, *Not by Theory Alone*.

44. Veblen, "Gustav Schmoller's Economics," 82, 89.

45. Veblen, "Arts and Crafts," 108–11; quotations are from pp. 9, 10, 11, emphasis added; Dorfman, *Thorstein Veblen and His America*, 204.

46. Sombart, *Der moderne Kapitalismus*, vol. 2, *Theorie der kapitalistischen Entwicklung*, 1st ed., 2:3–4. Note that there is very little discussion of Technik in the first volume of the 1902 edition, although the first volume of the revised edition of 1916 has a brief theoretical discussion of Technik in the opening chapter. Sombart, *Der moderne Kapitalismus*, 2nd ed., 1:4–7.

47. V. [Thorstein Veblen], "Review of *Der moderne Kapitalismus*"; quotations are from pp. 300, 303.

48. Ibid.; quotation is from p. 305, emphasis added; Sombart, *Der moderne Kapitalismus*, vol. 2, *Theorie der kapitalistischen Entwicklung*, 1st ed., 2:42–67.

49. Veblen, *The Theory of Business Enterprise*, 7, 302.

50. For an argument against viewing Veblen as an instrumentalist, in contrast to Dewey, see Tilman, *The Intellectual Legacy of Thorstein Veblen*, 124–28.

51. Cordes, "Veblen's 'Instinct of Workmanship,'" 2. See also Hobson, *Modern Sociologists: Veblen*, 193–95.

52. Veblen, "The Instinct of Workmanship and the Irksomeness of Labor," 189–90, 93–96, 200.

53. Veblen, "The Beginnings of Ownership," 353–54.

54. Veblen, "On the Nature of Capital [part 1]"; quotations are from pp. 18, 21.

55. Ibid., 525–26, 534.

56. In his critique of the Arts and Crafts movement, Veblen argued that the products of modern technology had no less cultural standing than the products of traditional crafts. Veblen, "Arts and Crafts."

57. Such discussions were common in German analyses of Technik, but they did not translate readily into English. The earliest explicit discussion of the relationship between science and technology that I have found is Vincent, "On Some Recent Processes for the Manufacture of Soda." Vincent's lecture had no discernible influence. See chapter 11 for more on the science-technology relationship.

58. Veblen, "The Place of Science in Modern Civilization"; quotations are from pp. 597–98.

59. Veblen, "On the Nature of Capital [part 2]," 110.

60. Veblen, "The Place of Science in Modern Civilization," 542. For a similar analysis, see Brette, "Thorstein Veblen's Theory of Institutional Change."

61. Veblen, *Absentee Ownership and Business Enterprise in Recent Times*, 62–63.

62. Jorgensen and Jorgensen, *Thorstein Veblen: Victorian Firebrand*, 157–58, 160; Edgell, *Veblen in Perspective*, 27; Veblen, *The Engineers and the Price System*, esp. 127–29.

63. Tilman, *The Intellectual Legacy of Thorstein Veblen*, 175–76.

64. Veblen, *The Engineers and the Price System*, 72, 82, 85; Knoedler and Mayhew, "Thorstein Veblen and the Engineers," 262–64; Edgell, *Veblen in Perspective*, 140–42, 151–58; Tilman, *The Intellectual Legacy of Thorstein Veblen*, 187–89. Veblen was clearly drawing from tendencies among rank-and-file engineers in his analysis, as Knoedler and Mayhew have demonstrated.

65. Oldenziel, *Making Technology Masculine*, 44–46; Bell, "Introduction to the Harbinger Edition."

66. Veblen, "On the Nature of Capital [part 2]," 519.

CHAPTER NINE

1. Dorfman, *Thorstein Veblen and His America*, 504–9; Oldenziel, *Making Technology Masculine*, 43; Ross, *The Origins of American Social Science*, 321, 372; Mumford, *Sketches from Life*, 220–21.

2. Ruth Oldenziel was, I believe, the first scholar to uncover Veblen's key role. Oldenziel, *Making Technology Masculine*, 42–46.

3. In addition to the examples below, see Mitchell, "Human Behavior and Economics," 25–29; R. Epstein, "Industrial Invention," 261–62; Harris, "Economic Evolution," 35; Ayres, "Moral Confusion in Economics," 189–90. Mitchell was a close friend of Veblen and a leading contributor to the Institutionalist school of American economics; Ayres drew heavily from Veblen to keep institutionalism alive after 1930. Stanfield and Stanfield, "The Significance of Clarence Ayres and the Texas School."

4. Davenport, "Capital as a Competitive Concept"; quotations are from p. 35. On Davenport, see Mitchell, "Thorstein Veblen."

5. Hansen, "The Technological Interpretation of History." For a convincing refutation of Hansen's analysis of Marx, see MacKenzie, "Marx and the Machine." Hansen was a student of John R. Commons at the University of Wisconsin, where he undoubtedly would have been exposed to Veblen. He went on to a prominent career as a leading American Keynesian. Brazelton, "Alvin Harvey Hansen: Economic Growth and a More Perfect Society"; Brazelton, "Alvin Harvey Hansen: A Note."

6. Hofstadter, *The Progressive Historians*, 179, 197–200, 285–87, 288; Dorfman, *Thorstein Veblen and His America*, 394, 449–51.

7. Hofstadter, *The Progressive Historians*, 200; Marcell, *Progress and Pragmatism*.

8. C. Beard, "Time, Technology, and the Creative Spirit in Political Science," 4–5.

9. L. Marx, *The Machine in the Garden*; Kasson, *Civilizing the Machine*; D. Nye, *Electrifying America*.

10. C. Beard, "Time, Technology, and the Creative Spirit in Political Science," 5.

11. E.g., Heilbroner, "Do Machines Make History?"

12. Merritt Smith, "Technological Determinism in American Culture," 2–13.

13. William Ogburn, for example, discussed this question in his idea of cultural lag, which I discuss further in chapter 10. Ogburn, *Social Change*, esp. 200–213.

14. Veblen, *Absentee Ownership and Business Enterprise in Recent Times*, e.g, 62–63.

15. In the next few years, Beard made similar sweeping statements about technology, perhaps most important in his introduction to the American edition of Bury's *The Idea of Progress* in 1932. C. Beard, introduction to Bury, *The Idea of Progress*, xx. Technology was also central in a book on American politics that Beard coauthored with his son William, an MIT graduate. Beard and Beard, *The American Leviathan*, vii.

16. E.g., "Exxon Mobil Meets Amazon.com."

17. Scholars in the 1920s dealing with subjects we would now classify as technological used the term only in passing. See Ogburn, *Social Change*, esp. 200–213; Gilfillan, "Who Invented It?"; Usher, *A History of Mechanical Inventions*, esp. vii, 1, 6. Stuart Chase used the term fairly often, in a

way clearly influenced by Veblen. Chase, *Men and Machines*, e.g., 12, 16, 70, 142, 247. Yet even as late as 1935, in his influential *The Sociology of Invention*, Gilfillan pointedly avoided *technology*, referring in his subtitle to "Technic Invention." Gilfillan, *The Sociology of Invention*.

18. See especially Beard's introductions to two widely reviewed essay collections. C. Beard, *Whither Mankind*, 1–25, esp. 14; C. Beard, *Toward Civilization*, 1–20.

19. Ayres, "Science: The False Messiah," 19; Ayres, "The Gospel of Technology"; McFarland, "Clarence Ayres and His Gospel of Technology"; Stanfield and Stanfield, "The Significance of Clarence Ayres and the Texas School."

20. Bix, *Inventing Ourselves Out of Jobs?*, 118–22; Oldenziel, *Making Technology Masculine*, 46–48.

21. For a concise summary of this period, see Edsforth, *The New Deal*, chaps. 2–4.

22. Based on a search for *technocracy* in the title of articles in the *New York Times, New York Herald Tribune, Wall Street Journal, Chicago Tribune*, and *Los Angeles Times* using the Proquest Historical Newspapers online database; Elsner, *The Technocrats*, 7.

23. For a more sympathetic assessment of technocracy, see Segal, introduction to Loeb, *Life in a Technocracy*, ix–xxxviii.

24. Raymond, *What Is Technocracy?*, 100–118.

25. Harrod, review of *Wealth, Virtual Wealth and Debt* by Frederick Soddy; Raymond, *What Is Technocracy?*, 135–44.

26. Ardzrooni, "Veblen and Technocracy"; Raymond, *What Is Technocracy?*, 120; Dorfman, *Thorstein Veblen and His America*, 510–14. Search on Google Books Ngram Viewer for *Engineers and the Price System* conducted 3/16/2016.

27. H. Scott, "Technology Smashes the Price System."

28. Ibid.; quotation is from p. 141. For similar determinist usage of *technology* in a discussion of technocracy, see L. Allen, "Technocracy—a Popular Summary."

29. H. Scott, "Technology Smashes the Price System," 135, 136, 139, 140.

30. Segal, introduction to Loeb, *Life in a Technocracy*.

31. Loeb, *Life in a Technocracy*, 3, 7 (emphasis removed), 192. These ideas about the clash between technology and capitalism clearly reference Veblen. Veblen, *The Engineers and the Price System*, 104–5, 120–21.

32. Veblen, *The Engineers and the Price System*, 129.

33. See chapter 7; Veblen, "The Place of Science in Modern Civilization."

34. Raymond, *What Is Technocracy?*, 16.

35. Rank-and-file engineers did play a role in the residual technocratic organizations that Scott continued to promote, including the geologist M. King Hubbert, who later became famous for his theory of "peak oil." Elsner, *The Technocrats*, 109.

36. Layton, *Revolt of the Engineers*, chap. 8.
37. *New York Herald Tribune*, "Engineering Council Assails Technocracy."
38. Frederick, *For and Against Technocracy.*
39. R. Williams, "Lewis Mumford as a Historian of Technology"; Molella, "Mumford in Historiographical Context"; R. Williams, "Lewis Mumford's Technics and Civilization (Classics Revisited)."
40. Mumford, *Technics and Civilization*, 6, 12.
41. For a similar focus on *the machine* over *technology*, see Chase, *Men and Machines.*
42. Mumford, *Technics and Civilization*, 4, 12.
43. Hughes, *American Genesis.*
44. Mumford quotation in Molella, "Mumford in Historiographical Context," 23.
45. Ibid., 41.
46. Mumford, *Technics and Civilization*, 7, 11, 12, 109, 110.
47. D. Miller, *Lewis Mumford*, 109–10; S. Long, "Lewis Mumford and Institutional Economics," 167–69; Mumford, *Sketches from Life*, 220–21.
48. Mumford, *Technics and Civilization*, 366, 472.
49. L. Marx, "Lewis Mumford," 173.
50. Most recently Renwick and Gunn, "Demythologizing the Machine," 59–76.
51. Geddes, "The Twofold Aspect of the Industrial Age," 177–78; Geddes, *Cities in Evolution*, 63–64; R. Williams, "Lewis Mumford as a Historian of Technology," 56.
52. I have found only one place where Geddes used *technics* alone, in a 1922 letter to Mumford: Mumford, Geddes, and Novak, *Lewis Mumford and Patrick Geddes*, 127.
53. R. Williams, "Lewis Mumford as a Historian of Technology," 50–62; quotations are from p. 50.
54. Mumford, *Technics and Civilization*, 6.
55. Webster, *An American Dictionary of the English Language*, s.v. "technics."
56. Eger, *Technologisches Wörterbuch in englischer und deutscher Sprache*, 2:814.
57. Spengler, *Man and Technics.*
58. Mumford, "The Decline of Spengler," 104.
59. Mumford, "The Drama of the Machines," 150. In this essay, Mumford mentioned *technics* once and *technic* twice, while *technology* and its variant forms appeared five times.
60. Mumford, "Personality, Partial Second Draft—10 June 1931," 34–42; Mumford Papers, box 105, folder 6590.
61. Mumford, *Sketches from Life*, 467; Mumford to Miller, 5 June 1932; Deutsches Museum Archives, VA 0357/1. Note that Mumford's letter to Miller, written in English, suggests that his work at the museum may have been more limited than he implied in his later reminiscences.
62. Mumford, Murray, and Novak, *"In Old Friendship,"* 75.

63. Note dated 12 November 1932; Mumford Papers, box 105, folder 6539; Mumford, Murray, and Novak, *"In Old Friendship,"* 83.
64. Mumford, *My Works and Days*, 176, 309.
65. Mumford, *Technics and Civilization*, 470.
66. Ibid., 4, 319, 324–25. Although *Technics and Civilization* contains few translations by Mumford from German works, he did translate *Technik* as "technics" at least once, in a quotation from the German economist Karl Bücher. Bücher, *Arbeit und Rhythmus*, 383; Mumford, *Technics and Civilization*, 344.
67. E.g., Mumford, "Authoritarian and Democratic Technics."
68. [Melvin Kranzberg], "Addenda to Addenda, [Jan. 1958]," Kranzberg Papers, SHOT records, I, series 1, box 1, folder 3. Thanks to Rosalind Williams for alerting me to this document. It contains Kranzberg's transcription of Mumford's remarks. For the original letter, see Mumford to Krantzberg [*sic*], 25 January 1958, SHOT records, box 1, folder 2.
69. Mumford's usage of *technics* did find some followers, e.g., Condit, "Modern Architecture."
70. E.g., Mumford, "An Appraisal of Lewis Mumford's 'Technics and Civilization' (1934)," 531; Mumford, "History."

CHAPTER TEN

1. See my discussion of Charles Beard in chapter 9.
2. Abbott, "Pragmatic Sociology and the Public Sphere," 339–40. In the social sciences, this trend toward "scientism" is well known. Bannister, *Sociology and Scientism*. On the clash of business and academic values, see Veblen, *The Higher Learning in America*.
3. See esp. Henderson, "Applied Sociology (Or Social Technology)." For a detailed study of social technology, see Derksen and Wierenga, "The History of 'Social Technology.'"
4. Small, "The Scope of Sociology," 809.
5. Derksen and Wierenga, "The History of Social Technology," 318–19.
6. Breslau, "The Scientific Appropriation of Social Research," 438.
7. Robert E. Park and Ernest W. Burgess, *Introduction to the Science of Sociology* (Chicago: University of Chicago Press, 1921), 339, as cited in Derksen and Wierenga, "The History of Social Technology," 319.
8. Gieryn, *Cultural Boundaries of Science*, as discussed in chapter 5 above.
9. Titchener, "Psychology: Science or Technology?," 39, 42 (emphasis removed), 44–45.
10. Ibid., 46, 48.
11. For an example of similar social-science rhetoric about pure science and technology, see G. Stanley Hall, "Clark University."
12. Titchener, "Psychology: Science or Technology?," 49. Titchener repeated these arguments fifteen years later in Titchener, *Systematic Psychology*, 65–69, 266–67.

13. Gieryn, *Cultural Boundaries of Science*, esp. 37–64.
14. Hayes, "Masters of Social Science," 669–70; Small, *The Cameralists*.
15. E.g., Henderson, *Modern Methods of Charity*, 609–702.
16. Post, *Arbeit statt Almosen*; "Dr. Jul. Post, Arbeit statt Almosen [review]," 283; "Arbeit statt Almosen von Dr. Jul. Post [review]"; Ofner, *Studien sozialer Jurisprudenz*, 1–30; Tönnies, "Natorp, Paul, Sozialpädagogik [review]," 448.
17. Frison, "Some German and Austrian Ideas," 106, 114.
18. Schumpeter, *History of Economic Analysis*, 571.
19. Berg, *The Machinery Question*, 43ff.
20. Jevons, *The Theory of Political Economy*, 239. On Jevons see Schabas, *A World Ruled by Number*.
21. Walras, *Éléments d'économie politique pure*, 313–14.
22. Carl Menger, the Austrian pioneer of marginalism, noted in 1871 that the ratio of intermediate products needed to produce consumer goods "depends on the current state of *Technik*." Menger, *Grundsätze der Volkswirthschaftslehre*, 40.
23. Böhm-Bawerk, *The Positive Theory of Capital*, 20; Böhm-Bawerk, *Positive Theorie des Kapitales*, 18.
24. Marshall, *Industry and Trade*.
25. Mitchell, "Human Behavior and Economics."
26. Bix, *Inventing Ourselves Out of Jobs?*
27. Gourvitch, *Survey of Economic Theory on Technological Change and Employment*, 20–33.
28. E.g., Ely, "The New Economic World and the New Economics," 349.
29. Morris, *The U.S. Department of Labor Bicentennial History of the American Worker*, chap. 5.
30. Slichter, "Market Shifts, Price Movements, and Employment," 5–7.
31. "Notes," 816.
32. *New York Times*, "Thomas Launches Socialist Campaign."
33. U.S. Senate, Committee on Education and Labor, *Unemployment in the United States*, 240–41.
34. "Report of Senate Committee on Causes and Relief of Unemployment," 68.
35. Ely, "The New Economic World and the New Economics," 349; "Our Accelerated Economic Life"; "Recent Economic Changes," 97, 103; Mitchell, "Forces That Make for American Prosperity."
36. Bix, *Inventing Ourselves Out of Jobs?*
37. E.g., Hunt, *Recent Economic Changes in the United States*. For articles about technological unemployment that do not mention *technology*, see Scheler, "Technological Unemployment"; Hansen, "Institutional Frictions and Technological Unemployment," 684–89.
38. Gill and Weintraub, *Unemployment and Technological Change*; Kaplan, *The Research Program of the National Research Project*; Gill, "WPA Studies 'Foe,' the Swift Machine." On the history of the concept of technological change, see Godin, "Technological Change."

39. See the work of Stiglitz and Sen, two future winners of the Nobel Prize in Economics: Atkinson and Stiglitz, "A New View of Technological Change"; Sen, *Choice of Techniques*. By the 1990s, some economists did distinguish between the two terms, with *technology* being the set of possible *techniques*. E.g., Gomulka, *The Theory of Technological Change and Economic Growth*, 5.

40. For an explicit example of an economist defining *technology* as a noneconomic factor, see O. Taylor, "Economic Theory, and Certain Noneconomic Elements in Social Life," 382–84.

41. U.S. Senate, Temporary National Economic Committee, *Technology in Our Economy*, xi, 87, 172, 195.

42. For a late-career exception to this neglect of technology, see Parsons, "The Impact of Technology on Culture and Emerging New Modes of Behaviour."

43. Camic, "Introduction: Talcott Parsons before *The Structure of Social Action*," xix–xxiii.

44. Ibid., xxiv–xxvi; Parsons, "'Capitalism' in Recent German Literature," 641–61.

45. Parsons, "'Capitalism' in Recent German Literature"; quotations are from pp. 644, 654–55 (my bracketed additions). In the last quotation, a more literal translation would be "the elimination of man in technical thought [*im technischen Denken*]." Sombart, *Der moderne Kapitalismus*, 2nd ed., 3:81.

46. Camic, "Introduction: Talcott Parsons before *The Structure of Social Action*," xxv; Parsons, "'Capitalism' in Recent German Literature," 654; Parsons, ""Capitalism" in Recent German Literature: Sombart and Weber (Concluded)," 48–49.

47. M. Weber, *The Protestant Ethic and the Spirit of Capitalism*, 24; M. Weber, *Gesammelte Aufsätze zur Religionssoziologie*, 1:10; Camic, "Introduction: Talcott Parsons before *The Structure of Social Action*," xxvii.

48. See chapter 9.

49. F. Knight, *Risk, Uncertainty and Profit*; M. Weber, *General Economic History*, 38, 180, 191; Weber, Hellmann, and Palyi, *Wirtschaftsgeschichte*, 50, 164, 173.

50. Levine, "The Continuing Challange of Weber's Theory of Rational Action," 104.

51. M. Weber, *Economy and Society*, 65–67.

52. Camic, "Introduction: Talcott Parsons before *The Structure of Social Action*," lvi–lix; Endres and Donoghue, "Defending Marshall's 'Masterpiece'."

53. Souter, "'The Nature and Significance of Economic Science' in Recent Discussion," 382–86.

54. Ibid., 381–82.

55. Parsons, "Some Reflections on 'The Nature and Significance of Economics,'" 525–26, 528.

56. Parsons continued his shift from *technique* to *technology* in a 1935 article that discussed Veblen and Sombart. Camic, "Introduction: Talcott Parsons before *The Structure of Social Action*," lix–lxi; Parsons, "Sociological Elements in Economic Thought."
57. Camic, "*Structure* after 50 Years."
58. Parsons, *The Structure of Social Action*, 233, 233n1, 234n1.
59. Ibid., 233–34, 655.
60. I counted roughly forty instances of *technological* and *technology* used in this narrow, instrumental sense.
61. Parsons, *The Structure of Social Action*, 466, 718, 780.
62. Ibid., 266.
63. M. Weber, *The Theory of Social and Economic Organization*, v, 160–62. Frison has also noted the problematic character of Weber's translation. Frison, "Linnaeus, Beckmann, Marx," 154n1. When Roth and Wittich revised Parsons's work in their 1968 translation of the full version of *Economy and Society*, they changed some occurrences of *technology* to the more appropriate *technique*, but still confused the terms. M. Weber, *Economy and Society*, xxv–xxvi, 65–67.
64. See, e.g., the reference to "research technology" in Gouldner, "Explorations in Applied Social Science," 176.
65. E.g., Lynn White, "Technology and Invention in the Middle Ages," 141.
66. Perrow, *Organizational Analysis*; Thompson, *Organizations in Action*.
67. Ellul, *The Technological Society*; Marcuse, *One Dimensional Man*.
68. McGee, "Making Up Mind."
69. Ibid., 773–79; Bannister, *Sociology and Scientism*, 169–72; quotation is from p. 161; Huff, "Theoretical Innovation in Science," 264; Westrum, *Technologies and Society*, 50–62; Volti, "Classics Revisited."
70. Tylor, *Primitive Culture*, 1:1; Ogburn, *Social Change*, 4.
71. Camic, "On Edge," 230.
72. Ogburn, *Social Change*, 200–201, 264, 280, 350; Volti, "Classics Revisited," 397–98.
73. Ogburn, *Social Change*, 213–36; quotation is from p. 218.
74. Huff, "Theoretical Innovation in Science," 265–67; Ogburn, *Social Change*, 4, 242.
75. Ogburn, *Social Change*, 343. See also Ogburn, "The Great Man versus Social Forces."
76. Mason's *The Origins of Invention* was well known, while Ogburn never absorbed the many insights of Gilfillan's work on invention. O. Mason, *The Origins of Invention*, esp. chaps. 1, 12; Gilfillan, *The Sociology of Invention*; Volti, "Classics Revisited," 403. McGee has provided a detailed analysis of the limitations of Ogburn's theory of invention, along with descriptions of the more sophisticated theories of Ogburn's contemporaries. McGee, "Making Up Mind," 777–89.
77. Ogburn, *Social Change*, 269; Volti, "Classics Revisited," 400.

78. Certainly, Veblen considered *technology* to be far broader than *invention*, as did most scholars writing on technological unemployment.
79. Ogburn, "The Future of Man," 295, 298–99.
80. Sorokin, "Recent Social Trends," 194 (quotation); Camic, "On Edge," 238. On *Recent Social Trends in the United States*, see also Bulmer, "The Methodology of Early Social Indicator Research"; Tobin, "Studying Society."
81. Research Committee on Social Trends, *Recent Social Trends in the United States*, 1:122.
82. Ibid., 1:xxv,122, 130.
83. Ogburn, "Technology and Sociology"; quotation is from p. 1. As David Edgerton has pointed out, the error of technological determinism lies in the idea that innovation directly shapes society, when in fact the social effects of technology result from its use. Edgerton, "From Innovation to Use," 120–21.
84. Volti, "Classics Revisited," 399.
85. Ogburn, "Technology and Governmental Change," 13.
86. Ibid., 13, 3–4; Ogburn, "The Future of Man," 299.
87. Camic, "On Edge," 245–72.
88. William F. Ogburn to John Merriam, 21 November 1935, Ogburn Papers, box 20, folder 5; Ogburn to Frederic A. Delano, 11 December 1935, Ogburn Papers, box 19, folder 7.
89. William F. Ogburn, "Technology and Planning: Memorandum on the organization and presentation of material on particular fields," 18 March 1935, Ogburn Papers, box 20, folder 14.
90. Elliott, who wrote a short chapter titled "The Interdependence of Science and Technology," admitted to Ogburn that drafting the Purdue university budget had prevented him from devoting adequate time to the project. Edward C. Elliott to William F. Ogburn, 31 March 1937, Ogburn Papers, box 19, folder 9.
91. United States, National Resources Committee, Science Committee, *Technological Trends and National Policy*, ix, 8; see also Ogburn, "The Influence of Inventions on American Social Institutions in the Future," 365, 367.
92. United States, National Resources Committee, Science Committee, *Technological Trends and National Policy*, 10; Godin, "Innovation without the Word," 294.
93. See e.g., Kelly, *What Technology Wants*.
94. Pelton, "Predictions on the Future of Inventions"; Crook, review of *Technological Trends and National Policy*; Brainerd, review of *Technological Trends and National Policy*; G. Meyer, "Note on Technological Trends and Social Planning"; Levy, review of *Technological Trends and National Policy*; "Technological Trends and National Policy."
95. William F. Ogburn to Waldemar Kaempffert, 25 April 1937, Ogburn Papers, box 20, folder 3; Laurence, "Inventions Survey Finds Major Changes

Imminent"; *New York Times*, "America's Future Studied in Light of Progressive Application of Technology."

96. Ogburn, Gilfillan, and Adams, *The Social Effects of Aviation*; Ogburn and Nimkoff, *Technology and the Changing Family*.
97. William F. Ogburn to Charles W. Eliot II, 5 June 1936, Ogburn Papers, box 20, folder 12; Rosen and Rosen, *Technology and Society*.
98. Ogburn and Nimkoff, *Sociology*, 846, 951.
99. Leslie White, "Energy and the Evolution of Culture"; Childe, "Archaeological Ages as Technological Stages."
100. W. Beard, "Technology and Political Boundaries"; Bain, "Technology and State Government." William Beard was Charles Beard's son. Bain was admittedly a sociologist, not a political scientist—these boundaries were not rigid.
101. For industrial relations, see E. Smith, *Technology and Labor*.
102. Rodgers, "Keywords," 691.

CHAPTER ELEVEN

1. Hughes, *American Genesis*, 47.
2. P. Long, *Artisan/Practitioners*, chap. 4.
3. On scientists in industry, see Shapin, *The Scientific Life*.
4. E.g., Neilson, *Webster's New International Dictionary of the English Language*, s.v. "technology." See the next chapter for a discussion of *technology* during the Cold War.
5. E.g., the 1973 Burndy Library Conference on Interaction of Science and Technology in the Industrial Age. See Reingold and Molella, "Introduction [to the Burndy Library Conference]." See also Layton, "Mirror-Image Twins."
6. Seely, "SHOT," 747–48.
7. See below for examples. The applied-science definition of *technology* is in some ways similar to the "linear model of innovation," which David Edgerton argues was more a straw man for academics to attack than a fully articulated and widely accepted model. Edgerton, "'The Linear Model' Did Not Exist." See also Godin, "The Linear Model of Innovation."
8. Mayr, "The Science-Technology Relationship as a Historiographic Problem."
9. Ibid., 670.
10. Other historians, particularly Ron Kline, have highlighted Mayr's critique. See Kline, "Construing 'Technology' as 'Applied Science,'" 194. For an overview of this issue, see McClellan, "What's Problematic about 'Applied Science.'"
11. For a problematic argument about the science-technology relationship rooted in the ideal of pure science, see Forman, "The Primacy of Science."
12. On the aristocratic roots of modern science, see Shapin, *A Social History of Truth*.

13. This issue comes up in critiques of the commercialization of academic research. For an analysis of this rhetoric, see Holden, "Lamenting the Golden Age."
14. Galison, *How Experiments End*; Shapin and Schaffer, *Leviathan and the Air-Pump*.
15. Roberts and Schaffer, preface to *The Mindful Hand*, ix.
16. Kline, "Construing 'Technology' as 'Applied Science,'" 201–2.
17. Layton, *Revolt of the Engineers*, esp. chap. 3.
18. Gooday, "'Vague and Artificial,'" 548; Babbage, *On the Economy of Machinery and Manufactures*, 307.
19. Gooday, "'Vague and Artificial,'" 549–52; Huxley, "Science and Culture," 155.
20. Bigelow, *Elements of Technology*, title page, iv, 4.
21. Bigelow, *The Useful Arts*.
22. Knapp, *Lehrbuch der chemischen Technologie*, vol. 1; Knapp et al., *Chemical Technology*, 1st American ed.
23. Kohlstedt, "A Step toward Scientific Self-Identity in the United States," 342.
24. Poinsett, *Discourse, on . . . the National Institution for the Promotion of Science*, 6, 7, 34. This same tension is present in George Wilson's lectures on technology, though Wilson did finesse the problem by defining *technology* as the collective term for all sciences "applicable to the industrial labours or utilitarian necessities of man." Wilson, "On the Physical Sciences Which Form the Basis of Technology," 64.
25. Bovey, "The Fundamental Conceptions Which Enter into Technology."
26. Ibid., 535, 543; Bacon, *The Works of Francis Bacon*, 4:114.
27. Bovey, "The Fundamental Conceptions Which Enter into Technology," 540, 542.
28. Ibid., 542. This is an early example of what Phillips describes as the divergence between the terms *Wissenschaft* and *science*, which were considered equivalent in the nineteenth century. Phillips, "Francis Bacon and the Germans," 389.
29. Bovey, "The Fundamental Conceptions Which Enter into Technology," 540–41.
30. Ibid., 549–51.
31. See chapters 4 and 6.
32. A 1901 article about the new London University in the *Times* of London also explicitly invoked *technology* as the application of science. "Science and the London University." The original article appeared here: *Times* (London), "The Organization of University Education in the Metropolis."
33. In fact, one of the principal organizers of the St. Louis congress was Albion Small (see chapter 10), who probably chose technology as the topic for Bovey's lecture. H. Rogers, *Congress of Arts and Science, Universal Exposition, St. Louis, 1904*, 6:1–15, 6:48.

34. I have not found a single explanation of the phrase, though some probably exist after World War II.
35. This ambiguity in the meanings of "science and technology" overlaps significantly with the ambiguity in "applied science." Kline, "Construing 'Technology' as 'Applied Science,'" 201–2.
36. *Eleventh Report of the Science and Art Department*, 1.
37. Armytage, "J. F. D. Donnelly," 7, 10; Robertson, "The South Kensington Museum," 4.
38. Wood, *A History of the Royal Society of Arts*, 17, 212.
39. "Proceedings of the Society: Annual Conference," 636–37; quotation is from p. 636, emphasis added; "Proceedings of the Society: Conference on Technological Education," 725–35.
40. "Proceedings of the Society: Annual Conference," 637.
41. *Times* (London), "Imperial College of Science and Technology."
42. *Report of the National Academy of Sciences for the Year 1919*, 22–23.
43. Edgerton, "From Innovation to Use," 125; Kline, "Construing 'Technology' as 'Applied Science,'" 201–2.
44. Kline gives only one example before the 1930s, by the engineer and NRC member Alfred D. Flinn, who equated *technology* and *applied science* in passing. Flinn, "The Relation of the Technical School to Industrial Research," 508; Kline, "Construing 'Technology' as 'Applied Science,'" 217.
45. Compton, "Inaugural Address," 595; S. Stratton, "Installation Address," 592, 593; J. Stratton, "Karl Taylor Compton," 47–48. When industrial researchers invoked *technology* before WWII, they also described it as distinct from *science*. For example, "technology and science consequently have their meeting point in industrial research." Weidlein and Hamor, *Science in Action*, 5.
46. Lawrence, "Science and Technology," 295, 297.
47. Ibid., 297–98. Lawrence drew his example from the philosopher Whitehead, who also wrote explicitly of an interactive relationship between *science* and *technology*. Whitehead, *Science and the Modern World*, 114–15. On Lawrence as engineer, see Hughes, *American Genesis*, 404–10. On Michelson's experiments in relation to Einstein's work, see Staley, *Einstein's Generation*.
48. *Science at the Cross Roads*; Freudenthal and McLaughlin, *The Social and Economic Roots of the Scientific Revolution*, 27–29.
49. Sarton, "The History of Science," 321.
50. Ibid., 336–38.
51. This statement, often repeated thirdhand, was supposedly made in 1948. M. Hall, "Recollections of a History of Science Guinea Pig," S72.
52. Sarton, "L'Histoire de la Technologie"; Sarton, "Bibliographie analytique," 177–78.
53. Pogo, review of *History of Technics*. See also Nikiforova, "The Concept of Technology and the Russian Cultural Research Tradition."

54. Freudenthal and McLaughlin, *The Social and Economic Roots of the Scientific Revolution*, 27.
55. See S. Cohen, *Bukharin and the Bolshevik Revolution*.
56. Ibid., 109–22.
57. Bukharin, *Historical Materialism*, 115, 121, 133, 136–44; Bukharin, *Teoriia istoricheskogo materializma*.
58. Bukharin, "Theory and Practice from the Standpoint of Dialectical Materialism," 11–33.
59. Bukharin, *Historical Materialism*, 161–69; quotations are from pp. 161, 163, 164, 165.
60. Joffe, "Physics and Technology," 37; Rubinstein, "Relations of Science, Technology, and Economics under Capitalism, and in the Soviet Union," 46, 48, 50.
61. I have used the new translation of Hessen's paper by Philippa Schimrat, based on a version of the paper published in 1933 in Russian. The new translation substitutes *technology* for all references to *technique*. Hessen, "The Social and Economic Roots of Newton's *Principia*," 45–52, 56, 59, 61, 84.
62. *New York Times*, "Soviet Scientists Discourse on Marx."
63. W. Adams, "The International Congress of the History of Science and Technology," 208, 211; "Societies and Academies," 177, 178.
64. On the reception of Hessen's paper, see Freudenthal and McLaughlin, "Classical Marxist Historiography of Science," 27–34; Schaffer, "Newton at the Crossroads."
65. Desch, "Pure and Applied Science," 496.
66. Freudenthal and McLaughlin, "Classical Marxist Historiography of Science," 30–31; Clark, *Science and Social Welfare in the Age of Newton*; Merton, "Science and the Economy of Seventeenth Century England," 3–7; Needham, "Capitalism and Science," 198–99.
67. Merton, "Science and Military Technique," 542.
68. Merton, "Science, Technology and Society in Seventeenth Century England," 510, 513–16, 562; Freudenthal and McLaughlin, "Classical Marxist Historiography of Science," 31–32.
69. Another example of the interactive framing of the science-technology relationship, one that denied explicitly that technology was applied science, is the influential history of science survey by Wolf, *A History of Science, Technology, and Philosophy in the 16th and 17th Centuries*, 7–8, 450–53.
70. Freudenthal and McLaughlin, "Classical Marxist Historiography of Science," 27.
71. Sarton, "The History of Science," 336–38; Trowbridge, "Pure Science and Engineering," 576. For Sarton's hostility toward Hessen, see Sarton, review of *Medicine and Health in the Soviet Union*, 203.
72. This is exactly the path that Merton took in his sociology of science. Schaffer, "Newton at the Crossroads," 25.

CHAPTER TWELVE

1. Merton is one exception, as were some Marxists. Merton, "Science and Military Technique"; Levy, review of *Technological Trends and National Policy*, 264–65.
2. I. Stewart, *Organizing Scientific Research for War*, 38.
3. Kaiser, "From Blackboards to Bombs," 523–24; Gelder, "Being Harvard's President"; Conant, "Chemists and the National Defense."
4. E.g., McMillen, "Relation of Physics to the War Effort," 272–73.
5. Kaiser, "From Blackboards to Bombs," 525. For an account of the Manhattan Project that stresses the work of engineers, see Hughes, *American Genesis*, chap. 8.
6. Kaiser, "From Blackboards to Bombs," 525.
7. E.g., Sarnoff, "'Science for Life or Death' Discussed by Sarnoff."
8. Edgerton, *Britain's War Machine*, 171.
9. Buckley, "Frank Baldwin Jewett, 1879–1949"; Reingold, "Physics and Engineering in the United States," 289.
10. Jewett, "The Promise of Technology," 2–3.
11. For nuanced discussions of the role of science in technological change, see Hughes, *American Genesis*; Mindell, *Between Human and Machine*; Vincenti, *What Engineers Know*.
12. Jewett, "The Promise of Technology," 2.
13. Lawrence, "Science and Technology," 295.
14. Forman, "Behind Quantum Electronics," 153.
15. I have used the 1960 reprint, which includes not just Bush's text but also the reports of the committees. Bush, *Science, the Endless Frontier*. See also Reingold, "Vannevar Bush's New Deal for Research."
16. For a careful and sophisticated analysis of the concept of basic research after World War II, see Schauz, "What Is Basic Research?," 298–313. Schauz shows that the concept of basic research underwent a shift in meaning toward pure science after World War II, a shift that cannot be fully attributed to the Bush report.
17. Bush, *Science, the Endless Frontier*, 19.
18. Goldberg, "Inventing a Climate of Opinion."
19. Edgerton, "'The Linear Model' Did Not Exist," 40.
20. Bush, *Science, the Endless Frontier*, 7, 18, 23, 24, 29. The phrase "science and technology" was even more common in the committee reports that Bush drew from; these reports are included in the 1960 reprint.
21. This threat to truth was not a recent phenomenon. Henrik Ibsen's 1882 play, *An Enemy of the People*, is about just such a threat. Conflict between engineers' analyses and employer demands were a sore point among engineers during the Progressive Era, especially among civil engineers. Ibsen, *An Enemy of the People*; Layton, *Revolt of the Engineers*.
22. Bush, *Science, the Endless Frontier*, 12, 19; Jewett, "The Promise of Technology," 3, 5.

23. Edgerton, "'The Linear Model' Did Not Exist," 48.
24. Edgerton, *Britain's War Machine*, esp. chap. 8; Scranton, "Technology-Led Innovation," 338–44.
25. The physicist was Lise Meitner, working with the chemists Otto Hahn and Fritz Strassmann. On this discovery, see Rhodes, *The Making of the Atomic Bomb*.
26. Boyer, *By the Bomb's Early Light*.
27. Dunn, "Utopia—1955."
28. "For the Future," *Newsweek*, 20 August 1945, 59–60; quoted in Boyer, *By the Bomb's Early Light*, 113.
29. E. Russell, *War and Nature*.
30. For examples, see Heimann, *50s: All-American Ads*.
31. Boyer, *By the Bomb's Early Light*, esp. chaps. 10, 11; e. g. Dietz, *Atomic Energy in the Coming Era*.
32. Ogburn, "Sociology and the Atom."
33. E.g., Lilienthal, "Science and the Spirit of Man."
34. Scranton, "Technology-Led Innovation," 340; Sherwin and Isenson, "First Interim Report on Project Hindsight (Summary)."
35. McGucken, "On Freedom and Planning in Science," 45. Polanyi fleshed out these ideas as a way to distinguish science from technology. Freedom of inquiry was essential for science, but less so for technology. Polanyi, "Pure and Applied Science." On Polanyi's life and work, see M. Nye, *Michael Polanyi and His Generation*.
36. Mayer, "Setting Up a Discipline, II," 58.
37. Merton, "A Note on Science and Democracy"; quotation is from p. 125. For discussion of the political context of Merton's norms of science, see S. Turner, "Merton's 'Norms' in Political and Intellectual Context"; Hollinger, "The Defense of Democracy."
38. Rossi, *Philosophy, Technology, and the Arts*. For more on the anti-Marxist animus toward the Zilsel thesis, see Freudenthal and McLaughlin, "Classical Marxist Historiography of Science."
39. For Koyré's dismissal of the Zilsel thesis, see Koyré, "Galileo and Plato," 401.
40. P. Long, *Artisan/Practitioners*, 24–25; A. Rupert Hall, "What Did the Industrial Revolution in Britain Owe to Science?"
41. Roberts, "Agency and Industry." For a nineteenth-century chemist's view of the Leblanc process as an interaction between science and technology, see Vincent, "On Some Recent Processes for the Manufacture of Soda."
42. *Isis* did, however, provide significant coverage of the history of technology in its book reviews.
43. *Chicago Daily Tribune*, "Warns Soviet Technology Is Threat to U.S."; Waterman, "The Science of Producing Good Scientists"; *New York Herald Tribune*, "The Threat of Soviet Technology." On Sputnik, see McQuade, "Sputnik Reconsidered."
44. Godin, "The Linear Model of Innovation"; Godin, "In the Shadow of Schumpeter"; Godin, *Narratives of Innovation*.

45. Maclaurin, *Invention and Innovation in the Radio Industry.*
46. Maclaurin, "The Sequence from Invention to Innovation," 98–100; Godin, "In the Shadow of Schumpeter," 346–49.
47. Maclaurin, "The Sequence from Invention to Innovation," 98, 102.
48. Maclaurin's Committee on Technological Change at MIT sponsored several other histories focused on technological change; e.g., Scoville, *Revolution in Glassmaking*; A. Bright, *The Electric-Lamp Industry.* For an overview of this early work in business history, see Cole, "Committee on Research in Economic History," 81–82.
49. Brady, "Yale Brozen 1917–1998"; Brozen, "Adapting to Technological Change"; Brozen, "Studies of Technological Change."
50. Passer, *The Electrical Manufacturers*; McCraw, *Prophet of Innovation,* 471–73.
51. Hunter, *Steamboats on the Western Rivers*; Waggoner, "Louis C. Hunter, 85, Is Dead"; Brown, "Louis C. Hunter, *Steamboats on the Western Rivers* (Classics Revisited)."
52. Regarding the influence of the pre–World War II German discourse of Technik on American intellectuals, see Greif, *The Age of the Crisis of Man,* esp. 47–51.
53. Special thanks for Amrys Williams for her research assistance on this topic.
54. Marcel, *Le déclin de la sagesse*; Marcel, *The Decline of Wisdom,* 3, 6, 12, 19.
55. Jünger, *The Failure of Technology.*
56. Ibid., 70, 71, 105–6, 204.
57. *New York Times,* "Pope Pius' Christmas Message." See also W. Norris Clarke, "Technology and Man," 433–36.
58. For the original Italian, see Pius XII, "Nuntius Radiophonicus."
59. Seely, "SHOT."
60. US Senate, Committee on Government Operations, *Create a Department of Science and Technology,* 1, 8–9.
61. Seely, "SHOT," 752–53. For more detail on Kranzberg's background, see R. Post, "Chance and Contingency."
62. Seely, "SHOT," 753; Kranzberg to Marie Boas, 16 October 1956, Kranzberg Papers, series 5, box 278, folder 7.
63. Seely, "SHOT," 754; Boas to Kranzberg, 29 October 1956, Kranzberg Papers, series 5, box 278, folder 7.
64. Seely, "SHOT," 758; Kranzberg to Thomas P. Hughes, 29 May 1957, SHOT Records, I, series 1, box 1, folder 1.
65. Seely, "SHOT," 758; Kranzberg to Guerlac, 28 June 1957, SHOT Records, I, series 1, box 1, folder 1.
66. Lewis Mumford, "Questionaire on Name of Quarterly Journal for the Society for the History of Technology," [February 1959], SHOT Records, II, Technology and Culture, series 1, box 1; R. Post, "Back at the Start," 973–78.
67. For White's early work in history of technology, see Lynn White, "Technology and Invention in the Middle Ages."

68. For an excellent discussion of this journal and its context, see Voskuhl, "Engineering Philosophy." See also Herf, "Reactionary Modernism Reconsidered," 157.
69. Kranzberg, "At the Start," 3.
70. On SHOT's deep and continuing links to engineering, see Seely, "SHOT," 760–68.
71. For an overview of the science-technology relationship as a theme in *Technology and Culture*, see Staudenmaier, *Technology's Storytellers*, 83–107.
72. Singer, Holmyard, and Hall, *A History of Technology*, viii. One early, highly critical reviewer commented on the inadequacy of this definition. Bickerman, review of *A History of Technology*, vol. 1.
73. Kranzberg, "At the Start," 8–9.
74. Drucker, "Work and Tools," 29–30, 37. On Drucker's long connection to SHOT, see R. Post, "Back at the Start," 981–85.
75. E.g., Feibleman, "Pure Science, Applied Science, Technology, Engineering."
76. An exception that proves the rule is Multhauf, "Some Observations on the State of the History of Technology," 1–4. Multhauf, who read broadly in European sources for the history of technology, glimpsed some of the broad outlines of the history of the concept of technology as presented in this book, yet his discussion remained little more than a list of relevant sources.
77. Here is a sampling of definitions: Drucker, "Work and Tools," 28–29; Multhauf, "The Scientist and the 'Improver' of Technology," 38; Mumford, "Tools and the Man," 322; Rae, "The 'Know-How' Tradition," 140–41; Woodbury, "Review: The Scholarly Future of the History of Technology," 347–48; Drucker, "The Technological Revolution," 342; Feibleman, "Pure Science, Applied Science, Technology, Engineering," 310–12; Ellul, "The Technological Order," 394–95; Skolimowski, "The Structure of Thinking in Technology," 372; Bunge, "Technology as Applied Science," 329.
78. Withey, "Public Opinion about Science and Scientists," 382–83; Pion and Lipsey, "Public Attitudes toward Science and Technology," 304.
79. SHOT scholars sometimes linked technology to the older concept of useful arts, but did not explore what this shift in meaning implied. E.g. Multhauf, "The Scientist and the 'Improver' of Technology," 38.
80. Ellul, "The Technological Order." See the following chapter for a more detailed discussion of Ellul.
81. Kirk, *Counterculture Green*.

CHAPTER THIRTEEN

1. A book that I happen to be writing, tentatively titled *Questioning Technology in the Long 1960s: Critics of Civilization versus Dissident Experts*.
2. Godin provides a list of dozens of different sequential models. Godin, *Narratives of Innovation*, 225–31, 51n10.

3. Sherwin and Isenson, "Project Hindsight," 1576; Godin, *Narratives of Innovation*, 103–4.
4. Godin, *Narratives of Innovation*, 125.
5. Ibid., 125–26.
6. Sherwin and Isenson, "First Interim Report on Project Hindsight (Summary)," 3.
7. Ruttan, "Usher and Schumpeter on Invention, Innovation and Technological Change: Reply," 154–55; Godin, *Narratives of Innovation*, 95–96.
8. This is, of course, a simplification, yet this distinction between *technique* and *technology* continued for decades. E.g., Gomulka, *The Theory of Technological Change and Economic Growth*, 4–5.
9. For critiques of mainstream economic theories of technology, see David, *Technical Choice, Innovation and Economic Growth*; Rosenberg, "Learning by Using"; David, "Path Dependence, Its Critics and the Quest for 'Historical Economics.'"
10. *New York Times*, "Rocket Race," 2.
11. *New York Times*, "Gaither Report Calls U. S. in Gravest Peril"; U.S. Office of Defense Mobilization, "Deterrence and Survival in the Nuclear Age." Occasional headlines also highlighted the Soviet technological threat in the 1950s. E.g., *New York Herald Tribune*, "The Threat of Soviet Technology."
12. Logsdon, *Exploring the Unknown*, 492.
13. Beck, "U.S. Faces 'Technology Gap' with Russ, GOP Forum Says."
14. B. Nelson, "Technological Innovation," 1229.
15. B. Nelson, "Technological Innovation."
16. J. Bright, "Opportunity and Threat in Technological Change," 76.
17. Schön, "Champions for Radical New Inventions," 83.
18. Schön, *Technology and Change*.
19. See chapter 10.
20. Michael Smith, "'Silence, Miss Carson!,'" 736.
21. United States Congress, *National Commission on Technology, Automation, and Economic Progress*, 116–17.
22. Wisnioski, *Engineers for Change*, 51–54; Harvard University Program on Technology and Society, *A Final Review, 1964–1972*.
23. Harvard University Program on Technology and Society, *A Final Review, 1964–1972*, 6–7; Cowan, "Looking Back in Order to Move Forward," 207–8.
24. Harvard University Program on Technology and Society, *A Final Review, 1964–1972*, 7.
25. McDermott, "Technology: The Opiate of the Intellectuals." For an excellent analysis and critique of McDermott's article, see Cowan, "Looking Back in Order to Move Forward."
26. Hughes and Hughes, *Lewis Mumford*; Greenman, Schuchardt, and Toly, *Understanding Jacques Ellul*; Mackey, "Herbert Marcuse."
27. Carlyle, "Signs of the Times," 444.

28. "The Encyclopaedia Britannica Conference on the Technological Order."
29. Ellul, *The Technological Society*, vi, xxvii, 14, 19, 80, 83, 418.
30. Ibid., xiii, 14, 78–79. For a more sympathetic analysis, see Greenman, Schuchardt, and Toly, *Understanding Jacques Ellul*, chap. 3.
31. This enthusiasm is captured brilliantly in Arthur Radebaugh's syndicated newspaper comic strip, "Closer Than We Think," which ran from 1958 to 1962. See http://arthur-radebaugh.blogspot.de/p/about-exhibit.html, accessed 2/11/18.
32. This tendency continued into postwar Soviet thought, with the concept of the "scientific-technological revolution." Buchholz, "The Scientific-Technological Revolution (STR) and Soviet Ideology."
33. For a thorough overview of Marcuse's life, see Mackey, "Herbert Marcuse." For a shorter introduction, see Feenberg, "Introduction to the Second Edition."
34. Feenberg, "Introduction to the Second Edition."
35. Marcuse, *One Dimensional Man*.
36. Ibid., 158–59.
37. Ibid., 231, 234.
38. Ibid., 256–57.
39. Lubar, "Do Not Fold, Spindle or Mutilate," 44–48.
40. E.g. "Marcuse on the Hippie Revolution."
41. Mumford, *Technics and Civilization*.
42. For essays on Mumford's changing ideas about technology, see Hughes and Hughes, *Lewis Mumford*.
43. Mumford, "Technics and the Nature of Man," 303–17.
44. Mumford, *The Myth of the Machine*, vol. 1, *Technics and Human Development*.
45. Ibid., vol. 2, *Pentagon of Power*.
46. See *New Yorker* 46 (10, 17, 24, and 31 October).
47. Gallagher, "Technology and Society."
48. *New York Times*, "Taming the Technological Monster."
49. Gooding, "Engineers Are Redesigning Their Own Profession."
50. E.g., Abelson, "Anxiety about Genetic Engineering."
51. Ramo, "Individual and the Good Technological Society"; Wisnioski, *Engineers for Change*, 55–59, 63. For a powerful argument about how technological determinism sustains structures of power, see Noble, *Forces of Production*.
52. E.g., Mathews, "Technology Peril Stirs Scientists"; Kantrowitz, "The Test," 21.
53. Fries, "Expertise against Politics"; Wisnioski, *Engineers for Change*, 52–57.
54. There is a large literature about the Office of Technology Assessment. For a recent discussion, see Sadowski, "Office of Technology Assessment."
55. Nader, *Unsafe at Any Speed*, ix; Carson, *Silent Spring*; Commoner, *The Closing Circle*; Commoner, *Science and Survival*.

56. Bookchin, "Towards a Liberatory Technology," 86, 91, 94.
57. Schumacher, *Small Is Beautiful*; Schumacher, *This I Believe, and Other Essays*.
58. Hoffman, *Revolution for the Hell of It*, 86.
59. Kirk, *Counterculture Green*; F. Turner, *From Counterculture to Cyberculture*.
60. Wisnioski, *Engineers for Change*, 36–39, 47–49; quotation is from p. 49; Horgan, "Technology and Human Values." Wisnioski claims that engineers like Horgan "were the exception rather than the rule" (49).
61. Florman, *The Existential Pleasures of Engineering*, esp. 56–73.
62. Preece, "A Report of the Discussion," 468.
63. Staudenmaier, *Technology's Storytellers*.
64. Reagan, "Remarks at a White House Luncheon Honoring the Astronauts of the Space Shuttle Columbia, May 19, 1981"; Katz, "US Energy Policy"; R. Smith, "Environmental Policies Attacked."
65. Markoff, *What the Dormouse Said*.
66. In the early 1980s, Gingrich supported tax credits allowing families to buy personal computers to make it easier for mothers to work from home. Gingrich later became a leading digital utopian in the 1990s. Christensen, "Cottage Industry and Women."
67. See, e.g., Mumford's attack on McLuhan's electronic utopianism. Mumford, *Pentagon of Power*, 292–99.
68. Kline, "Cybernetics, Management Science, and Technology Policy"; quotation is from p. 535.
69. "Exxon Mobil Meets Amazon.com."
70. Kelly, *What Technology Wants*; Friedman, "U.S.G. and P.T.A."
71. In the late 1950s, the idea of technology as technique spread to management theorists, who began writing about "organizational technology" in the sense of management methods. E.g., Goelz, "Toward a Concept of Education for Administration," 62.
72. Foucault, *The Foucault Reader*, 341–42; Foucault, "About the Beginning of the Hermeneutics of the Self," 203, 223n4; Foucault, *Hermeneutics of the Subject*, 46–47; Foucault, *L'herméneutique du sujet*, 48; Foucault, "Technologies of the Self," 18, 545.
73. E.g., Tell, "Rhetoric and Power."
74. For a critique of misuse of language by academics, see Billig, *Learn to Write Badly*.
75. Pielke, ""Basic Research" as a Political Symbol," 356–58.
76. Schauz, "What Is Basic Research?," 303.
77. Liu and Liu, *The Three-Body Problem*.

Bibliography

ARCHIVAL SOURCES

Deutsches Museum Archives, Munich
Melvin Kranzberg Papers, 1934–88, Archives Center, National
    Museum of American History, Smithsonian Institution,
    Washington, DC
Lewis Mumford Papers, Rare Book and Manuscript Library, Uni-
    versity of Pennsylvania, Philadelphia
William Fielding Ogburn Papers, Department of Special Collec-
    tions, Regenstein Library, University of Chicago

PUBLISHED SOURCES

Abbott, Andrew. "Pragmatic Sociology and the Public Sphere:
    The Case of Charles Richmond Henderson." *Social Science
    History* 34, no. 3 (2010): 337–71.
Abelson, Philip H. "Anxiety about Genetic Engineering." *Science*
    173, no. 3994 (1971): 285.
Adams, Henry. "The Dynamo and the Virgin." In *The Education
    of Henry Adams*, 379–90. New York: Modern Library, 1931.
Adams, W. "The International Congress of the History of Science
    and Technology." *History* 16, no. 63 (1931): 202–13.
Adas, Michael. *Machines as the Measure of Men: Science, Technol-
    ogy, and Ideologies of Western Dominance*. Ithaca, NY: Cornell
    University Press, 1989.
Agassi, Joseph. *The Very Idea of Modern Science: Francis Bacon and
    Robert Boyle*. Dordrecht: Springer, 2012.
Aitken, Hugh G. J. *Scientific Management in Action: Taylorism at
    Watertown Arsenal, 1908–1915*. Princeton, NJ: Princeton
    University Press, 1985. First published 1960.

Albertson, David. *Mathematical Theologies: Nicholas of Cusa and the Legacy of Thierry of Chartres*. New York: Oxford University Press, 2014.

Alder, Ken. *Engineering the Revolution: Arms and Enlightenment in France, 1763–1815*. Princeton, NJ: Princeton University Press, 1997.

Alembert, Jean Le Rond d'. *Preliminary Discourse to the Encyclopedia of Diderot*. Translated by Richard N. Schwab. Chicago: University of Chicago Press, 1995. First published 1751.

Allard, Guy H. "Les arts mécaniques aux yeux de l'idéologie médiévale." In *Les arts mécaniques au moyen âge*, edited by G. H. Alllard and S. Lusignan, 13–31. Montreal: Bellarmin, 1982.

Allen, Glen Scott. *Master Mechanics and Wicked Wizards: Images of the American Scientist as Hero and Villain from Colonial Times to the Present*. Amherst: University of Massachusetts Press, 2009.

Allen, Leroy. "Technocracy—a Popular Summary." *Social Science* 8, no. 2 (1933): 175–88.

Allgemeine deutsche Real-Encyklopädie für die gebildeten Stände: Conversations-Lexikon. 10th ed. 15 vols. Vol. 14, Leipzig: F. A. Brockhaus, 1854.

Allgemeine deutsche Real-Encyklopädie für die gebildeten Stände: Conversations-Lexikon. 7th ed. 12 vols. Vol. 8, Leipzig: F. A. Brockhaus, 1827.

Ames, William. *Technometry*. Translated by Lee W. Gibbs. Philadelphia: University of Pennsylvania Press, 1979.

Anderson, R. G. W. "'What Is Technology': Education through Museums in the Mid-19th Century." *British Journal for the History of Science* 25 (1992): 169–84.

Angulo, A. J. *William Barton Rogers and the Idea of MIT*. Baltimore: Johns Hopkins University Press, 2009.

Angus, Ian H. "Disenchantment and Modernity: The Mirror of Technique." *Human Studies* 6 (1983): 141–66.

"Arbeit statt Almosen von Dr. Jul. Post" [review]. *Preußische Jahrbücher* 52 (1883): 410.

Ardzrooni, Leon. "Veblen and Technocracy." *Living Age*, March 1933, 39–42.

Aristotle. *Aristotle's Nicomachean Ethics*. Translated by Robert C. Bartlett and Susan D. Collins. Chicago: University of Chicago Press, 2011.

———. *The Metaphysics*. Translated by Hugh Tredennick. Cambridge, MA: Harvard University Press, 1933.

———. *Nicomachean Ethics*. Translated by Martin Ostwald. Indianapolis: Bobbs-Merrill, 1962.

Armytage, W. H. G. "J. F. D. Donnelly: Pioneer in Vocational Education." *Vocational Aspect of Education* 2, no. 4 (1950): 6–21.

Ash, Eric H. *Power, Knowledge, and Expertise in Elizabethan England*. Baltimore: Johns Hopkins University Press, 2004.

Ashworth, William J. "The Ghost of Rostow: Science, Culture and the British Industrial Revolution." *History of Science* 46, no. 3 (2008): 249–74.

Atkinson, Anthony B., and Joseph E. Stiglitz. "A New View of Technological Change." *Economic Journal* 79, no. 315 (1969): 573–78.

Atwill, Janet. *Rhetoric Reclaimed: Aristotle and the Liberal Arts Tradition.* Ithaca, NY: Cornell University Press, 1998.

Ayres, C. E. *Science: The False Messiah.* Indianapolis: Bobbs-Merrill, 1927.

———. "The Gospel of Technology." In *American Philosophy Today and Tomorrow,* edited by Horace M. Kallen and Sidney Hook, 25–42. New York: Lee Furman, Inc., 1935.

———. "Moral Confusion in Economics." *International Journal of Ethics* 45, no. 2 (1935): 170–99.

Babbage, Charles. *On the Economy of Machinery and Manufactures.* London: Charles Knight, 1832.

Bacon, Francis. *The Advancement of Learning.* Edited by William Aldis Wright. 5th ed. Oxford: Clarendon, 1900.

———. *The New Organon.* Edited by Lisa Jardine and Michael Silverthorne. Cambridge: Cambridge University Press, 2000.

———. *The Works of Francis Bacon.* Edited by Douglas Denon Heath, James Spedding, and Robert Leslie Ellis. 7 vols. Vol. 4, London: Longmans, 1858.

Bain, Read. "Technology and State Government." *American Sociological Review* 2 (1937): 860–74.

Balabkins, Nicholas W. *Not by Theory Alone . . . The Economics of Gustav von Schmoller and Its Legacy to America.* Berlin: Duncker and Humblot, 1988.

Ball, Terence. *Transforming Political Discourse: Political Theory and Conceptual History.* Oxford: Basil Blackwell, 1988.

Bannister, Robert C. *Sociology and Scientism: The American Quest for Objectivity, 1880–1940.* Chapel Hill: University of North Carolina Press, 1987.

Banse, Gerhard, and Hans-Peter Müller. *Johann Beckmann und die Folgen: Erfindungen, Versuch der historischen, theoretischen und empirischen Annäherung an einen vielschichtigen Begriff.* Münster: Waxmann, 2001.

Beard, Charles Austin. Introduction to *The Idea of Progress,* by J. B. Bury, ix–xl. New York: Dover Publications, 1955. First published 1932.

———. "Time, Technology, and the Creative Spirit in Political Science." *American Political Science Review* 21, no. 1 (1927): 1–11.

———, ed. *Toward Civilization.* London: Longmans, Green, 1930.

———, ed. *Whither Mankind: A Panorama of Modern Civilization.* New York: Longmans, Green, 1928.

Beard, Charles Austin, and William Beard. *The American Leviathan: The Republic in the Machine Age.* New York: Macmillan, 1931.

Beard, William. "Technology and Political Boundaries." *American Political Science Review* 25, no. 3 (1931): 557–72.

Beaune, Jean-Claude. *La technologie introuvable: Recherche sur la définition et l'unité de la technologie à partir de quelques modèles du XVIIIe et XIXe siècles.* Paris: Librairie Philosophique J. Vrin, 1980.

Beck, Paul. "U.S. Faces 'Technology Gap' with Russ, GOP Forum Says." *Los Angeles Times,* 1 April 1964.

Beckert, Manfred. *Johann Beckmann.* Leipzig: B. G. Teubner, 1983.

Beckmann, Johann. *A History of Inventions and Discoveries*. Translated by William Johnston. 3 vols. London: Printed for J. Bell, 1797.

———. *Anleitung zur Technologie; oder, zur Kentniß der Handwerke, Fabriken und Manufacturen: vornehmlich derer, die mit der Landwirthschaft, Polizey und Cameralwissenschaft in nächster Verbindung stehn—nebst Beyträgen zur Kunstgeschichte*. Göttingen: Wittwe Vandenhoeck, 1777.

———. *Anleitung zur Technologie; oder, zur Kentniß der Handwerke, Fabriken und Manufacturen*. 2nd ed. Göttingen: Wittwe Vandenhoeck, 1780.

Behrent, Michael C. "Foucault and Technology." *History and Technology* 29, no. 1 (2013): 54–104.

Bell, Daniel. "Introduction to the Harbinger Edition." In *The Engineers and the Price System*, by Thorstein Veblen, 1–35. New York: Harcourt, Brace and World, 1963.

Bennett, Jim. "The Mechanical Arts." Chap. 27 in *The Cambridge History of Science: Early Modern Science*, edited by Katharine Park and Lorraine Daston, 673–95. Cambridge: Cambridge University Press, 2006.

Bentham, Jeremy. *Chrestomathia*. Oxford: Clarendon Press, 1983. Posthumous ed., 1843; orig. pub. between 1815 and 1817.

Berg, Maxine. *The Age of Manufactures, 1700–1820*. 2nd ed. London: Routledge, 1994.

———. "The Genesis of 'Useful Knowledge.'" *History of Science* 45 (2007): 123–33.

———. *The Machinery Question and the Making of Political Economy*. Cambridge: Cambridge University Press, 1980.

Bernstein, Eduard. *The Preconditions of Socialism*. Translated by Henry Tudor. Cambridge: Cambridge University Press, 1993.

Bernstein, Richard J. *Beyond Objectivism and Relativism: Science, Hermeneutics, and Praxis*. Philadelphia: University of Pennsylvania Press, 1988.

———. "Heidegger's Silence? *Ēthos* and Technology." Chap. 4 in *The New Constellation: The Ethical-Political Horizons of Modernity/Postmodernity*, 79–141. Cambridge, MA: MIT Press, 1992.

Bevir, Mark. "Review: Begriffsgeschichte." *History and Theory* 39 (2000): 273–84.

Bickerman, Elias J. Review of *A History of Technology*, vol 1. *American Journal of Philology* 77, no. 1 (1956): 96–100.

Bigelow, Jacob. *An Address on the Limits of Education, Read before the Massachusetts Institute of Technology, November 16, 1865*. Boston: E. P. Dutton, 1865.

———. *Elements of Technology: Taken Chiefly from a Course of Lectures Delivered at Cambridge, on the Application of the Sciences to the Useful Arts; Now Published for the Use of Seminaries and Students*. Boston: Hilliard Gray Little and Wilkins, 1829.

———. *The Useful Arts: Considered in Connexion with the Applications of Science*. Boston: Marsh, Capen, Lyon, and Webb, 1840. Reprint, New York: Arno Press, 1972.

Bijker, Wiebe E., Thomas Parke Hughes, and T. J. Pinch, eds. *The Social Construction of Technological Systems: New Directions in the Sociology and History of Technology.* Cambridge, MA: MIT Press, 1987.

Billig, Michael. *Learn to Write Badly: How to Succeed in the Social Sciences.* Cambridge: Cambridge University Press, 2013.

Bix, Amy Sue. *Inventing Ourselves Out of Jobs? America's Debate over Technological Unemployment 1929–1981.* Baltimore: Johns Hopkins University Press, 2000.

Blount, Thomas. *Glossographia; or, A dictionary interpreting all such hard words of whatsoever language now used in our refined English tongue with etymologies, definitions and historical observations on the same: Also the terms of divinity, law, physick, mathematicks and other arts and sciences explicated.* 2nd ed. London: Printed by Tho. Newcombe for George Sawbridge, 1661.

Böhm-Bawerk, Eugen von. *Positive Theorie des Kapitales.* Innsbruck: Wagner, 1889.

———. *The Positive Theory of Capital.* Translated by William Smart. New York: G. E. Stechert, 1891.

The Book of Knowledge, for All Classes: With Familiar Illustrations. Hartford: S. Andrus, 1848.

Bookchin, Murray. "Towards a Liberatory Technology." In *Post-Scarcity Anarchism,* 85–139. Berkeley, CA: Ramparts Press, 1971.

Bovey, Henry Taylor. "The Fundamental Conceptions Which Enter into Technology." In *Congress of Arts and Science, Universal Exposition, St. Louis, 1904,* edited by Howard Jason Rogers, 6:535–51. Boston: Houghton Mifflin, 1905.

Boyer, Paul S. *By the Bomb's Early Light: American Thought and Culture at the Dawn of the Atomic Age.* Chapel Hill: University of North Carolina Press, 1994.

Brady, Mark. "Yale Brozen 1917–1998." *Economic Affairs* 18, no. 3 (1 September 1998): 59.

Brainerd, J. G. Review of *Technological Trends and National Policy. Annals of the American Academy of Political and Social Science* 196 (1938): 234–35.

Braun, Hans-Joachim. "Allgemeine Fragen der Technik an der Wende zum 20. Jahrhundert: Zum Werk P. K. von Engelmeyers." *Technikgeschichte* 42 (1975): 306–26.

———. "Technik als 'Kulturhebel' und 'Kulturfactor': Zum Verhältnis von Technik und Kultur bei Franz Reuleaux." In *Technische Intelligenz und "Kulturfaktor Technik,"* edited by Burkhard Dietz, Michael Fessner, and Helmut Maier, 35–43. Münster: Waxmann, 1996.

Brazelton, W. Robert. "Alvin Harvey Hansen: A Note on His Analysis of Keynes, Hayek, and Commons." *Journal of Economic Issues* 27, no. 3 (1993): 940–48.

———. "Alvin Harvey Hansen: Economic Growth and a More Perfect Society; The Economist's Role in Defining the Stagnation Thesis and in Popularizing Keynesianism." *American Journal of Economics and Sociology* 48, no. 4 (1989): 427–40.

Breslau, Daniel. "The Scientific Appropriation of Social Research: Robert Park's Human Ecology and American Sociology." *Theory and Society* 19, no. 4 (1990): 417–46.

Brette, Olivier. "Thorstein Veblen's Theory of Institutional Change: Beyond Technological Determinism." *European Journal of the History of Economic Thought* 10, no. 3 (2003): 455–77.

Breul, Karl. *Cassell's New German and English Dictionary.* Rev. and enl. ed. New York: Funk and Wagnalls, 1936, 1939.

Brey, Philip. "Theorizing Modernity and Technology." Chap. 2 in *Modernity and Technology,* edited by Thomas J. Misa, Philip Brey, and Andrew Feenberg, 33–71. Cambridge, MA: MIT Press, 2003.

Bright, Arthur A. *The Electric-Lamp Industry: Technological Change and Economic Development from 1800 to 1947.* New York: Macmillan, 1949.

Bright, James R. "Opportunity and Threat in Technological Change." *Harvard Business Review* 41, no. 6 (1963): 76–86.

Brockhaus Enzyklopädie. Vol. 18, Wiesbaden: F. A. Brockhaus, 1973.

Brockhaus' Konversations-Lexikon. 14th ed. Vol. 15, Leipzig: F. A. Brockhaus, 1903.

Brown, John K. "Louis C. Hunter, Steamboats on the Western Rivers (Classics Revisited)." *Technology and Culture* 44, no. 4 (2003): 786–93.

Brozen, Yale. "Adapting to Technological Change." *Journal of Business of the University of Chicago* 24, no. 2 (1951): 114–26.

———. "Studies of Technological Change." *Southern Economic Journal* 17, no. 4 (1951): 438–50.

Buchanan, W. M. *A Technological Dictionary: Explaining the Terms of the Arts, Sciences, Literature, Professions, and Trades.* London: W. Tegg, 1846.

Bücher, Karl. *Arbeit und Rhythmus.* 2nd ed. Leipzig: B. G. Teubner, 1899.

———. *Die Entstehung der Volkswirtschaft: Vorträge und Versuche.* 3rd ed. Tübingen: Verlag der H. Laupp'schen Buchhandlung, 1901.

———. *Industrial Evolution.* Translated by S. Morley Wickett. New York: H. Holt, 1901.

Buchholz, Arnold. "The Scientific-Technological Revolution (STR) and Soviet Ideology." *Studies in Soviet Thought* 30, no. 4 (1985): 337–46.

Buckley, Oliver E. "Frank Baldwin Jewett, 1879–1949." *Biographical Memoirs of the National Academy of Sciences* 27 (1952): 239–63.

Bud, Robert. "'Applied Science' in Nineteenth-Century Britain: Public Discourse and the Creation of Meaning, 1817–1876." *History and Technology* 30, no. 1/2 (2014): 3–36.

———. "'Applied Science': A Phrase in Search of a Meaning." *Isis* 103, no. 3 (2012): 537–45.

Bud, Robert, and Gerrylynn K. Roberts. *Science versus Practice.* Manchester: Manchester University Press, 1984.

Bukharin, Nikolai Ivanovich. *Historical Materialism: A System of Sociology.* New York: International Publishers, 1925. Reprint, New York: Russell and Russell, 1965. Page references are to the 1965 edition.

———. *Teoriia istoricheskogo materializma: Populiarnyi uchebnik marksistskoi sotsiologii.* [Moscow]: Gosudarstvennoe izdatel'stvo, 1921.

———. "Theory and Practice from the Standpoint of Dialectical Materialism." In *Science at the Cross Roads,* 11–33.

Bulmer, Martin. "The Methodology of Early Social Indicator Research: William Fielding Ogburn and 'Recent Social Trends,' 1933." *Social Indicators Research* 13, no. 2 (1983): 109–30.

Bunge, Mario. "Technology as Applied Science." *Technology and Culture* 7, no. 3 (1966): 329–47.

Burford, Alison. *Craftsmen in Greek and Roman Society.* [London]: Thames and Hudson, 1972.

Burkitt, Ian. "Technologies of the Self: Habitus and Capacities." *Journal for the Theory of Social Behaviour* 32, no. 2 (June 2002): 219–37.

Burton, Richard Francis. "The Lake Regions of Central Equatorial Africa, with Notices of the Lunar Mountains and the Sources of the White Nile; Being the Results of an Expedition Undertaken under the Patronage of Her Majesty's Government and the Royal Geographical Society of London, in the Years 1857–1859." *Journal of the Royal Geographical Society of London* 29 (1859): 1–454.

———. *Mission to Gelele, King of Dahome; with Notices of the So-called "Amazons," the Grand Customs, the Yearly Customs, the Human Sacrifices, the Present State of the Slave Trade, and the Negro's Place in Nature.* 2 vols. Vol. 2, London: Tinsley Bros., 1864.

Bury, J. B. *The Idea of Progress.* New York: Macmillan, 1932. Reprint, New York: Dover Publications, 1960.

Bush, Vannevar. *Science, the Endless Frontier.* Washington, DC: Government Printing Office, 1945. Reprint, Washington, DC: National Science Foundation, 1960. Page references are to the 1960 edition.

Butterworth, Benjamin. *The Growth of Industrial Art.* Washington, DC: Government Printing Office, 1888.

Calhoun, Daniel Hovey. *The American Civil Engineer: Origins and Conflict.* Cambridge, MA: Harvard University Press, 1960.

Camic, Charles. "Alexander's Antisociology." *Sociological Theory* 14, no. 2 (July 1996): 172–86.

———. "Introduction: Talcott Parsons before *The Structure of Social Action.*" In *Talcott Parsons: The Early Essays,* ix–lxix. Chicago: University of Chicago Press, 1991.

———. "On Edge: Sociology during the Great Depression and New Deal." Chap. 7 in *Sociology in America: A History,* edited by Craig Calhoun, 225–80. Chicago: University of Chicago Press, 2007.

————. "Reputation and Predecessor Selection: Parsons and the Institutionalists." *American Sociological Review* 57, no. 4 (1992): 421–45.

————. "*Structure* after 50 Years: The Anatomy of a Charter." *American Journal of Sociology* 95 (July 1989): 38–107.

————. "Veblen's Apprenticeship: On the Translation of Gustav Cohn's System der Finanzwissenschaft." *History of Political Economy* 42 (2010): 679–721.

Camic, Charles, Philip S. Gorski, and David M. Trubek, eds. *Max Weber's Economy and Society: A Critical Companion.* Stanford, CA: Stanford University Press, 2005.

Carlyle, Thomas. *Past and Present.* London: Chapman and Hall, 1870. First published 1843.

————. "Signs of the Times." *Edinburgh Review* 49, no. 98 (June 1829): 439–59.

Carson, Rachel. *Silent Spring.* New York: Houghton Mifflin, 1962.

Casini, Lorenzo. "Juan Luis Vives [Joannes Ludovicus Vives]." In *The Stanford Encyclopedia of Philosophy*, edited by Edward N. Zalta, 2012. http://plato .stanford.edu/archives/win2012/entries/vives/.

Castells, Manuel. *The Rise of the Network Society.* Cambridge, MA: Blackwell, 1996.

Chambers, Ephraim. *Cyclopædia; or, An Universal Dictionary of Arts and Sciences.* 2 vols. Vol. 1, London: Printed for J. and J. Knapton, 1728.

Channell, David F. "The Harmony of Theory and Practice: The Engineering Science of W. J. M. Rankine." *Technology and Culture* 23, no. 1 (1982): 39–52.

Chase, Stuart. *Men and Machines.* New York: Macmillan, 1929.

Chaussard, Pierre Jean-Baptiste. *Ode philosophique sur les arts industriels.* Paris: De l'Imprimerie des sciences et arts . . . 1799.

Chicago Daily Tribune. "Warns Soviet Technology Is Threat to U.S." 2 April 1959.

Childe, V. Gordon. "Archaeological Ages as Technological Stages." *Journal of the Royal Anthropological Institute of Great Britain and Ireland* 74, no. 1/2 (1944): 7–24.

Chisholm, Hugh, ed. *The Encyclopædia Britannica: A Dictionary of Arts, Sciences, Literature and General Information.* 11th ed. 29 vols. New York: Encyclopædia Britannica, 1910.

Christensen, Kathleen. "Cottage Industry and Women." *Chicago Tribune*, 22 September 1983.

Cicero, Marcus Tullius. *On Obligations.* Translated by P. G. Walsh. Oxford: Oxford University Press, 2000.

Clark, G. N. *Science and Social Welfare in the Age of Newton.* Oxford: Clarendon Press, 1937.

Clarke, W. Norris, SJ. "Technology and Man: A Christian Vision." *Technology and Culture* 3, no. 4 (1962): 422–42.

Cohen, H. Floris. *The Scientific Revolution: A Historiographical Inquiry.* Chicago: University of Chicago Press, 1994.

Cohen, Stephen F. *Bukharin and the Bolshevik Revolution: A Political Biography, 1888–1938.* New York: Alfred A. Knopf, 1973.

Cohn, Gustav. *The Science of Finance.* Translated by Thorstein Veblen. Chicago: University of Chicago Press, 1895.

———. *System der Finanzwissenschaft: Ein Lesebuch für Studierende.* Stuttgart: F. Enke, 1889.

Cole, Arthur H. "Committee on Research in Economic History: A Description of Its Purposes, Activities, and Organization." *Journal of Economic History* 13, no. 1 (1953): 79–87.

Collins, H. M. *Tacit and Explicit Knowledge.* Chicago: University of Chicago Press, 2013.

Collins, Roger. "Making Sense of the Early Middle Ages." *English Historical Review* 124, no. 508 (June 1, 2009): 641–65.

Colvin, Sidney. "Art." In *Encyclopædia Britannica: A Dictionary of Arts, Sciences, Literature and General Information,* edited by Hugh Chisholm, 2:657–60. 11th ed. New York: Encyclopædia Britannica, 1910.

Committee of Associated Institutions of Science and Arts. *Objects and Plan of an Institute of Technology: Including a Society of Arts, a Museum of Arts, and a School of Industrial Science, Proposed to be Established in Boston.* [Boston?]: s.n., 1860.

Commoner, Barry. *The Closing Circle: Nature, Man, and Technology.* New York: Knopf, 1971.

———. *Science and Survival.* New York: Viking, 1966.

Commons, John R., and Selig Perlman. "Review of *Der moderne Kapitalismus,* by Werner Sombart." *American Economic Review* 19, no. 1 (1929): 78–88.

Compton, Karl T. "Inaugural Address." *Science* 71, no. 1850 (1930): 593–96.

Conant, James B. "Chemists and the National Defense." *Science* 94, no. 2450 (12 December 1941): 563–64.

Condit, Carl W. "Modern Architecture: A New Technical-Aesthetic Synthesis." *Journal of Aesthetics and Art Criticism* 6, no. 1 (1947): 45–54.

Conversations-Lexikon: Allgemeine deutsche Real-Encyklopädie. 12th ed. 15 vols. Vol. 14, Leipzig: Brockhaus, 1879.

Cordes, Christian. "Veblen's 'Instinct of Workmanship,' Its Cognitive Foundations, and Some Implications for Economic Theory." *Journal of Economic Issues* 39, no. 1 (2005): 1–20.

Cowan, Ruth Schwartz. "Looking Back in Order to Move Forward: John McDermott, 'Technology: The Opiate of the Intellectuals.'" *Technology and Culture* 51, no. 1 (2010): 199–215.

Crabb, George. *Universal Technological Dictionary; or, Familiar Explanation of the Terms Used in All Arts and Sciences.* 2 vols. London: Baldwin, Cradock, and Joy, 1823.

Crook, Wilfrid H. Review of *Technological Trends and National Policy. American Sociological Review* 2, no. 6 (1937): 957.

Cuomo, S. *Technology and Culture in Greek and Roman Antiquity.* Cambridge: Cambridge University Press, 2007.

Darnton, Robert. "Philosophers Trim the Tree of Knowledge: The Epistemological Strategy of the *Encyclopédie.*" Chap. 5 in *The Great Cat Massacre and*

Other Episodes in French Cultural History, 191–214. New York: Vintage Books, 1984.

Davenport, H. J. "Capital as a Competitive Concept." *Journal of Political Economy* 13, no. 1 (December 1904): 31–47.

David, Paul A. "Path Dependence, Its Critics and the Quest for 'Historical Economics.'" Chap. 2 in *Evolution and Path Dependence in Economic Ideas: Past and Present*, edited by Pierre Garrouste and Stavros Ioannides, 15–40. Cheltenham, UK: Edward Elger, 2001.

———. *Technical Choice, Innovation and Economic Growth: Essays on American and British Experience in the Nineteenth Century*. London: Cambridge University Press, 1975.

Davis, Gregory H. *Means without End: A Critical Survey of the Ideological Genealogy of Technology without Limits, from Apollonian Techne to Postmodern Technoculture*. Lanham, MD: University Press of America, 2006.

Dear, Peter. "Mixed Mathematics." Chap. 6 in *Wrestling with Nature: From Omens to Science*, edited by Peter Harrison, Ronald L. Numbers, and Michael H. Shank, 149–72. Chicago: University of Chicago Press, 2011.

———. *Revolutionizing the Sciences: European Knowledge and Its Ambitions, 1500–1700*. 2nd ed. Princeton, NJ: Princeton University Press, 2009.

Dennis, Michael Aaron. "Accounting for Research: New Histories of Corporate Laboratories and the Social History of American Science." *Social Studies of Science* 17, no. 3 (1987): 479–518.

Der große Brockhaus. 15th ed. Vol. 18, Leipzig: Brockhaus, 1934.

Derksen, Maarten, and Tjardie Wierenga. "The History of 'Social Technology,' 1898–1930." *History and Technology* 29, no. 4 (2013): 311–30.

Desch, Cecil H. "Pure and Applied Science." *Science* 74, no. 1925 (1931): 495–502.

Dessauer, Friedrich. *Streit um die Technik*. Frankfurt: Verlag Josef Knecht, 1958.

Detienne, Marcel, and Jean-Pierre Vernant. *Cunning Intelligence in Greek Culture and Society*. Hassocks, UK: Harvester Press, 1978.

Dewey, John. *The Quest for Certainty: A Study of the Relation of Knowledge and Action*. New York: Minton, Balch, 1929.

Diderot, Denis. "Art (Applied Natural History)." In *The Encyclopedia of Diderot and d'Alembert Collaborative Translation Project*. Ann Arbor: Michigan Publishing, University of Michigan Library, 2003. http://hdl.handle.net/2027/spo.did2222.0000.139.

Diderot, Denis, and Jean Le Rond d'Alembert. *Encyclopédie, ou dictionnaire raisonné des sciences, des arts et des métiers, etc.* Edited by Robert Morrissey. Spring 2013 ed. Chicago: ARTFL Encyclopédie Project, 1751–1772. http://encyclopedie.uchicago.edu/.

Dietz, Burkhard, Michael Fessner, and Helmut Maier, eds. *Technische Intelligenz und "Kulturfaktor Technik": Kulturvorstellungen von Technikern und Ingenieuren zwischen Kaiserreich und früher Bundesrepublik Deutschland*. Münster: Waxmann, 1996.

Dietz, David. *Atomic Energy in the Coming Era.* New York: Dodd, Mead, 1945.

Diggins, John P. *The Bard of Savagery: Thorstein Veblen and Modern Social Theory.* New York: Seabury Press, 1978.

Dorfman, Joseph. "The Role of the German Historical School in American Economic Thought." *American Economic Review* 45, no. 2 (1955): 17–28.

———. *Thorstein Veblen and His America.* New York: Viking Press, 1934.

"Dr. Jul. Post, Arbeit statt Almosen [review]." *Die neue Zeit: Revue des geistigen und öffentlichen Lebens* 3, no. 6 (1886): 283.

Drucker, Peter F. "The Technological Revolution: Notes on the Relationship of Technology, Science, and Culture." *Technology and Culture* 2, no. 4 (Autumn 1961): 342–51.

———. "Work and Tools." *Technology and Culture* 1, no. 1 (1959): 28–37.

Du Gay, Paul. *Doing Cultural Studies: The Story of the Sony Walkman.* London: Sage, in association with the Open University, 1997.

Dunn, Alan. "Utopia—1955." *Saturday Evening Post,* 30 December 1944, 36–37.

Durbin, Paul T. "Philosophy of Technology: In Search of Discourse Synthesis." *Techné: Research in Philosophy and Technology* 10, no. 2 (2006): 1–288.

Eco, Umberto. *The Aesthetics of Thomas Aquinas.* Cambridge, MA: Harvard University Press, 1988.

———. "In Praise of Thomas Aquinas." *Wilson Quarterly* 10, no. 4 (1986): 78–87.

Edebohls, George M. "The Technique of Vaginal Hysterectomy." *American Journal of the Medical Sciences* 109 (1895): 42–49.

Edgell, Stephen. *Veblen in Perspective: His Life and Thought.* Armonk, NY: M.E. Sharpe, 2001.

Edgell, Stephen, and Rick Tilman. "The Intellectual Antecedents of Thorstein Veblen: A Reappraisal." *Journal of Economic Issues* 23, no. 4 (1989): 1003–26.

Edgerton, David. *Britain's War Machine: Weapons, Resources, and Experts in the Second World War.* London: Penguin, 2012.

———. "From Innovation to Use: Ten Eclectic Theses on the Historiography of Technology." *History and Technology* 16, no. 2 (1999): 111–36.

———. "'The Linear Model' Did Not Exist: Reflections on the History and Historiography of Science and Research in Industry in the Twentieth Century." In *The Science-Industry Nexus: History, Policy, Implications,* edited by Karl Grandin, Nina Wormbs, and Sven Widmalm, 31–57. Sagamore Beach, MA: Science History Publications/USA, 2004.

———. *The Shock of the Old: Technology and Global History since 1900.* Oxford: Oxford University Press, 2007.

Edsforth, Ronald. *The New Deal: America's Response to the Great Depression.* Malden, MA: Wiley, 2000.

Edwards, Paul N. "Infrastructure and Modernity: Force, Time, and Social Organization in the History of Sociotechnical Systems." In *Modernity and Technology,* edited by Thomas J. Misa, Philip Brey, and Andrew Feenberg, 185–225. Cambridge, MA: MIT Press, 2003.

Edwards, Steve. "Factory and Fantasy in Andrew Ure." *Journal of Design History* 14, no. 1 (2001): 17–33.

Eger, Gustav. *Technologisches Wörterbuch in englischer und deutscher Sprache.* Braunschweig: F. Vieweg und Sohn, 1882.

Great Britain. Department of Science and Art. *Eleventh Report of the Science and Art Department of the Committee of Council on Education.* London: G. E. Eyre and W. Spottiswoode, 1864.

Ellul, Jacques. "The Technological Order." *Technology and Culture* 3, no. 4 (1962): 394–421.

———. *The Technological Society.* Translated by John Wilkinson. New York: Knopf, 1964.

Elsner, Henry. *The Technocrats: Prophets of Automation.* [Syracuse, NY]: Syracuse University Press, 1967.

Ely, Richard T. "The New Economic World and the New Economics." *Journal of Land and Public Utility Economics* 5, no. 4 (1929): 341–53.

Emerson, G. B. "[Review of Bigelow, *Elements of Technology*]." *North American Review* 30 (April 1830): 337–60.

"The Encyclopaedia Britannica Conference on the Technological Order." Special issue, edited by Carl F. Stover. *Technology and Culture* 3, no. 4 (Fall 1962).

Endres, Anthony M., and M. Donoghue. "Defending Marshall's 'Masterpiece': Ralph Souter's Critique of Robbins' Essay." *Cambridge Journal of Economics* 34 (2010): 547–68.

Engelhardt, Viktor. "Technik und soziale Ethik." *Sozialistische Monatshefte* 27 (1921): 481–85.

Engelmeyer, Peter Klimentisch von. "Allgemeine Fragen der Technik." *Dinglers Polytechnisches Journal* 311 (1899): 21–22, 69–71, 101–3, 133–34, 149–51; 312 (1899): 1–2, 65–67, 97–99, 129–30, 145–47; 313 (1899): 17–19, 65–67.

Engels, Friedrich. "Engels an W. Borgius." In Marx and Engels, *Werke,* 39:205–7.

Epstein, Ralph C. "Industrial Invention: Heroic, or Systematic?" *Quarterly Journal of Economics* 40, no. 2 (1926): 232–72.

Epstein, S. R. "Craft Guilds in the Pre-Modern Economy: A Discussion." *Economic History Review* 61 (2008): 155–74.

Epstein, Steven A. "Urban Society." Chap. 1(b) in *The New Cambridge Medieval History,* edited by David Abulafia, 5:26–37. Cambridge: Cambridge University Press, 1999.

Evans, Richard J. *Altered Pasts: Counterfactuals in History.* Menahem Stern Jerusalem Lectures. Waltham, MA: Brandeis University Press / Historical Society of Israel, 2013.

"Exxon Mobil Meets Amazon.com." *Business Week,* 14 December 1998, 178.

Farrar, W. V. "Andrew Ure, F.R.S., and the Philosophy of Manufactures." *Notes and Records of the Royal Society of London* 27, no. 2 (February 1973): 299–324.

Feenberg, Andrew. "Introduction to the Second Edition." In *One-Dimensional Man: Studies in the Ideology of Advanced Industrial Society,* xi–xxix. Boston: Beacon Press, 1991.

———. "Modernity Theory and Technology Studies: Reflections on Bridging the Gap." Chap. 7 in *Between Reason and Experience: Essays in Technology and Modernity*, 129–56. Cambridge, MA: MIT Press, 2010.

———. *Questioning Technology*. London: Routledge, 1999.

Feibleman, James K. "Pure Science, Applied Science, Technology, Engineering: An Attempt at Definitions." *Technology and Culture* 2, no. 4 (Autumn 1961): 305–17.

Fellowes, F. Wayland. "What Technique Does for a Picture." *New Englander and Yale Review* 55 (1890): 393–95.

Ferguson, Eugene S. *Bibliography of the History of Technology*. Cambridge, MA: MIT Press, 1968.

Flinn, Alfred D. "The Relation of the Technical School to Industrial Research." *Science* 54, no. 1404 (1921): 508–10.

Florman, Samuel C. *The Existential Pleasures of Engineering*. New York: St. Martin's Press, 1976.

Flower, J. W. "On the Relative Ages of the Stone Implement Periods in England." *Journal of the Anthropological Institute of Great Britain and Ireland* 1 (1872): 274–95.

Forman, Paul. "Behind Quantum Electronics: National Security as Basis for Physical Research in the United States, 1940–1960." *Historical Studies in the Physical and Biological Sciences* 18, no. 1 (1987): 149–229.

———. "The Primacy of Science in Modernity, of Technology in Postmodernity, and of Ideology in the History of Technology." *History and Technology* 23, no. 1/2 (March/June 2007): 1–152.

Formigari, Lia. *A History of Language Philosophies*. Translated by Gabriel Poole. Amsterdam: Benjamins, 2004.

Foster, Patrick N. "The Founders of Industrial Arts in the US." *Journal of Technology Education* 7, no. 1 (Fall 1995): 6–21.

———. "Industrial Arts/Technology Education as a Social Study: The Original Intent?" *Journal of Technology Education* 6, no. 2 (Spring 1995): 4–18.

Foucault, Michel. "About the Beginning of the Hermeneutics of the Self: Two Lectures at Dartmouth." *Political Theory* 21, no. 2 (1993): 198–227.

———. *The Foucault Reader*. Edited by Paul Rabinow. New York: Pantheon Books, 1984.

———. *The Hermeneutics of the Subject: Lectures at the Collège de France, 1981–82*. New York: Palgrave Macmillan, 2005.

———. *The History of Sexuality*. Translated by Robert Hurley. 3 vols. New York: Pantheon Books, 1978–88.

———. *L'herméneutique du sujet: Cours au Collège de France (1981–1982)*. [Paris?]: Gallimard, 2001.

———. "Nietzsche, Genealogy, History." In *The Foucault Reader*, edited by Paul Rabinow, 76–100. New York: Pantheon Books, 1984.

———. "Technologies of the Self." In *Technologies of the Self: A Seminar with Michel Foucault*, edited by Luther H. Martin, Huck Gutman, and Patrick H. Hutton, 16–49. Amherst: University of Massachusetts Press, 1988.

Frederick, J. George, ed. *For and Against Technocracy.* New York: Business Bourse, 1933.

Freudenthal, Gideon, and Peter McLaughlin. "Classical Marxist Historiography of Science: The Hessen-Grossmann-Thesis." In Freudenthal and McLaughlin, *The Social and Economic Roots of the Scientific Revolution,* 1–38.

———, eds. *The Social and Economic Roots of the Scientific Revolution: Texts by Boris Hessen and Henryk Grossmann.* Boston Studies in the Philosophy of Science, vol. 278. Dordrecht: Springer, 2009.

Friedel, Robert. *A Culture of Improvement: Technology and the Western Millennium.* Cambridge, MA: MIT Press, 2007.

Friedman, Thomas L. "U.S.G. and P.T.A." *New York Times,* 23 November 2010.

———. *The World Is Flat: A Brief History of the Twenty-First Century.* New York: Farrar, Straus and Giroux, 2005.

Fries, Sylvia Doughty. "Expertise against Politics: Technology as Ideology on Capitol Hill, 1966–1972." *Science, Technology, and Human Values* 8, no. 2 (1983): 6–15.

Frison, Guido. "Linnaeus, Beckmann, Marx and the Foundations of Technology: Between Natural and Social Sciences; A Hypothesis of an Ideal Type." *History and Technology* 10 (1993): 139–60, 161–73.

———. "Smith, Marx and Beckmann: Division of Labour, Technology and Innovation." In *Technologie zwischen Fortschritt und Tradition: Beiträge zum internationalen Johan Beckmann-Symposium Göttingen 1989,* edited by Hans-Peter Müller and Ulrich Troitzsch, 17–40. Frankfurt am Main: Peter Lang, 1992.

———. "Some German and Austrian Ideas on *Technologie* and *Technik* between the End of the Eighteenth Century and the Beginning of the Twentieth." *History of Economic Ideas* 6, no. 1 (1998): 107–33.

———. "Technical and Technological Innovation in Marx." *History and Technology* 6 (1988): 299–324.

Gadamer, Hans-Georg. *Truth and Method.* New York: Crossroad, 1990.

Galen. "Exhortation to the Study of the Arts Especially Medicine: To Menodotus." *Medical Life* 37 (1930): 507–29. Published electronically April 1996. http://www.ucl.ac.uk/~ucgajpd/medicina antiqua/tr_GalExhort.html.

Galison, Peter. *How Experiments End.* Chicago: University of Chicago Press, 1987.

Gallagher, Cornelius Edward. "Technology and Society." *Vital Speeches of the Day* 35 (15 June 1969): 528–33.

Galluzzi, Paolo. *Renaissance Engineers: From Brunelleschi to Leonardo da Vinci.* Florence: Istituto e Museo di Storia della Scienza, 1996.

Geddes, Patrick. *Cities in Evolution: An Introduction to the Town Planning Movement and to the Study of Civics.* London: Williams and Norgate, 1915.

———. "The Twofold Aspect of the Industrial Age: Paleotechnic and Neotechnic." *Town Planning Review* 3, no. 3 (1912): 176–87.

Gelder, Robert van. "Being Harvard's President, Aiding National Defense and Acting as Private Citizen Keep One Man Busy." *New York Times,* 20 July 1941.

Gerrie, Jim. "Was Foucault a Philosopher of Technology?" *Techné: Research in Philosophy and Technology* 7, no. 2 (2003): 66–73.

Ghosh, Peter. "Max Weber, Werner Sombart and the Archiv für Sozialwissenschaft: The Authorship of the 'Geleitwort' (1904)." *History of European Ideas* 36, no. 1 (2010): 71–100.

Gieryn, Thomas F. *Cultural Boundaries of Science.* Chicago: University of Chicago Press, 1999.

Gies, Frances, and Joseph Gies. *Cathedral, Forge, and Waterwheel: Technology and Invention in the Middle Ages.* New York: HarperCollins, 1994.

Gilbert, Humphrey. *Queene Elizabethes Achademy.* Edited by Frederick James Furnivall. London: N. Trübner, 1869.

Gilfillan, S. Colum. *The Sociology of Invention: An Essay in the Social Causes of Technic Invention and Some of Its Social Results.* Chicago: Follett, 1935.

———. "Who Invented It?" *Scientific Monthly* 25, no. 6 (December 1927): 529–34.

Gill, Corrington. "WPA Studies 'Foe,' the Swift Machine." *New York Times,* May 31, 1936.

Gill, Corrington, and David Weintraub. *Unemployment and Technological Change.* Philadelphia: Work Projects Administration, 1940.

Gille, Bertrand. *Histoire des techniques: Technique et civilisations, technique et sciences.* [Paris]: Gallimard, 1978.

———. *History of Techniques.* New York: Gordon and Breach, 1986.

Gimpel, Jean. *The Medieval Machine: The Industrial Revolution of the Middle Ages.* New York: Holt, Rinehart and Winston, 1976.

Gispen, Kees. *New Profession, Old Order: Engineers and German Society, 1815–1914.* Cambridge: Cambridge University Press, 1989.

Glick, Thomas F. "Technology." In *Islamic and Christian Spain in the Early Middle Ages,* 247–94. Princeton, NJ: Princeton University Press, 1979.

Godin, Benoît. "In the Shadow of Schumpeter: W. Rupert Maclaurin and the Study of Technological Innovation." *Minerva* 46, no. 3 (2008): 343–60.

———. "Innovation without the Word: William F. Ogburn's Contribution to the Study of Technological Innovation." *Minerva* 48, no. 3 (2010): 277–307.

———. "The Linear Model of Innovation: The Historical Construction of an Analytical Framework." *Science, Technology and Human Values* 31, no. 6 (2006): 639–67.

———. *Narratives of Innovation: The Invention of Models, 1920–1980.* Cambridge, MA: MIT Press, 2017.

———. "Technological Change: What Do Technology and Change Stand For?" Project on the Intellectual History of Innovation, Working Paper no. 24, 2015, accessed 2/28/18, http://www.csiic.ca/wp-content/uploads/2015/12/Paper24_TC.pdf.

Goelz, Paul C. "Toward a Concept of Education for Administration." *Journal of the Academy of Management* 1, no. 1 (1958): 62–63.

Goldberg, Stanley. "Inventing a Climate of Opinion: Vannevar Bush and the Decision to Build the Bomb." *Isis* 83, no. 3 (1992): 429–52.

Gomulka, Stanislaw. *The Theory of Technological Change and Economic Growth.* Hoboken, NJ: Taylor and Francis, 1990.

Good, John Mason. "On Medical Technology." *Transactions of the Medical Society of London* 1, pt. 1 (1810): 1–50.

Gooday, Graeme. "'Vague and Artificial': The Historically Elusive Distinction between Pure and Applied Science." *Isis* 103, no. 3 (2012): 546–54.

Gooding, Judson. "Engineers Are Redesigning Their Own Profession." *Fortune,* June 1971, 72.

Gordon, Robert J. *The Rise and Fall of American Growth: The U.S. Standard of Living since the Civil War.* Princeton, NJ: Princeton University Press, 2016.

Götz, Nikolaus [Ernst Niekisch]. "Menschenfresser Technik (1931)." In *Widerstand,* edited by Uwe Sauermann, 56–65. Krefeld: SINUS-Verlag, 1982.

Gouldner, Alvin W. "Explorations in Applied Social Science." *Social Problems* 3, no. 3 (1956): 169–81.

Gourvitch, Alexander. *Survey of Economic Theory on Technological Change and Employment.* New York: A. M. Kelley, 1966. First published 1940.

Graham, Loren R. *The Ghost of the Executed Engineer: Technology and the Fall of the Soviet Union.* Cambridge, MA: Harvard University Press, 1993.

Grant, Alexander. *The Story of the University of Edinburgh during Its First Three Hundred Years.* 2 vols. London: Longmans, Green, 1884.

Greene, Benjamin Franklin. *The Rensselaer Polytechnic Institute: Its Reorganization in 1849–50, Its Condition at the Present Time, Its Plans and Hopes for the Future.* Troy, NY: D. H. Jones, 1855.

Greenman, Jeffrey P., Read Mercer Schuchardt, and Noah Toly. *Understanding Jacques Ellul.* Cambridge: James Clarke, 2012.

Greif, Mark. *The Age of the Crisis of Man: Thought and Fiction in America, 1933–1973.* Princeton, NJ: Princeton University Press, 2015.

Grignon, Pierre Clément. *Mémoires de physique sur l'art de fabriquer le fer.* Paris: Librairie chez Delalain, 1775.

Grimmer-Solem, Erik. *The Rise of Historical Economics and Social Reform in Germany, 1864–1894.* Oxford: Clarendon Press, 2003.

Griscom, J. "Foreign Literature and Science, Extracted and Translated." *American Journal of Science and Arts* 13 (1828): 393–400.

Grundmann, Reiner, and Nico Stehr. "Why Is Werner Sombart Not Part of the Core of Classical Sociology?" *Journal of Classical Sociology* 2 (2001): 257–87.

Guillerme, J., and J. Sebestik. "Les commencements de la technologie." *Thalès* 12 (1966): 1–72.

Hacking, Ian. *Representing and Intervening: Introductory Topics in the Philosophy of Natural Science.* Cambridge: Cambridge University Press, 1983.

"Half Yearly Retrospect of German Literature." *Monthly Magazine* 18 (1804): 623–45.

Hall, A. Rupert. "What Did the Industrial Revolution in Britain Owe to Science?" In *Historical Perspectives: Studies in English Thought and Society in Honour of J. H. Plumb,* edited by Neil McKendrick, 129–51. London: Europa, 1974.

Hall, G. Stanley. "Clark University." *Science* 15, no. 362 (January 10, 1890): 18–22.

Hall, Marie Boas. "Recollections of a History of Science Guinea Pig." *Isis* 90 (1999): S68–S83.

Halle, Johann Samuel. *Werkstäte der heutigen Künste; oder, Die neue Kunsthistorie.* 6 vols. Vol. 1, Brandenburg, 1761.

Halliwell, Stephen. *Aristotle's Poetics.* Chapel Hill: University of North Carolina Press, 1986.

Hammond, William A. "[Review of *Century Dictionary*]." *North American Review* 114 (January 1892): 112.

Hansen, Alvin H. "Institutional Frictions and Technological Unemployment." *Quarterly Journal of Economics* 45, no. 4 (1931): 684–97.

———. "The Technological Interpretation of History." *Quarterly Journal of Economics* 36, no. 1 (November 1921): 72–83.

Hård, Mikael. "German Regulation: The Integration of Modern Technology into National Culture." In *The Intellectual Appropriation of Technology: Discourses on Modernity, 1900–1939,* edited by Mikael Hård and Andrew Jamison, 33–67. Cambridge, MA: MIT Press, 1998.

Harkness, Deborah E. *The Jewel House: Elizabethan London and the Scientific Revolution.* New Haven, CT: Yale University Press, 2007.

Harris, Abram L. "Economic Evolution: Dialectical and Darwinian." *Journal of Political Economy* 42, no. 1 (1934): 34–79.

Harrison, Peter. "'Science' and 'Religion': Constructing the Boundaries." *Journal of Religion* 86 (2006): 81–106.

Harrod, R. F. Review of *Wealth, Virtual Wealth and Debt* by Frederick Soddy. *Economic Journal* 37, no. 146 (1927): 271–73.

Harvard University Program on Technology and Society. *A Final Review, 1964–1972.* Cambridge, MA, 1972.

Hayes, Edward Cary. "Masters of Social Science: Albion Woodbury Small." *Social Forces* 4, no. 4 (1926): 669–77.

Heidegger, Martin. *Die Technik und die Kehre.* Stuttgart: Neske, 1988.

———. *The Question concerning Technology and Other Essays.* Translated by William Lovitt. New York: Harper and Row, 1977.

Heilbroner, Robert L. "Do Machines Make History?" *Technology and Culture* 8, no. 3 (1967): 335–45.

Heimann, Jim. *50s: All-American Ads.* Köln: Taschen, 2001.

Henderson, Charles Richmond. "Applied Sociology (or Social Technology)." *American Journal of Sociology* 18, no. 2 (1912): 215–21.

———. *Modern Methods of Charity: An Account of the Systems of Relief, Public and Private, in the Principal Countries Having Modern Methods.* New York: MacMillan, 1904.

Herbst, Jurgen. *The German Historical School in American Scholarship.* Ithaca, NY: Cornell University Press, 1965.

Herf, Jeffrey. "Reactionary Modernism Reconsidered: Modernity, the West and the Nazis." In *The Intellectual Revolt against Liberal Democracy 1870–1945*, edited by Zeev Sternhell, 131–58. Jerusalem: Israel Academy of Science and Humanities, 1996.

———. *Reactionary Modernism: Technology, Culture, and Politics in Weimar and the Third Reich*. Cambridge: Cambridge University Press, 1984.

Hessen, Boris. "The Social and Economic Roots of Newton's *Principia*." In Freudenthal and McLaughlin, *The Social and Economic Roots of the Scientific Revolution*, 41–101.

Hickman, Larry. *Philosophical Tools for Technological Culture: Putting Pragmatism to Work*. Bloomington: Indiana University Press, 2001.

Hicks, Andrew. "Martianus Capella and the Liberal Arts." In *The Oxford Handbook of Medieval Latin Literature*, edited by David Townsend and Ralph J. Hexter, 307–34. Oxford: Oxford University Press, 2012.

Hilaire-Perez, Liliane. "Technology as a Public Culture in the Eighteenth Century: The Artisans' Legacy." *History of Science* 45 (2007): 135–53.

Hindle, Brooke. *Emulation and Invention*. New York: New York University Press, 1981.

Hindle, Brooke, and Steven D. Lubar. *Engines of Change: The American Industrial Revolution, 1790–1860*. Washington, DC: Smithsonian Institution Press, 1986.

Hobson, J. A. *The Evolution of Modern Capitalism: A Study of Machine Production*. London: Charles Scribner's Sons, 1894.

———. *Modern Sociologists: Veblen*. New York: John Wiley and Sons, 1937.

Hoffman, Abbie. *Revolution for the Hell of It*. New York: Dial Press, 1968.

Hofstadter, Richard. *The Progressive Historians: Turner, Beard, Parrington*. New York: Knopf, 1968.

Holden, Kerry. "Lamenting the Golden Age: Love, Labour and Loss in the Collective Memory of Scientists." *Science as Culture* 24, no. 1 (2014): 24–45.

Hollinger, David A. "The Defense of Democracy and Robert K. Merton's Formulation of the Scientific Ethos." *Knowledge and Society* 4 (1983): 1–15.

Horgan, J. D. "Technology and Human Values: The 'Circle of Action.'" *Mechanical Engineering* 95, no. 8 (August 1973): 19–22.

Horn, Jeff. "The Privilege of Liberty: Challenging the Society of Orders." *Proceedings of the Western Society for French History* 35 (2007): 171–83.

Hortleder, Gerd. *Das Gesellschaftsbild des Ingenieurs: Zum politschen Verhalten der technischen Intelligenz in Deutschland*. Frankfurt: Suhrkamp Verlag, 1970.

Hounshell, David A. "Edison and the Pure Science Ideal in 19th-Century America." *Science* 207, no. 4431 (1980): 612–17.

Hoven, Birgit van den. *Work in Ancient and Medieval Thought: Ancient Philosophers, Medieval Monks and Theologians and Their Concept of Work, Occupations and Technology*. Amsterdam: J. C. Gieben, 1996.

Huff, Toby E. "Theoretical Innovation in Science: The Case of William F. Ogburn." *American Journal of Sociology* 79, no. 2 (September 1973): 261–77.

Hugh of Saint-Victor. *The Didascalicon of Hugh of St. Victor: A Medieval Guide to the Arts*. Translated by Jerome Taylor. New York: Columbia University Press, 1961; reprint, 1991. Page references are to the 1991 edition.

Hughes, Thomas P. *American Genesis: A Century of Invention and Technological Enthusiasm*. New York: Viking, 1989.

———. "Emerging Themes in the History of Technology." *Technology and Culture* 20, no. 4 (1979): 697–711.

Hughes, Thomas P., and Agatha C. Hughes, eds. *Lewis Mumford: Public Intellectual*. New York: Oxford University Press, 1990.

Hunt, Edward Eyre, ed. *Recent Economic Changes in the United States*. 2 vols. New York: McGraw-Hill, 1929.

Hunter, Louis C. *Steamboats on the Western Rivers: An Economic and Technological History*. Cambridge, MA: Harvard University Press, 1949.

Huxley, Thomas H. "Science and Culture." In *Collected Essays*, 134–59. London: Methuen, 1898. First published 1880.

Ibsen, Henrik. *An Enemy of the People*. Translated by Christopher Hampton. London: Faber, 1997.

Ingold, Tim. "Materials against Materiality." *Archaeological Dialogues* 14, no. 1 (2007): 1–16.

J. [Thomas P. Jones]. "Bigelow's Elements of Technology." *Journal of the Franklin Institute* 4, no. 3 (September 1892): 215–16.

James, Ryan, and Andrew Weiss. "An Assessment of Google Books' Metadata." *Journal of Library Metadata* 12, no. 1 (2012): 15–22.

Jarves, James Jackson. *The Art Idea: Sculpture, Painting and Architecture in America*. Boston: Houghton Mifflin, 1864.

Jeremy, David J. *Transatlantic Industrial Revolution: The Diffusion of Textile Technologies between Britain and America, 1790–1830s*. North Andover, MA: Merrimack Valley Textile Museum, 1981.

Jevons, W. Stanley. *The Theory of Political Economy*. London: Macmillan, 1871.

Jewett, Frank Baldwin. "The Promise of Technology." *Science* 99 (January 7, 1944): 1–6.

Joerges, Bernward. "Soziologie und Maschinerie." In *Technik als sozialer Prozess*, edited by Peter Weingart, 44–89. Frankfurt am Main: Suhrkamp, 1989.

Joffe, A. "Physics and Technology." In *Science at the Cross Roads*, 37–40.

Johnson, Samuel. *A Dictionary of the English Language*. 3rd ed. Dublin: Jones, 1768.

Jorgensen, Elizabeth Watkins, and Henry Irvin Jorgensen. *Thorstein Veblen: Victorian Firebrand*. Armonk, London: M. E. Sharpe, 1999.

Joshel, Sandra R. *Work, Identity, and Legal Status at Rome: A Study of the Occupational Inscriptions*. Norman: University of Oklahoma Press, 1992.

Jünger, Friedrich Georg. *The Failure of Technology: Perfection without Purpose*. Hinsdale, IL: H. Regnery, 1956. First published 1949.

Justi, Johann Heinrich Gottlob von, Daniel Gottfried Schreber, Johann Conrad Harrepeter, B. C. Rosenthal, C. G. D. Müller, and Johann Samuel

Halle. *Schauplatz der Künste und Handwerke; oder, Vollständige Beschreibung derselben*. 21 vols. Berlin, Stettin, and Leipzig: Johann Heinrich Rüdigern, 1762–1805.

Kaiser, David. "From Blackboards to Bombs." *Nature* 523 (30 July 2015): 523–25.

Kames, Henry Home. *Elements of Criticism*. Edinburgh: A. Miller, 1762.

Kant, Immanuel. *Kant's Critique of Judgement*. Translated by J. H. Bernard. 2nd ed. London: Macmillan, 1914.

Kantrowitz, Arthur. "The Test: Meeting the Challenge of New Technology." *Bulletin of the Atomic Scientists* 25 (November 1969): 20–22, 48.

Kaplan, Irving. *The Research Program of the National Research Project*. Philadelphia: Works Progress Administration, 1937.

Karmarsch, Karl. *Die polytechnische Schule zu Hannover*. Hannover: Verlage der Hahn'chen Hofbuchhandlung, 1848.

———. *Geschichte der Technologie seit der Mitte des achzehnten Jahrhunderts*. New York: Johnson Reprint, 1965. First published 1872.

Kasson, John F. *Civilizing the Machine: Technology and Republican Values in America 1776–1900*. Middlesex, UK: Penguin Books, 1976.

Katz, James Everett. "US Energy Policy: Impact of the Reagan Administration." *Energy Policy* 12, no. 2 (June 1984): 135–45.

Kautsky, Karl. "Was will und kann die materialistische Geschichtsauffassung leisten?" *Die Neue Zeit* 15, no. 1 (1897): 228–38.

Kelly, Kevin. *What Technology Wants*. New York: Viking, 2010.

Kevles, Daniel J. *The Physicists: The History of a Scientific Community in Modern America*. New York: Knopf, 1978.

Kirk, Andrew G. *Counterculture Green: The Whole Earth Catalog and American Environmentalism*. Lawrence: University Press of Kansas, 2007.

Klein, Ursula. "Artisanal-Scientific Experts in Eighteenth-Century France and Germany." *Annals of Science* 69, no. 3 (2012): 303–6.

Klein, Ursula, and E. C. Spary, eds. *Materials and Expertise in Early Modern Europe: Between Market and Laboratory*. Chicago: University of Chicago Press, 2010.

Kline, Ronald R. "Construing 'Technology' as 'Applied Science.'" *Isis* 86 (1995): 194–221.

———. "Cybernetics, Management Science, and Technology Policy: The Emergence of 'Information Technology' as a Keyword, 1948–1985." *Technology and Culture* 47, no. 3 (2006): 513–35.

———. "Forman's Lament." *History and Technology* 24, no. 1/2 (2007): 160–66.

———. "Science and Technology." Chap. 9 in *Wrestling with Nature: From Omens to Science*, edited by Peter Harrison, Ronald L. Numbers, and Michael H. Shank, 225–52. Chicago: University of Chicago Press, 2011.

Knapp, Friedrich Ludwig. *Lehrbuch der chemischen Technologie: Zum Unterricht und Selbststudium*. 2 vols. Vol. 1, Braunschweig: Vieweg, 1847.

Knapp, Friedrich Ludwig, Edmund Ronalds, Thomas Richardson, and Walter R. Johnson. *Chemical Technology; or, Chemistry, Applied to the Arts and to Manufactures*. 1st American ed. Philadelphia: Lea and Blanchard, 1848.

Knight, Edward H. *Knight's New Mechanical Dictionary: A Description of Tools, Instruments, Machines, Processes, and Engineering, with Indexical References to Technical Journals (1876–1880)*. Boston: Houghton, Mifflin, 1884.

Knight, Frank H. *Risk, Uncertainty and Profit*. Boston: Hougthon Mifflin, 1921.

Knoedler, Janet, and Anne Mayhew. "Thorstein Veblen and the Engineers: A Reinterpretation." *History of Political Economy* 31, no. 2 (Summer 1999): 255–72.

Kohlstedt, Sally. "A Step toward Scientific Self-Identity in the United States: The Failure of the National Institute, 1844." *Isis* 62, no. 3 (1971): 339–62.

Koyré, Alexandre. "Galileo and Plato." *Journal of the History of Ideas* 4, no. 4 (1943): 400–428.

Kranzberg, Melvin. "At the Start." *Technology and Culture* 1, no. 1 (1959): 1–10.

———. "Technology and History: 'Kranzberg's Laws.'" *Technology and Culture* 27, no. 3 (1986): 544–60.

Kristeller, Paul Oskar. "The Modern System of the Arts: A Study in the History of Aesthetics Part I." *Journal of the History of Ideas* 12, no. 4 (October 1951): 496–527.

———. "The Modern System of the Arts: A Study in the History of Aesthetics (II)." *Journal of the History of Ideas* 13, no. 1 (January 1952): 17–46.

Kundert, Joshua. "German Engineers and Bildung during the Nineteenth Century." MA thesis, History of Science, University of Wisconsin–Madison, 2001.

LaFollette, Marcel C. *Making Science Our Own: Public Images of Science 1910–1955*. Chicago: University of Chicago Press, 1990.

Landes, David S. *Revolution in Time: Clocks and the Making of the Modern World*. Cambridge, MA: Harvard University Press, 1983.

———. *Unbound Prometheus*. London: Cambridge University Press, 1969.

Lash, Scott. "Technological Forms of Life." *Theory, Culture and Society* 18 (2001): 105–20.

Laudan, Rachel. "Natural Alliance or Forced Marriage? Changing Relations between the Histories of Science and Technology." *Technology and Culture* 36, no. 2 (1995): S17–S30.

Laurence, William L. "Inventions Survey Finds Major Changes Imminent; Urges Labor Safeguards." *New York Times*, 18 July 1937.

Law, John. *Aircraft Stories: Decentering the Object in Technoscience*. Durham, NC: Duke University Press, 2002.

Lawrence, Ernest O. "Science and Technology." *Science* 86, no. 2231 (October 1, 1937): 295–98.

Layton, Edwin. "Mirror-Image Twins: The Communities of Science and Technology in 19th-Century America." *Technology and Culture* 12, no. 4 (1971): 562–80.

———. *Revolt of the Engineers*. Cleveland: Case Western Reserve Press, 1971.

Levenson, Alan T. "The German Peace Movement and the Jews: An Unexplored Nexus." *Leo Baeck Institute Year Book* 46 (2001): 277–301.

Levine, Donald N. "The Continuing Challenge of Weber's Theory of Rational Action." Chap. 4 in *Max Weber's Economy and Society: A Critical Companion*, edited by Charles Camic, Philip S. Gorski, and David M. Trubek, 101–26. Stanford, CA: Stanford University Press, 2005.

Levy, H. Review of *Technological Trends and National Policy. Science and Society* 2 (1938): 262–65.

Lienhard, John H. *The Engines of Our Ingenuity.* Oxford: Oxford University Press, 2000.

Lilienthal, David E. "Science and the Spirit of Man." *Bulletin of the Atomic Scientists* 5, no. 4 (1949): 98–100.

Lindenfeld, David F. *The Practical Imagination: The German Sciences of State in the Nineteenth Century.* Chicago: University of Chicago Press, 1997.

Lippmann, Edmund O. von. *Beiträge zur Geschichte der Naturwissenschaften und der Technik.* Berlin: Springer, 1923.

Liu, Cixin, and Ken Liu. *The Three-Body Problem.* New York: Tor Books, 2014.

Lobkowicz, Nicholas. *Theory and Practice: History of a Concept from Aristotle to Marx.* Notre Dame, IN: University of Notre Dame Press, 1967.

Loeb, Harold. *Life in a Technocracy: What It Might Be Like.* Syracuse, NY: Syracuse University Press, 1996. First published 1933.

Logsdon, John M., ed. *Exploring the Unknown: Selected Documents in the History of the U.S. Civil Space Program* Vol. 7. Washington, DC: NASA, 1995.

Long, Pamela O. *Artisan/Practitioners and the Rise of the New Sciences, 1400–1600.* Corvallis: Oregon State University Press, 2011.

———. *Openness, Secrecy, Authorship: Technical Arts and the Culture of Knowledge from Antiquity to the Renaissance.* Baltimore: Johns Hopkins University Press, 2001.

Long, Stewart. "Lewis Mumford and Institutional Economics." *Journal of Economic Issues* 36, no. 1 (2002): 167–82.

Lubar, Steven. "'Do Not Fold, Spindle or Mutilate': A Cultural History of the Punch Card." *Journal of American Culture* 15, no. 4 (1992): 43–55.

Lucier, Paul. "The Origins of Pure and Applied Science in Gilded Age America." *Isis* 103, no. 3 (2012): 527–36.

———. *Scientists and Swindlers: Consulting on Coal and Oil in America, 1820–1890.* Baltimore: Johns Hopkins University Press, 2008.

MacIntyre, Alasdair C. *After Virtue: A Study in Moral Theory.* 2nd ed. Notre Dame, IN: University of Notre Dame Press, 1984.

———. *A Short History of Ethics.* New York: Macmillan, 1966.

MacKenzie, Donald A. "Marx and the Machine." *Technology and Culture* 25 (July 1984): 473–502.

MacKenzie, Donald A., and Judy Wajcman. *The Social Shaping of Technology.* 2nd ed. Buckingham, UK: Open University Press, 1999.

Mackey, Theresa M. "Herbert Marcuse." In *Twentieth-Century European Cultural Theorists: First Series*, edited by Paul Hansom, 315–29. Dictionary of Literary Biography, vol. 242. Detroit: Gale, 2001.

Maclaurin, William Rupert. *Invention and Innovation in the Radio Industry*. New York: Macmillan, 1949.

———. "The Sequence from Invention to Innovation and Its Relation to Economic Growth." *Quarterly Journal of Economics* 67, no. 1 (1953): 97–111.

MacLeod, Christine. *Heroes of Invention: Technology, Liberalism and British Identity, 1750–1914*. Cambridge: Cambridge University Press, 2007.

Manegold, Karl-Heinz. *Universität, Technische Hochschule und Industrie*. Berlin: Duncker and Humblot, 1971.

Marcel, Gabriel. *The Decline of Wisdom*. New York: Philosophical Library, 1955.

———. *Le déclin de la sagesse*. Paris: Librairie Plon, 1954.

Marcell, David W. *Progress and Pragmatism: James, Dewey, Beard, and the American Idea of Progress*. Westport, CT: Greenwood Press, 1974.

Marcuse, Herbert. *One Dimensional Man: Studies in the Ideology of Advanced Industrial Society*. Boston: Beacon Press, 1964.

"Marcuse on the Hippie Revolution." *Berkeley (CA) Barb*, 4–10 August 1967.

Maritain, Jacques. *Art and Scholasticism*. Translated by James Scanlan. New York: C. Scribner's Sons, 1930.

Markoff, John. *What the Dormouse Said: How the Sixties Counterculture Shaped the Personal Computer Industry*. New York: Viking Penguin, 2005.

Marshall, Alfred. *Industry and Trade: A Study of Industrial Technique and Business Organization*. London: Macmillan, 1919, 1920.

———. *Principles of Economics*. London: Macmillan, 1890.

Marx, Karl. *Capital*. Translated by Samuel Moore and Edward Aveling. 3rd ed. Vol. 1, Moscow: Progress Publishers, 1954. First published 1887.

———. *Capital: A Critique of Political Economy*. Translated by Ben Fowkes. 3 vols. Vol. 1, Harmondsworth, UK: Penguin, 1976.

———. *Capital: A Critique of Political Economy*. Translated by Samuel Moore and Edward B. Aveling. 3 vols. Vol. 1, Chicago: C. H. Kerr, 1906.

———. *Das Kapital: Kritik der politischen Oekonomie*. 1st ed. Vol. 1, Hamburg: O. Meissner, 1867.

———. *Das Kapital: Kritik der politischen Oekonomie*. 2nd ed. Hamburg: O. Meissner, 1872.

———. *Die technologisch-historischen Exzerpte: Historisch-kritische Ausgabe*. Edited by Hans-Peter Müller. Frankfurt: Ullstein Materialien, 1981.

———. Preface to *A Contribution to the Critique of Political Economy*. Moscow: Progress Publishers, 1977. https://www.marxists.org/archive/marx/works/1859/critique-pol-economy/preface.htm.

Marx, Karl, and Friedrich Engels. *Werke*. Vol. 23, Berlin: Dietz Verlag, 1962.

———. *Werke*. Vol. 30, Berlin: Dietz Verlag, 1964.

———. *Werke*. Vol. 39, Berlin: Dietz Verlag, 1968.

Marx, Leo. "Lewis Mumford: Prophet of Organicism." In Hughes and Hughes, *Lewis Mumford: Public Intellectual*, 164–80.

———. *The Machine in the Garden: Technology and the Pastoral Ideal in America*. London: Oxford University Press, 1964.

———. "*Technology*: The Emergence of a Hazardous Concept." *Social Research* 64, no. 3 (Fall 1997): 965–88.

———. "*Technology*: The Emergence of a Hazardous Concept." *Technology and Culture* 51, no. 3 (2010): 561–77.

Maskell, William. *The Industrial Arts: Historical Sketches.* [London]: [Chapman and Hall], 1876.

Mason, John Hope. *The Value of Creativity: The Origins and Emergence of a Modern Belief.* Aldershot, UK: Ashgate, 2003.

Mason, Otis T. *The Origins of Invention: A Study of Industry among Primitive People.* London: Walter Scott, 1895.

Mathews, Cleve. "Technology Peril Stirs Scientists." *New York Times,* 31 August 1969.

Matschoss, Conrad. *Geschichte der Dampfmaschine: Ihre kulturelle Bedeutung, technische Entwicklung und ihre grossen Männer.* Berlin: Springer, 1901.

Mayer, Anna K. "Setting Up a Discipline, II: British History of Science and 'The End of Ideology,' 1931–1948." *Studies in History and Philosophy of Science Part A* 35, no. 1 (2004): 41–72.

Mayr, Otto. "The Science-Technology Relationship as a Historiographic Problem." *Technology and Culture* 17 (1976): 663–73.

McClellan, James E. "What's Problematic about 'Applied Science.'" In *The Applied-Science Problem,* edited by James E. McClellan, 1–36. Jersey City, NJ: Jensen/Daniels, 2008.

McClelland, Charles E. *The German Experience of Professionalization: Modern Learned Professions and Their Organizations from the Early Nineteenth Century to the Hitler Era.* Cambridge: Cambridge University Press, 1991.

McCraw, Thomas K. *Prophet of Innovation: Joseph Schumpeter and Creative Destruction.* Cambridge, MA: Belknap Press of Harvard University Press, 2007.

McDermott, John. "Technology: The Opiate of the Intellectuals." *New York Review of Books* 13 (1969): 25–35.

McFarland, Floyd B. "Clarence Ayres and His Gospel of Technology." *History of Political Economy* 18, no. 4 (December 1, 1986): 617–37.

McGee, David. "Making Up Mind: The Early Sociology of Invention." *Technology and Culture* 36, no. 4 (October 1995): 773–801.

McGucken, William. "On Freedom and Planning in Science: The Society for Freedom in Science, 1940–46." *Minerva* 16, no. 1 (1978): 42–72.

McMillen, J. Howard. "Relation of Physics to the War Effort." *Transactions of the Kansas Academy of Science* 46 (1943): 272–75.

McNeill, William H. *The Pursuit of Power: Technology, Armed Force, and Society since A.D. 1000.* Chicago: University of Chicago Press, 1982.

McQuade, Kim. "Sputnik Reconsidered: Image and Reality in the Early Space Age." *Canadian Review of American Studies* 37, no. 3 (2007): 371–401.

Meier, Hugo Arthur. "The Technological Concept in American Social History, 1750–1860." PhD diss., University of Wisconsin–Madison, 1950.

———. "Technology and Democracy, 1800–1860." *Mississippi Valley Historical Review* 43, no. 4 (1957): 618–40.

Meijers, Anthonie, ed. *Philosophy of Technology and Engineering Sciences.* Handbook of the Philosophy of Science. Vol. 9, Amsterdam: Elsevier/North Holland, 2009.

Menger, Carl. *Grundsätze der Volkswirthschaftslehre.* Vienna: W. Braumüller, 1871.

Mertens, Joost. "Technology as the Science of the Industrial Arts: Louis Sébastien Lenormand (1757–1837) and the Popularization of Technology." *History and Technology* 18 (2002): 203–31.

Merton, Robert K. "A Note on Science and Democracy." *Journal of Legal and Political Sociology* 1 (1942): 115–26.

———. "Science and the Economy of Seventeenth Century England." *Science and Society* 3, no. 1 (1939): 3–27.

———. "Science and Military Technique." *Scientific Monthly* 41, no. 6 (1935): 542–45.

———. "Science, Technology and Society in Seventeenth Century England." *Osiris* 4 (1938): 360–632.

Messinger, Heinz, and Werner Rüdenberg. *Langenscheidt's New College German Dictionary.* New ed. Berlin: Langenscheidt, 1973.

Meyer, Gerhard. "Note on Technological Trends and Social Planning." *American Journal of Sociology* 43, no. 6 (1938): 951–63.

Meyer, Torsten. "Zwischen Ideologie und Wissenschaft: 'Technik und Kultur' im Werk Werner Sombarts." In *Technische Intelligenz und "Kulturfaktor Technik,"* edited by Burkhard Dietz, Michael Fessner, and Helmut Maier, 67–86. Münster: Waxmann, 1996.

Mill, J. S. *A System of Logic, Ratiocinative and Inductive, Being a Connected View of the Principles, and the Methods of Scientific Investigation.* 4th ed. Vol. 2, London: J. W. Parker, 1856.

Miller, Donald L. *Lewis Mumford: A Life.* New York: Weidenfeld and Nicolson, 1989.

Miller, Perry. *The Life of the Mind in America from the Revolution to the Civil War: Books One through Three.* New York: Harcourt, Brace and World, 1965.

———. *The New England Mind: The Seventeenth Century.* 3rd ed. Cambridge, MA: Harvard University Press, 1939.

Mindell, David A. *Between Human and Machine: Feedback, Control, and Computing before Cybernetics.* Baltimore: Johns Hopkins University Press, 2002.

Mitcham, Carl. *Thinking Through Technology: The Path between Engineering and Philosophy.* Chicago: University of Chicago Press, 1994.

Mitcham, Carl, and Eric Schatzberg. "Defining Technology and the Engineering Sciences." In *Philosophy of Technology and Engineering Sciences,* edited by Anthonie Meijers, 9:27–63. Amsterdam: Elsevier/North Holland, 2009.

Mitchell, Wesley C. "Forces That Make for American Prosperity." *New York Times,* 12 May 1929.

———. "Human Behavior and Economics: A Survey of Recent Literature." *Quarterly Journal of Economics* 29, no. 1 (1914): 1–47.

———. "Sombart's Hochkapitalismus." *Quarterly Journal of Economics* 43, no. 2 (1929): 303–23.

———. "Thorstein Veblen." In *What Veblen Taught: Selected Writings of Thorstein Veblen.* Edited with an introduction by Wesley C. Mitchell, vii–xlix. New York: A. M. Kelley, 1964.

Mokyr, Joel. *The Gifts of Athena: Historical Origins of the Knowledge Economy.* Princeton, NJ: Princeton University Press, 2002.

Molella, Arthur P. "Mumford in Historiographical Context." In Hughes and Hughes, *Lewis Mumford: Public Intellectual,* 21–42.

Morris, Richard B., ed. *The U.S. Department of Labor Bicentennial History of the American Worker.* Washington, DC: Government Printing Office, 1976.

Mueller-Vollmer, Kurt. "How Brockhaus' *Conversations-Lexicon* Became the *Encyclopaedia Americana.*" In *Der Mnemosyne Träume: Festschrift zum 80. Geburtstag von Joseph P. Strelka,* edited by Ilona Slawinski, 209–23. Tübingen: Francke Verlag, 2007.

Müller, Hans-Peter, and Ulrich Troitzsch, eds. *Technologie zwischen Fortschritt und Tradition: Beiträge zum internationalen Johan Beckmann-Symposium Göttingen 1989.* Frankfurt am Main: Peter Lang, 1992.

Multhauf, Robert P. "The Scientist and the 'Improver' of Technology." *Technology and Culture* 1, no. 1 (1959): 38–47.

———. "Some Observations on the State of the History of Technology." *Technology and Culture* 15, no. 1 (1974): 1–12.

Mumford, Lewis. "An Appraisal of Lewis Mumford's 'Technics and Civilization' (1934)." *Daedalus* 88, no. 3 (1959): 527–36.

———. "Authoritarian and Democratic Technics." *Technology and Culture* 5, no. 1 (Winter 1964): 1–8.

———. "The Decline of Spengler." *New Republic,* 9 March 1932, 104.

———. "The Drama of the Machines." *Scribner's Magazine,* August 1930, 150–61.

———. "History: Neglected Clue to Technological Change." *Technology and Culture* 2 (1961): 230–36.

———. *My Works and Days: A Personal Chronicle.* 2nd ed. New York: Harcourt Brace Jovanovich, 1979.

———. *The Myth of the Machine.* Vol. 1, *Technics and Human Development.* New York: Harcourt, Brace, 1967.

———. *The Myth of the Machine.* Vol. 2, *The Pentagon of Power.* New York: Harcourt Brace Jovanovich, 1970.

———. *Sketches from Life: The Autobiography of Lewis Mumford, the Early Years.* New York: Dial Press, 1982.

———. *Technics and Civilization.* New York: Harcourt Brace Jovanovich, 1963. First published 1934.

———. "Technics and the Nature of Man." *Technology and Culture* 7, no. 3 (1966): 303–17.

———. "Tools and the Man." *Technology and Culture* 1, no. 4 (Autumn 1960): 320–34.

Mumford, Lewis, Patrick Geddes, and Frank G. Novak. *Lewis Mumford and Patrick Geddes: The Correspondence.* London: Routledge, 1995.

Mumford, Lewis, Henry Alexander Murray, and Frank G. Novak. *"In Old Friendship": The Correspondence of Lewis Mumford and Henry A. Murray, 1928–1981.* Syracuse, NY: Syracuse University Press, 2007.

Murray, Hugh, William Wallace, and T. G. Bradford. *The Encyclopædia of Geography.* Rev. ed. 3 vols. Philadelphia: Carey, Lea and Blanchard, 1837.

Nader, Ralph. *Unsafe at Any Speed: The Designed-in Dangers of the American Automobile.* New York: Grossman, 1965.

Nanni, Romano. "Technical Knowledge and the Advancement of Learning: Some Questions About 'Perfectibility' and 'Invention.'" In *Philosophies of Technology: Francis Bacon and His Contemporaries*, edited by Claus Zittel, Gisela Engel, Romano Nanni, and Nicole C. Karafyllis, 51–65. Leiden: Brill, 2008.

National Academy of Sciences (U.S.). Committee on Science and Public Policy. *Basic Research and National Goals: A Report to the Committee on Science and Astronautics, U.S. House of Representatives.* [Washington, DC], 1965.

Needham, Joseph. "Capitalism and Science." *Economic History Review* 8, no. 2 (1938): 198–99.

Neilson, William Allan. *Webster's New International Dictionary of the English Language.* 2nd ed. Springfield, MA: G. and C. Merriam, 1939.

Nelson, Bryce. "Technological Innovation: Panel Stresses Role of Small Firms." *Science* 155, no. 3767 (1967): 1229–31.

Nelson, Daniel. *Frederick W. Taylor and the Rise of Scientific Management.* Madison: University of Wisconsin Press, 1980.

New York Herald Tribune. "Engineering Council Assails Technocracy." 15 January 1933.

———. "Gaither Report Calls U. S. in Gravest Peril." 20 December 1957.

———. "The Threat of Soviet Technology." 16 March 1956.

New York Times. "America's Future Studied in Light of Progressive Application of Technology." 20 July 1937.

———. "Pope Pius' Christmas Message." 25 December 1953.

———. "Rocket Race." 20 October 1957.

———. "Soviet Scientists Discourse on Marx." 1 July 1931.

———. "Taming the Technological Monster." 4 November 1970.

———. "Thomas Launches Socialist Campaign." 5 August 1928.

Nietzsche, Friedrich. *On the Genealogy of Morality.* Translated by Alan J. Swensen and Maudemarie Clark. Indianapolis: Hackett, 1998.

Nikiforova, Natalia. "The Concept of Technology and the Russian Cultural Research Tradition." *Technology and Culture* 56, no. 1 (2015): 184–203.

Noble, David F. *Forces of Production: A Social History of Industrial Automation.* New York: Knopf, 1984.

"Notes." *American Economic Review* 18, no. 4 (1928): 816–28.

Nussbaum, Frederick L. *A History of the Economic Institutions of Modern Europe.* New York: F. S. Crofts, 1933.

Nussbaum, Martha Craven. *The Fragility of Goodness: Luck and Ethics in Greek Tragedy and Philosophy.* Cambridge: Cambridge University Press, 1986.

Nye, David E. *Electrifying America: Social Meanings of a New Technology, 1880–1940.* Cambridge, MA: MIT Press, 1990.

Nye, Mary Jo. *Michael Polanyi and His Generation: Origins of the Social Construction of Science.* Chicago: University of Chicago Press, 2011.

Ofner, Julius. *Studien sozialer Jurisprudenz.* Vienna: Alfred Hölder, 1894.

Ogburn, William Fielding. "The Future of Man in Light of His Past: The Viewpoint of a Sociologist." *Scientific Monthly* 32, no. 4 (April 1931): 294–300.

———. "The Great Man versus Social Forces." *Social Forces* 5, no. 2 (1926): 225–31.

———. "The Influence of Inventions on American Social Institutions in the Future." *American Journal of Sociology* 43, no. 3 (1937): 365–76.

———. *Social Change with Respect to Culture and Original Nature.* New York: B. W. Huebsch, 1922.

———. "Sociology and the Atom." *American Journal of Sociology* 51, no. 4 (1946): 267–75.

———. "Technology and Governmental Change." *Journal of Business of the University of Chicago* 9, no. 1 (1936): 1–13.

———. "Technology and Sociology." *Social Forces* 17, no. 1 (1938): 1–8.

Ogburn, William Fielding, S. Colum Gilfillan, and Jean Adams. *The Social Effects of Aviation.* Boston: Houghton Mifflin, 1946.

Ogburn, William Fielding, and Meyer Francis Nimkoff. *Sociology.* Boston: Houghton Mifflin, 1940.

———. *Technology and the Changing Family.* Boston: Houghton Mifflin, 1955.

Oldenziel, Ruth. "Gender and the Meanings of Technology: Engineering in the US, 1880–1945." PhD diss., Yale University, 1992.

———. *Making Technology Masculine.* Amsterdam: Amsterdam University Press, 1999.

Olson, Richard. "Science, Technology, and the Industrial Revolution: The Conflation of Science and Technics." In *Science Deified, Science Defied: The Historical Significance of Science in Western Culture*, 316–44. Berkeley: University of California Press, 1982.

Ong, Walter J. *Ramus: Method, and the Decay of Dialogue; from the Art of Discourse to the Art of Reason.* Cambridge, MA: Harvard University Press, 1958.

Onions, C. T., and James Augustus Henry Murray. *A New English Dictionary on Historical Principles; Founded Mainly on the Materials Collected by the Philological Society.* 10 vols. Vol. 9, Oxford: Clarendon Press, 1919.

"Our Accelerated Economic Life." *Science News-Letter* 15, no. 425 (1929): 341.

Ovitt, George, Jr. "The Cultural Context of Western Technology: Early Christian Attitudes toward Manual Labor." *Technology and Culture* 27, no. 3 (1986): 477–500.

———. *The Restoration of Perfection: Labor and Technology in Medieval Culture.* New Brunswick, NJ: Rutgers University Press, 1987.

Parsons, Talcott. "'Capitalism' in Recent German Literature: Sombart and Weber." *Journal of Political Economy* 36, no. 6 (1928): 641–61.

———. "'Capitalism' in Recent German Literature: Sombart and Weber (Concluded)." *Journal of Political Economy* 37, no. 1 (1929): 31–51.

———. "The Impact of Technology on Culture and Emerging New Modes of Behaviour." *International Social Science Journal* 22 (1970): 607–27.

———. "Sociological Elements in Economic Thought." *Quarterly Journal of Economics* 49 (1935): 414–53.

———. "Some Reflections on 'The Nature and Significance of Economics.'" *Quarterly Journal of Economics* 48 (1934): 511–45.

———. *The Structure of Social Action: A Study in Social Theory with Special Reference to a Group of Recent European Writers.* New York: McGraw-Hill, 1937.

Partington, Charles F. *The British Cyclopaedia of the Arts and Sciences; including Treatises on the Various Branches of Natural and Experimental Philosophy, the Useful and Fine Arts, Mathematics, Commerce, &c.* 2 vols. London: Orr and Smith, 1835.

Passer, Harold C. *The Electrical Manufacturers, 1875–1900: A Study in Competition, Entrepreneurship, Technical Change, and Economic Growth.* Cambridge, MA: Harvard University Press, 1953.

Pelton, Frank M. "Predictions on the Future of Inventions and Changing Social Customs, and Their Effect on Education." *Clearing House* 13, no. 3 (1938): 131–34.

Perrin, Porter G. "Possible Sources of Technologia at Early Harvard." *New England Quarterly* 7, no. 4 (1934): 718–24.

Perrow, Charles. *Organizational Analysis: A Sociological View.* Belmont, CA: Wadsworth, 1970.

Phillips, Denise. *Acolytes of Nature: Defining Natural Science in Germany, 1770–1850.* Chicago: University of Chicago Press, 2012.

———. "Francis Bacon and the Germans: Stories from When 'Science' Meant 'Wissenschaft.'" *History of Science* 53, no. 4 (2015): 378–94.

Pickstone, John V. *Ways of Knowing: A New History of Science, Technology, and Medicine.* Chicago: University of Chicago Press, 2001.

Pielke, Roger. "'Basic Research' as a Political Symbol." *Minerva* 50, no. 3 (2012): 339–61.

Pion, Georgine M., and Mark W. Lipsey. "Public Attitudes toward Science and Technology: What Have the Surveys Told Us?" *Public Opinion Quarterly* 45, no. 3 (1981): 303–16.

Pius XII. "Nuntius Radiophonicus." *Acta Apostolicae Sedis* 46 (1954): 5–16.

Plato. *Republic.* Translated by Joe Sachs. Newburyport, MA: Focus Publishing/R. Pullins, 2011.

Plato, and Aristotle. *Plato, Gorgias, and Aristotle, Rhetoric.* Translated by Joe Sachs. Newburyport, MA: Focus Publishing/R. Pullins, 2009.

Playfair, Lyon. *Industrial Instruction on the Continent: Being the Introductory Lecture of the Session 1852–1853.* London: H. M. Stationery Office, 1852.

Pogo, Alexander. Review of *History of Technics. Isis* 26, no. 2 (1937): 486–87.

Poinsett, Joel Roberts. *Discourse, on the Objects and Importance of the National Institution for the Promotion of Science, Established at Washington, 1840, Delivered at the First Anniversary.* Washington, DC: P. Force, 1841.

Polanyi, Michael. "Pure and Applied Science and Their Appropriate Forms of Organization." *Dialectica* 10, no. 3 (1956): 231–42.

Poni, Carlo. "The Worlds of Work: Formal Knowledge and Practical Abilities in Diderot's Encyclopédie." *Jahrbuch für Wirtschaftsgeschichte* 50, no. 1 (2009): 135–50.

Poppe, Johann Heinrich Moritz von. *Handbuch der Technologie: Vornehmlich zum Gebrauch auf Schulen und Universitäten.* Frankfurt am Main: Mohr, 1806.

Post, Julius. *Arbeit statt Almosen: Beitrag zur Social-Technik.* Bremen: Roussell, 1881.

Post, Robert C. "Back at the Start: History and Technology and Culture." *Technology and Culture* 51, no. 4 (2010): 961–94.

———. "Chance and Contingency: Putting Mel Kranzberg in Context." *Technology and Culture* 50, no. 4 (2009): 839–72.

Pot, Johan Hendrik Jacob van der. *Die Bewertung des technischen Fortschritts: Eine systematische Übersicht der Theorien.* 2 vols. Assen: Van Gorcum, 1985.

Powell, John Wesley. "The Evolution of Religion." *Monist* 8, no. 2 (1898): 183–204.

———. "Human Evolution: Annual Address of the President, J. W. Powell, Delivered November 6, 1883." *Transactions of the Anthropological Society of Washington* 2 (1883): 176–208.

———. "Technology, or the Science of Industries." *American Anthropologist* 1, no. 2 (April 1899): 319–49.

Powell, John Wesley, and Franz Boas. "Museums of Ethnology and Their Classification." *Science* 9, no. 229 (June 24, 1887): 612–14.

Preece, Warren E. "A Report of the Discussion." *Technology and Culture* 3, no. 4 (1962): 466–85.

"Proceedings of the Society: Annual Conference." *Journal of the Society of Arts* 21, no. 1076 (1873): 635–60.

"Proceedings of the Society: Conference on Technological Education." *Journal of the Society of Arts* 20, no. 1027 (1872): 725–35.

Quintilian. *The Institutio Oratoria of Quintilian.* Translated by Harold Edgeworth Butler. 4 vols. London: W. Heinemann, 1921.

Rae, John B. "The 'Know-How' Tradition: Technology in American History." *Technology and Culture* 1, no. 2 (Spring 1960): 139–50.

Ramo, Simon. "Individual and the Good Technological Society." *Vital Speeches of the Day* 33 (15 August 1967): 646–48.

Rankine, William John Macquorn. *A Manual of Applied Mechanics.* 5th ed. London: Charles Griffin, 1869.

Raymond, Allen. *What Is Technocracy?* New York: Whittlesey House, McGraw-Hill, 1933.

Reagan, Ronald. "Remarks at a White House Luncheon Honoring the Astronauts of the Space Shuttle Columbia, May 19, 1981." https://www.reaganlibrary.archives.gov/archives/speeches/1981/51981c.htm.

"Recent Economic Changes." *Monthly Labor Review* 28, no. 6 (1929): 96–107.

Redondi, Pietro. "History and Technology: Research on the Borderline." *History and Technology* 1 (October 1983): 1–6.

Reingold, Nathan. "Physics and Engineering in the United States, 1945–1965: A Study of Pride and Prejudice." In *The Michelson Era in American Science 1870–1930*, edited by Stanley Goldberg and Roger H. Stuewer, 288–99. New York: American Institute of Physics, 1987.

———. "Vannevar Bush's New Deal for Research; or, The Triumph of the Old Order." In *Science, American Style*, 284–333. New Brunswick, NJ: Rutgers University Press, 1991.

Reingold, Nathan, and Arthur Molella. "Introduction [to the Burndy Library Conference]." *Technology and Culture* 17, no. 4 (1976): 624–33.

Remsen, Ira. "The Age of Science." *Science* 20, no. 498 (15 July 1904): 65–73.

Renwick, Chris, and Richard C. Gunn. "Demythologizing the Machine: Patrick Geddes, Lewis Mumford, and Classical Sociological Theory." *Journal of the History of the Behavioral Sciences* 44, no. 1 (2008): 59–76.

"Report of Senate Committee on Causes and Relief of Unemployment." *Monthly Labor Review* 28, no. 5 (1929): 65–78.

Report of the National Academy of Sciences for the Year 1919. Washington, DC: Government Printing Office, 1920.

Research Committee on Social Trends. *Recent Social Trends in the United States: Report of the President's Research Committee on Social Trends*. 2 vols. Vol. 1, New York: McGraw-Hill, 1933.

Reuleaux, Franz. "Cultur und Technik." *Zeitschrift des Vereines Deutscher Ingenieure* 29 (1885): 24–28, 41–46.

———. "Technology and Civilization." *Annual Report of the Board of Regents of the Smithsonian Institution* 45 (1890): 705–19.

Reynolds, Terry S. "The Education of Engineers in America before the Morrill Act of 1862." *History of Education Quarterly* 32, no. 4 (1992): 459–82.

Rhode Island General Assembly, House of Representatives, Committee on Education. *The Industrial Arts in the Public Schools*. Providence: Angell, Burlingame, 1877.

Rhodes, Richard. *The Making of the Atomic Bomb*. New York: Simon and Schuster, 1986.

Rice, Stephen P. *Minding the Machine: Languages of Class in Early Industrial America*. Berkeley: University of California Press, 2004.

Richter, Melvin. "Begriffsgeschichte and the History of Ideas." *Journal of the History of Ideas* 48 (1987): 247–63.

Rieger, Bernhard. *Technology and the Culture of Modernity in Britain and Germany, 1890–1945*. Cambridge: Cambridge University Press, 2005.

Ringer, Fritz K. *The Decline of the German Mandarins: The German Academic Community, 1890–1933*. Cambridge, MA: Harvard University Press, 1969.

Roberts, Lissa. "Agency and Industry: Charles C. Gillispie's 'The Natural History of Industry,' Then and Now." *Technology and Culture* 54, no. 4 (2013): 922–41.

———. Introduction to section 3. In Roberts, Schaffer, and Dear, *The Mindful Hand*, 189–95.

Roberts, Lissa, and Simon Schaffer. Preface to Roberts, Schaffer, and Dear, *The Mindful Hand*, xiii–xxvii.

Roberts, Lissa, Simon Schaffer, and Peter Dear, eds. *The Mindful Hand: Inquiry and Invention from the Late Renaissance to Early Industrialisation*. Amsterdam: Royal Netherlands Academy of Arts and Sciences, 2007.

Robertson, Bruce. "The South Kensington Museum in Context: An Alternative History." *Museum and Society* 2, no. 1 (2015): 1–14.

Rodgers, Daniel T. "Keywords: A Reply." *Journal of the History of Ideas* 49 (1988): 669–76.

Rogers, Emma, ed. *Life and Letters of William Barton Rogers*. 2 vols. Boston: Houghton, Mifflin, 1896.

Rogers, Howard Jason, ed. *Congress of Arts and Science, Universal Exposition, St. Louis, 1904*. 8 vols. Vol. 1, Boston: Houghton, Mifflin, 1905.

Rogers, William Barton. "A Plan for a Polytechnic School in Boston." In *Life and Letters of William Barton Rogers*, edited by Emma Rogers, 1:420–27. Boston: Houghton, Mifflin, 1896.

Rohkrämer, Thomas. "Antimodernism, Reactionary Modernism and National Socialism: Technocratic Tendencies in Germany, 1890–1945." *Contemporary European History* 8, no. 1 (1999): 29–50.

———. *Eine andere Moderne? Zivilisationskritik, Natur und Technik in Deutschland 1880–1933*. Paderborn: Schöningh, 1999.

Roochnik, David. "Is Rhetoric an Art?" *Rhetorica: A Journal of the History of Rhetoric* 12, no. 2 (1994): 127–54.

———. *Of Art and Wisdom: Plato's Understanding of Techne*. University Park: Pennsylvania State University Press, 1996.

Ropohl, Günter. "Prolegomena zu einem neuen Entwurf der allgemeine Technologie." In *Techne Technik Technologie*, 152–72. Munich: Hans Lenk und Simon Moser, 1992.

Rosen, S. McKee, and Laura F. Rosen. *Technology and Society: The Influence of Machines in the United States*. New York: Macmillan, 1941.

Rosenberg, Nathan. "Learning by Using." In *Inside the Black Box: Technology and Economics*, 120–40. Cambridge: Cambridge University Press, 1982.

Ross, Dorothy. *The Origins of American Social Science*. Cambridge: Cambridge University Press, 1991.

Rossi, Paolo. *Francis Bacon: From Magic to Science*. Translated by Sacha Rabinovitch. 2nd ed. Chicago: University of Chicago Press, 1978.

———. *Philosophy, Technology, and the Arts in the Early Modern Era.* Translated by Salvator Attanasio. 2nd ed. New York: Harper Torchbooks, 1970.

Rowland, H. A. "A Plea for Pure Science." *Science* 2, no. 29 (24 August 1883): 242–50.

Rubinstein, M. "Relations of Science, Technology, and Economics under Capitalism, and in the Soviet Union." In *Science at the Cross Roads*, 42–66.

Russell, Bertrand. *Sceptical Essays.* New York: W. W. Norton, 1928.

———. *Unpopular Essays.* New York: Simon and Schuster, 1950.

Russell, Edmund. *War and Nature: Fighting Humans and Insects with Chemicals, from World War I to Silent Spring.* Cambridge: Cambridge University Press, 2001.

Ruttan, Vernon W. "Usher and Schumpeter on Invention, Innovation and Technological Change: Reply." *Quarterly Journal of Economics* 75, no. 1 (1961): 154–56.

Sabel, Charles F., and Jonathan Zeitlin, eds. *World of Possibilities: Flexibility and Mass Production in Western Industrialization.* Cambridge: Cambridge University Press, 1997.

Sadowski, Jathan. "Office of Technology Assessment: History, Implementation, and Participatory Critique." *Technology in Society* 42 (1 August 2015): 9–20.

Salomon, Jean-Jacques. "What Is Technology? The Issue of Its Origins and Definitions." *History and Technology* 1 (1984): 113–56.

Sarnoff, David. "'Science for Life or Death' Discussed by Sarnoff." *New York Times*, 10 August 1945.

Sarton, George."Bibliographie analytique des publications relatives à l'histoire de la science parues depuis le 1er janvier 1912." *Isis* 1 (1913): 136–42.

———. "The History of Science." *The Monist* 26, no. 3 (1916): 321–65.

———. "L'Histoire de la Technologie." *Revue générale des sciences pures et appliquées* 23 (1912): 421.

———. Review of *Medicine and Health in the Soviet Union. Isis* 39, no. 3 (1948): 202–3.

Schabas, Margaret. *The Natural Origins of Economics.* Chicago: University of Chicago Press, 2005.

———. *A World Ruled by Number: William Stanley Jevons and the Rise of Mathematical Economics.* Princeton, NJ: Princeton University Press, 1990.

Schaffer, Simon. "Newton at the Crossroads." *Radical Philosophy* 37 (1984): 23–28.

Schatzberg, Eric. "From Art to Applied Science." *Isis* 103, no. 3 (2012): 555–63.

———. "*Technik* Comes to America: Changing Meanings of *Technology* before 1930." *Technology and Culture* 47 (2006): 486–512.

Schauz, Désirée. "What Is Basic Research? Insights from Historical Semantics." *Minerva* 52 (2014): 273–328.

Scheler, Michael B. "Technological Unemployment." *Annals of the American Academy of Political and Social Science* 154 (1931): 17–27.

Schick, Kathy Diane, and Nicholas Patrick Toth. *Making Silent Stones Speak: Human Evolution and the Dawn of Technology.* New York: Simon and Schuster, 1993.

Schivelbusch, Wolfgang. *Disenchanted Night: The Industrialization of Light in the Nineteenth Century.* Berkeley: University of California Press, 1988.

Schmoller, Gustav von. *Grundriss der allgemeinen Volkswirtschaftslehre.* 2 vols. Leipzig: Duncker and Humblot, 1900.

Schneider, Max. "Uber Technik, technisches Denken und technische Wirkungen." Dissertation, Friedrich-Alexanders-Universität, Erlangen, 1912.

Schön, Donald A. "Champions for Radical New Inventions." *Harvard Business Review* 41, no. 2 (1963): 77–86.

———. *Technology and Change: The New Heraclitus.* New York: Delacorte Press, 1967.

Schumacher, E. F. *Small Is Beautiful: A Study of Economics as if People Mattered.* London: Blond and Briggs, 1973.

———. *This I Believe, and Other Essays.* Totnes, UK: Green Books, 1997.

Schumpeter, Joseph A. *History of Economic Analysis.* New York: Oxford University Press, 1954.

Schweber, Howard. "The 'Science' of Legal Science: The Model of the Natural Sciences in Nineteenth-Century American Legal Education." *Law and History Review* 17, no. 3 (1999): 421–66.

"Science and the London University." *Science* 13, no. 339 (1901): 1021–24.

Science at the Cross Roads: Papers Presented to the International Congress of the History of Science and Technology Held in London from June 29th to July 3rd, 1931 by the Delegates of the U.S.S.R. 2nd ed. London: F. Cass, 1971.

"Science, Art, Discovery." *Christian Advocate* 57, no. 30 (27 July 1882): 10.

Scott, Howard. "Technology Smashes the Price System." *Harper's,* January 1933, 129–42.

Scott, James C. *Seeing Like a State: How Certain Schemes to Improve the Human Condition Have Failed.* New Haven, CT: Yale University Press, 1998.

Scott, Robert, and Henry George Liddell. *An Intermediate Greek-English Lexicon, Founded upon the Seventh Edition of Liddell and Scott's Greek-English Lexicon.* Oxford: Clarendon Press, 1900.

Scoville, Warren Candler. *Revolution in Glassmaking: Entrepreneurship and Technological Change in the American Industry, 1880–1920.* Cambridge, MA: Harvard University Press, 1948.

Scranton, Philip. "Technology-Led Innovation: The Non-linearity of US Jet Propulsion Development." *History and Technology* 22, no. 4 (2006): 337–67.

Seely, Bruce E. "SHOT, the History of Technology and Engineering Education." *Technology and Culture* 36, no. 4 (October 1995): 739–72.

Segal, Howard P. Introduction to Loeb, *Life in a Technocracy,* ix–xxxviii.

———. *Technological Utopianism in American Culture.* 2nd ed. Syracuse: Syracuse University Press, 2005.

Seibicke, Wilfried. "'Technica aut Technologia'; Christian Wolffs Anteil an der Herausbildung des modernen Technikbegriffs." *Festschrift für Friedrich von Zahn* 2 (1971): 179–99.

———. *Technik: Versuch einer Geschichte der Wortfamilie um* τέχνη *in Deutschland vom 16. Jahrhundert bis etwa 1830.* Dusseldorf: VDI-Verlag, 1968.

Seligman, Edwin R. A. "The Economic Interpretation of History. I." *Political Science Quarterly* 16, no. 4 (1901): 612–40.

———. "The Economic Interpretation of History. II." *Political Science Quarterly* 17, no. 1 (1902): 71–98.

Sellberg, Erland. "Petrus Ramus." In *The Stanford Encyclopedia of Philosophy,* edited by Edward N. Zalta, Summer 2016, https://plato.stanford.edu/archives/sum2016/entries/ramus/.

Sen, Amartya. *Choice of Techniques: An Aspect of the Theory of Planned Economic Development.* Oxford: B. Blackwell, 1962.

Shapin, Steven. *The Scientific Life: A Moral History of a Late Modern Vocation.* Chicago: University of Chicago Press, 2008.

———. *A Social History of Truth: Civility and Science in Seventeenth-Century England.* Chicago: University of Chicago Press, 1994.

———. "The Virtue of Scientific Thinking." *Boston Review,* 20 January 2015. http://bostonreview.net/steven-shapin-scientism-virtue.

Shapin, Steven, and Simon Schaffer. *Leviathan and the Air-Pump: Hobbes, Boyle, and the Experimental Life.* Princeton, NJ: Princeton University Press, 1985.

Sheridan, Geraldine. "Recording Technology in France: The *Descriptions des arts*, Methodological Innovation and Lost Opportunities at the Turn of the Eighteenth Century." *Cultural and Social History* 5, no. 3 (2008): 329–54.

Sherwin, Chalmers W., and Raymond S. Isenson. "First Interim Report on Project Hindsight (Summary)." Washington, DC: Office of the Director of Defense Research and Engineering, 1966. http://www.dtic.mil/get-tr-doc/pdf?AD=AD0642400.

———. "Project Hindsight." *Science* 156, no. 3782 (1967): 1571–77.

Shields, Mark A. "Reinventing Technology in Social Theory." *Current Perspectives in Social Theory* 17 (1997): 187–216.

Shiner, Larry. *The Invention of Art: A Cultural History.* Chicago: University of Chicago Press, 2001.

Sieferle, Rolf Peter. *Fortschrittsfeinde? Opposition gegen Technik und Industrie von der Romantik bis zur Gegenwart.* Munich: C. H. Beck, 1984.

Siemens, C. William. "Science in Relation to the Arts." *Journal of the Franklin Institute* 115 (1883): 48–66, 127–34, 208–19.

Singer, Charles, E. J. Holmyard, and A. R. Hall, eds. *A History of Technology.* Vol. 1, *From Early Times to Fall of Ancient Empires.* New York: Oxford University Press, 1954.

Skinner, Quentin. "Meaning and Understanding in the History of Ideas." *History and Theory* (1969): 3–53.

Skolimowski, Henryk. "The Structure of Thinking in Technology." *Technology and Culture* 7, no. 3 (1966): 371–83.

Slichter, Sumner H. "Market Shifts, Price Movements, and Employment." *American Economic Review* 19, no. 1 (1929): 5–22.

Small, Albion W. *The Cameralists: The Pioneers of German Social Polity.* New York: B. Franklin, 1909.

———. "The Scope of Sociology. III. The Problems of Sociology." *American Journal of Sociology* 5, no. 6 (1900): 778–813.

Smith, Cyril Stanley, and Martha Teach Gnudi. Introduction to *The Pirotechnia of Vannoccio Biringuccio*, ix–xxv. New York: American Institute of Mining and Metallurgical Engineers, 1942.

Smith, Elliott Dunlap. *Technology and Labor: A Study of the Human Problems of Labor Saving.* New Haven, CT: Yale University Press, 1939.

Smith, Merritt Roe. "Technological Determinism in American Culture." Chap. 1 in *Does Technology Drive History? The Dilemma of Technological Determinism*, edited by Merritt Roe Smith and Leo Marx, 2–35. Cambridge, MA: MIT Press, 1994.

Smith, Michael B. "'Silence, Miss Carson!' Science, Gender, and the Reception of *Silent Spring.*" *Feminist Studies* 27, no. 3 (2001): 733–52.

Smith, Pamela H. *The Body of the Artisan: Art and Experience in the Scientific Revolution.* Chicago: University of Chicago Press, 2004.

Smith, R. Jeffrey. "Environmental Policies Attacked." *Science* 216, no. 4553 (25 June 1982): 1394.

"Societies and Academies." *Science* 74, no. 1911 (1931): 175–79.

Sombart, Werner. *Der moderne Kapitalismus.* 2nd ed. Vol. 1, Munich: Duncker and Humblot, 1919.

———. *Der moderne Kapitalismus.* Vol. 2, *Theorie der kapitalistischen Entwicklung.* 1st ed. Munich: Duncker and Humblot, 1902.

———. *Der moderne Kapitalismus: Historisch-systematische Darstellung des gesamteuropäischen Wirtschaftslebens von seinen Anfängen bis zur Gegenwart.* 2nd ed. 3 vols. in 6. Munich: Duncker and Humblot, 1924–28.

———. "Technik und Kultur." Chap. 1 in *Verhandlungen des ersten Deutschen Soziologentages*, 63–110. Tübingen: J. C. B. Mohr (Paul Siebeck), 1911.

———. "Technik und Kultur." *Archiv für Sozialwissenschaft und Sozialpolitik* 33 (1911): 305–47.

———. "Technology and Culture." In *Sociological Beginnings: The First Conference of the German Society for Sociology*, edited by Christopher Adair-Toteff, 94–109. Liverpool: Liverpool University Press, 2005.

Sorokin, Pitirim A. "Recent Social Trends: A Criticism." *Journal of Political Economy* 41, no. 2 (1933): 194–210.

Sorrenson, Richard. "George Graham, Visible Technician." *British Journal for the History of Science* 32, no. 2 (1999): 203–21.

Souter, R. W. "'The Nature and Significance of Economic Science' in Recent Discussion." *Quarterly Journal of Economics* 47 (1933): 377–413.

Spade, Paul Vincent. "Medieval Philosophy." In *The Stanford Encyclopedia of Philosophy*, edited by Edward N. Zalta, 2013. http://plato.stanford.edu/archives/spr2013/entries/medieval-philosophy.

Spencer, Herbert. "The Genesis of Science." *British Quarterly Review* 20 (1854): 108–62.

Spengler, Oswald. *Man and Technics: A Contribution to a Philosphy of Life*. Translated by Charles Francis Atkinson. New York: Knopf, 1932.

Sperber, Jonathan. *Karl Marx: A Nineteenth-Century Life*. New York: Liveright, 2013.

Sprondel, Walter M., and Constans Seyfarth. *Max Weber und die Rationalisierung sozialen Handelns*. Stuttgart: Ferdinand Enke Verlag, 1981.

Stahl, William Harris, Richard Johnson, and E. L. Burge. *Martianus Capella and the Seven Liberal Arts*. New York: Columbia University Press, 1971.

Staiti, Paul J. *Samuel F. B. Morse*. Cambridge: Cambridge University Press, 1989.

Staley, Richard. *Einstein's Generation: The Origins of the Relativity Revolution*. Chicago: University of Chicago Press, 2008.

Stanfield, James R., and Jacqueline B. Stanfield. "The Significance of Clarence Ayres and the Texas School." In *Is Economics an Evolutionary Science? The Legacy of Thorstein Velben*, edited by Francisco Louçã and Mark Perlman, 83–94. Cheltenham, UK: Edward Elgar, 2000.

"Statut des Vereins deutscher Ingenieure." *Zeitschrift des Vereins deutscher Ingenieure* 1, no. 1 (January 1857): 4–6.

Staudenmaier, John M. *Technology's Storytellers: Reweaving the Human Fabric*. Cambridge, MA: MIT Press, 1985.

Stehr, Nico, and Reiner Grundmann. "Introduction: Werner Sombart." In *Economic Life in the Modern Age*, edited by Nico Stehr and Reiner Grundmann, ix–lxii. New Brunswick, NJ: Transaction, 2001.

Stewart, Irvin. *Organizing Scientific Research for War: The Administrative History of the Office of Scientific Research and Development*. Boston: Little, Brown, 1948.

Stewart, Larry R. *The Rise of Public Science: Rhetoric, Technology, and Natural Philosophy in Newtonian Britain, 1660–1750*. Cambridge: Cambridge University Press, 1992.

Stratton, Julius Adams. "Karl Taylor Compton: September 14, 1887–June 22, 1954." *Biographical Memoirs of the National Academy of Sciences* 91 (1992): 39–57.

Stratton, Julius Adams, and Loretta H. Mannix. *Mind and Hand: The Birth of MIT*. Cambridge, MA: MIT Press, 2005.

Stratton, Samuel W. "Installation Address." *Science* 71, no. 1850 (1930): 591–93.

Struik, Dirk J. *Yankee Science in the Making*. Boston: Little, Brown, 1948.

Swedberg, Richard. *Max Weber and the Idea of Economic Sociology*. Princeton, NJ: Princeton University Press, 1998.

Tatarkiewicz, W. "Classification of Arts in Antiquity." *Journal of the History of Ideas* 24, no. 2 (1963): 231–40.

Taylor, Frederick Winslow. *Scientific Management, Comprising Shop Management, the Principles of Scientific Management [and] Testimony before the Special House Committee*. New York: Harper, 1947.

Taylor, Jerome. Introduction to *The Didascalicon of Hugh of St. Victor: A Medieval Guide to the Arts*, translated by Jerome Taylor, 3–39. New York: Columbia University Press, 1961; reprint, 1991. Page references are to the 1991 edition.

Taylor, O. H. "Economic Theory, and Certain Non-economic Elements in Social Life." In *Explorations in Economics: Notes and Essays Contributed in Honor of F. W. Taussig*, 380–90. New York: McGraw-Hill, 1936.

"Technological Trends and National Policy." *Science* 86, no. 2221 (1937): 69–71.

"Technologie." In *Karmarsch und Heeren's Technisches Wörterbuch*, edited by Friedrich Kick and Wilhelm Gintl, 8:780–83. 3rd ed. Prague: A. Haase, 1885.

"Technologie." In *Historisches Wörterbuch der Philosophie*, edited by Joachim Ritter and Karlfried Gründer. 10:958–62. Basel: Schwabe, 1998.

"Technology, n." In *OED Online*, Oxford University Press, 2012. http://www.oed.com/view/Entry/198469.

Tell, Dave. "Rhetoric and Power: An Inquiry into Foucault's Critique of Confession." *Philosophy and Rhetoric* 43, no. 2 (2010): 95–117.

Thomas Aquinas. *Commentary on the Metaphysics of Aristotle*. 2 vols. Vol. 1, Chicago: H. Regnery, 1961.

———. *New English Translation of St. Thomas Aquinas's Summa Theologiae (Summa Theologica)*. Translated by Alfred J. Freddoso. Notre Dame, Indiana, 2013. http://www3.nd.edu/~afreddos/summa-translation/TOC.htm.

Thomas, Grace Powers. *Where to Educate, 1898–1899: A Guide to the Best Private Schools, Higher Institutions of Learning, etc., in the United States*. Boston: Brown, 1898.

Thompson, James D. *Organizations in Action: Social Science Bases of Administrative Theory*. New York: McGraw-Hill, 1967.

Tilman, Rick. *The Intellectual Legacy of Thorstein Veblen: Unresolved Issues*. Westport, CT: Greenwood Press, 1996.

Times (London). "Imperial College of Science and Technology." 11 May 1907.

———. "The Organization of University Education in the Metropolis." 4 June 1901.

Titchener, Edward Bradford. "Psychology: Science or Technology?" *Popular Science Monthly* 84 (1914): 39–51.

———. *Systematic Psychology: Prolegomena*. Ithaca, NY: Cornell University Press, 1929.

Tobin, William A. "Studying Society: The Making of 'Recent Social Trends in the United States, 1929–1933.'" *Theory and Society* 24, no. 4 (1995): 537–65.

Tönnies, Ferdinand. "Natorp, Paul, Sozialpädagogik [review]." *Archiv für Sozialwissenschaft und Sozialpolitik* 14 (1899): 445–62.

Treadwell, Donald. "[Review of] Elements of Technology . . . by Jacob Bigelow." *Christian Examiner* (November 1829): 187–202.

Tribe, Keith. "Cameralism and the Science of Government." *Journal of Modern History* 56, no. 2 (1984): 263–84.

Trowbridge, Augustus. "Pure Science and Engineering." *Science* 68, no. 1772 (1928): 575–79.

Turner, Denys. *Thomas Aquinas: A Portrait.* New Haven, CT: Yale University Press, 2013.

Turner, Fred. *From Counterculture to Cyberculture: Stewart Brand, the Whole Earth Network, and the Rise of Digital Utopianism.* Chicago: University of Chicago Press, 2006.

Turner, James. "Le concept de science dans l'Amérique du XIXe siècle." *Annales: Histoire, Sciences Sociales* 57, no. 3 (2002): 753–72.

Turner, Stephen. "Merton's 'Norms' in Political and Intellectual Context." *Journal of Classical Sociology* 7, no. 2 (2007): 161–78.

Tylor, Edward B. *Primitive Culture: Researches into the Development of Mythology, Philosophy, Religion, Art, and Custom.* 2 vols. London: J. Murray, 1871.

United States. National Resources Committee. Science Committee. *Technological Trends and National Policy Including the Social Implications of New Inventions.* Washington, DC: Government Printing Office, 1937.

Ure, Andrew. *The Philosophy of Manufactures; or, An Exposition of the Scientific, Moral, and Commercial Economy of the Factory System of Great Britain.* London: Charles Knight, 1835.

———. Preface to *A Dictionary of Arts, Manufactures, and Mines; Containing a Clear Exposition of Their Principles and Practice,* 1:iii–xiv. 4th ed. New York: D. Appleton, 1853.

U.S. Congress. House Committee on Education and Labor, Select Subcommittee on Labor. *National Commission on Technology, Automation, and Economic Progress: Hearings.* Washington, DC: Government Printing Office, 1964.

U.S. Office of Defense Mobilization. "Deterrence and Survival in the Nuclear Age." 1957. Available from the National Security Archive, George Washington University, http://nsarchive.gwu.edu/NSAEBB/NSAEBB139/nitze02.pdf.

U.S. Senate. Committee on Education and Labor. *Unemployment in the United States,* 70th Cong., 2nd sess., 1929.

———. Committee on Government Operations. *Hearings on S.676, A Bill to Create a Department of Science and Technology, and to Transfer Certain Agencies and Functions to Such Department.* 86th Cong., 1st Sess. Vol. 1, Washington, DC: Government Printing Office, 1959.

———. Temporary National Economic Committee. *Technology in Our Economy.* Washington, DC: Government Printing Office, 1941.

Usher, Abbott Payson. "The Genesis of Modern Capitalism." *Quarterly Journal of Economics* 36, no. 3 (1922): 525–35.

———. *A History of Mechanical Inventions.* New York: McGraw-Hill, 1929.

V. [Thorstein Veblen]. "Review of *Der moderne Kapitalismus.*" *Journal of Political Economy* 11, no. 2 (1903): 300–305.

Veblen, Thorstein. *Absentee Ownership and Business Enterprise in Recent Times: The Case of America.* Boston: Beacon Press, 1923.

———. "Arts and Crafts." *Journal of Political Economy* 11, no. 1 (1902): 108–11.

———. "The Beginnings of Ownership." *American Journal of Sociology* 4, no. 3 (1898): 352–65.

————. *The Engineers and the Price System*. New York: B. W. Huebsch, 1921. Reprint, New York: Harcourt, Brace and World, 1963. Page references are to the 1963 edition.

————. "Gustav Schmoller's Economics." *Quarterly Journal of Economics* 16, no. 1 (1901): 69–93.

————. *The Higher Learning in America: A Memorandum on the Conduct of Universities by Business Men*. New York: B. W. Huebsch, 1918.

————. "Industrial and Pecuniary Employments." *Publications of the American Economic Association* 2, no. 1 (1901): 190–235.

————. "The Instinct of Workmanship and the Irksomeness of Labor." *American Journal of Sociology* 4, no. 2 (1898): 187–201.

————. *The Instinct of Workmanship and the State of Industrial Arts*. New York: Macmillan, 1914.

————. "On the Nature of Capital [part 1]." *Quarterly Journal of Economics* 22, no. 4 (August 1908): 517–42.

————. "On the Nature of Capital: Investment, Intangible Assets, and the Pecuniary Magnate." *Quarterly Journal of Economics* 23, no. 1 (November 1908): 104–36.

————. "The Place of Science in Modern Civilization." *American Journal of Sociology* 11, no. 5 (March 1906): 585–609.

————. "The Preconceptions of Economic Science, III." *Quarterly Journal of Economics* 14, no. 2 (1900): 240–69.

————. "Review of *Einführung in der Socialismus* by Richard Calwer." *Journal of Political Economy* 5, no. 2 (1897): 270–72.

————. "The Socialist Economics of Karl Marx and His Followers, I." *Quarterly Journal of Economics* 20, no. 4 (1906): 575–95.

————. *The Theory of Business Enterprise*. New York: C. Scribner's Sons, 1904.

————. *The Theory of the Leisure Class*. New York: Dover Publications, 1994. First published 1899.

————. *The Vested Interests and the State of the Industrial Arts*. New York: Viking Press, 1919.

————. *What Veblen Taught: Selected Writings of Thorstein Veblen*. Edited with an introduction by Wesley C. Mitchell. New York: A. M. Kelley, 1964.

————. "Why Is Economics Not an Evolutionary Science?" *Quarterly Journal of Economics* 12 (1898): 373–97.

Vincent, C. W. "On Some Recent Processes for the Manufacture of Soda." *Journal of the Society of Arts* 22, no. 1117 (1874): 470–77.

Vincenti, Walter G. *What Engineers Know and How They Know It: Analytical Studies from Aeronautical History*. Baltimore: Johns Hopkins University Press, 1990.

Volti, Rudi. "Classics Revisited: William F. Ogburn, *Social Change with Respect to Culture and Original Nature*." *Technology and Culture* 45, no. 2 (2004): 396–405.

Voskuhl, Adelheid. *Androids in the Enlightenment: Mechanics, Artisans, and Cultures of the Self*. Chicago: University of Chicago Press, 2013.

———. "Engineering Philosophy: Theories of Technology, German Idealism, and Social Order in High-Industrial Germany." *Technology and Culture* 57 (2016): 721–52.

Waggoner, Walter H. "Louis C. Hunter, 85, Is Dead; A Specialist on U.S. Industry." *New York Times*, 24 March 1984.

Wagner, Johannes Rudolf, and William Crookes. *A Handbook of Chemical Technology.* New York: D. Appleton, 1872.

Walras, Léon. *Éléments d'économie politique pure; ou, Théorie de la richesse sociale.* Lausanne: L. Corbaz, 1874.

Waterman, Alan T. "The Science of Producing Good Scientists: Soviet Technology Is Making Notable Strides." *New York Times*, 31 July 1955.

Watkins, Mel. "Rodney Dangerfield, Comic Seeking Respect, Dies at 82." *New York Times*, 6 October 2004.

Weber, Max. *Economy and Society: An Outline of Interpretive Sociology.* New York: Bedminster Press, 1968.

———. *General Economic History.* New Brunswick, NJ: Transaction Books, 1981. First published 1927.

———. *Gesammelte Aufsätze zur Religionssoziologie.* 3 vols. Tübingen: Mohr, 1922.

———. "'Objectivity' in Social Science and Social Policy." In *The Methodology of the Social Sciences,* edited by Edward A. Shils and Henry A. Finch, 49–112. New York: Free Press, 1949.

———. *The Protestant Ethic and the Spirit of Capitalism.* Translated by Talcott Parsons. London: G. Allen and Unwin, 1930.

———. "Remarks on Technology and Culture." *Theory, Culture and Society* 22, no. 4 (2005): 23–38.

———. *The Theory of Social and Economic Organization.* Translated by A. M. Henderson and Talcott Parsons. New York: Oxford University Press, 1947.

———. *Wirtschaft und Gesellschaft.* Vol. 3, Tübingen: J. C. B. Mohr (Paul Siebeck), 1922.

Weber, Max, Sigmund Hellmann, and Melchior Palyi. *Wirtschaftsgeschichte.* 2nd ed. Munich: Duncker and Humblot, 1924.

Weber, Wolfhard. "Grosse Technologen und ihre berufspädagogische Bedeutung: Johann Beckmann (1739–1811)." In *Berufsausbildung und sozialer Wandel: 150 Jahre Preussische allgemeine Gewerbeordnung von 1845,* edited by Wolf-Dietrich Greinert, 225–43. Berlin: Bundesinstitut für Berufsbildung, 1996.

Webster, Noah. *An American Dictionary of the English Language.* 2 vols. New York: S. Converse, 1828.

Weeks, Sophie. "Francis Bacon and the Art–Nature Distinction." *Ambix* 54, no. 2 (2007): 101–29.

———. "The Role of Mechanics in Francis Bacon's *Great Instauration.*" In *Philosophies of Technology: Francis Bacon and His Contemporaries,* edited by Claus Zittel, Gisela Engel, Romano Nanni, and Nicole C. Karafyllis, 133–95. Leiden: Brill, 2008.

Weidlein, Edward R., and William Allen Hamor. *Science in Action: A Sketch of the Value of Scientific Research in American Industries.* New York: McGraw-Hill, 1931.

Wendling, Amy E. *Karl Marx on Technology and Alienation.* Basingstoke, UK: Palgrave Macmillan, 2009.

Westrum, Ron. *Technologies and Society: The Shaping of People and Things.* Belmont, CA: Wadsworth, 1991.

Whimster, Sam. *Understanding Weber.* London: Routledge, 2007.

White, Leslie A. "Energy and the Evolution of Culture." *American Anthropologist* 45, no. 3 (1943): 335–56.

White, Lynn. "Cultural Climates and Technological Advance in the Middle Ages." Chap. 14 in *Medieval Religion and Technology: Collected Essays,* 217–53. Berkeley: University of California Press, 1978.

———. *Medieval Religion and Technology: Collected Essays.* Berkeley: University of California Press, 1978.

———. "Technology and Invention in the Middle Ages." *Speculum* 15 (1940): 141–59.

Whitehead, Alfred North. *Science and the Modern World.* 2nd ed. New York: Free Press, 1925.

Whitney, Elspeth. "Paradise Restored: The Mechanical Arts from Antiquity through the Thirteenth Century." *Transactions of the American Philosophical Society* 80 (1990): 1–169.

Whitney, William Dwight, ed. *The Century Dictionary: A Work of Universal Reference in All Departments of Knowledge.* 2nd ed. 13 vols. Vol. 8, New York: Century, 1897.

Whitney, William Dwight, and Benjamin E. Smith. *The Century Dictionary and Cyclopedia, with a New Atlas of the World.* Rev. and enl. ed. 12 vols. New York: Century, 1911.

Wickham, Chris. *The Inheritance of Rome: A History of Europe from 400 to 1000.* New York: Viking, 2009.

Williams, Henry Smith. *A History of Science.* 11 vols. New York: Harper and Brothers, 1904–10.

Williams, Rosalind H. "Lewis Mumford as a Historian of Technology in *Technics and Civilization.*" In Hughes and Hughes, *Lewis Mumford: Public Intellectual,* 43–65.

———. "Lewis Mumford's Technics and Civilization (Classics Revisited)." *Technology and Culture* 43 (2002): 139–49.

———. *The Triumph of Human Empire: Verne, Morris, and Stevenson at the End of the World.* Chicago: University of Chicago Press, 2013.

Wilson, George. "On the Physical Sciences Which Form the Basis of Technology; Being the Introductory Prelection for 1856." *Edinburgh New Philosophical Journal* 5 (1857): 64–101.

———. "On the Relations of Technology to Agriculture." *Transactions of the Highland and Agricultural Society of Scotland* (1857): 254–67.

———. *What Is Technology? An Inaugural Lecture Delivered in the University of Edinburgh, on November 7, 1855.* Edinburgh: Sutherland and Knox, 1855.

Winchester, Simon. *The Meaning of Everything: The Story of the Oxford English Dictionary.* Oxford: Oxford University Press, 2003.

Winner, Langdon. "Do Artifacts Have Politics?" *Daedalus* 109, no. 1 (1980): 121–36.

———. "Technologies as Forms of Life." Chap. 1 in *The Whale and the Reactor: A Search for Limits in an Age of High Technology,* 3–18. Chicago: University of Chicago Press, 1986.

Winterer, Caroline. *The Culture of Classicism: Ancient Greece and Rome in American Intellectual Life, 1780–1910.* Baltimore: Johns Hopkins University Press, 2002.

Wisnioski, Matthew H. *Engineers for Change: Competing Visions of Technology in 1960s America.* Cambridge, MA: MIT Press, 2012.

Withey, Stephen B. "Public Opinion about Science and Scientists." *Public Opinion Quarterly* 23, no. 3 (1959): 382–88.

Wolf, A. *A History of Science, Technology, and Philosophy in the 16th and 17th Centuries.* London: George Allen and Unwin, 1935.

Wolff, Christian. *Philosophia rationalis sive logica.* Francofurti & Lipsiæ: Prostat in Officina libraria Rengeriana, 1740.

Wood, Henry Trueman. *A History of the Royal Society of Arts.* London: J. Murray, 1913.

Woodbury, Robert S. "Review: The Scholarly Future of the History of Technology." *Technology and Culture* 1, no. 4 (1960): 345–48.

Wyrwa, Ulrich. "Richard Nikolaus Graf Coudenhove-Kalergi (1894–1972) und die Paneuropa-Bewegung in den zwanziger Jahren." *Historische Zeitschrift* 283, no. 1 (2006): 103–22.

Yeo, Richard. *Encyclopaedic Visions: Scientific Dictionaries and Enlightenment Culture.* Cambridge: Cambridge University Press, 2001.

Zilsel, Edgar. "The Sociological Roots of Science." *American Journal of Sociology* 47, no. 4 (1942): 544–62.

Index

lone genius, 61; manual labor, 31–32; and mathematics, 38; and medicine, 34; medieval concept of, 30–37; and modernity, 49; moral value of, 36–37; natural philosophy, relationship between, 54; and physics, 38; praise of, 50; and technology, 73–74, 96; as term, 32; and theatrics, 34; as virtuous, 39, 58–59

Meier, Hugo, 84, 258n63

Meitner, Lise, 283n25

Menger, Carl, 157, 268n40, 274n22

Merriam, John C., 171

Merton, Robert K., 191–93, 202, 223

Mesthene, Emmanuel G., 219–21

Metaphysics (Aristotle), 245n24

Meyer, Torsten, 112

Michael of Rhodes, 45

Michelson, Albert A., 186, 196

Mill, James, 157

Mill, John Stuart, 60–61, 81

Miller, Perry, 76, 84, 272n61

Mindful Hand, The (Roberts, Schaffer, and Dear), 178

Mitcham, Carl, 105, 243n36

Mitchell, Wesley C., 270n3

Mobil Oil, 231

Moderne Kapitalismus (Sombart), 150, 163

modernity, 113, 120, 202, 232; agency, removal of, 72; mechanical arts, 49; and technology, 2–4, 90–91, 165, 194, 214; unease with, 206

Mokyr, Joel, industrial enlightenment, 55

Mongols, 37

Moore, Samuel, 102, 261n21

moral values, 10, 17, 21–24, 26, 30, 36–37, 47

Morris, William, 61

Multhauf, Robert P., 285n76

Mumford, Lewis, 6, 31, 137, 145–46, 153, 208–9, 221–22, 225–27, 230, 272n52, 272n59, 272n61, 273n66, 273n68; megamachine, 2; technics, use of, 3, 147–51

Myth of the Machine, The (Mumford), 226

Nader, Ralph, 221, 228–29

Nagasaki, 196

Nanni, Romano, 50–51

Narratives of Innovation (Godin), 215

National Academy of Sciences, 185

National Association of Science Writers, 212

National Defense Research Committee, 195

National Institution for the Promotion of Science, 181

National Research Council (NRC), 172, 185–86

National Research Project on Reemployment Opportunities and Recent Changes in Industrial Technique, 161

National Science Foundation, 203

National Socialism, 109, 122, 202, 224, 263n51

natural philosophy, 11, 44, 47–51, 85, 86, 89; as spiritual enterprise, 54

Needham, Joseph, 191

Needles, Enoch R., 207

New Atlantis (Bacon), 51

New Deal, 171

New England, 77, 85

New Left, 224–26

New Organon (Bacon), 50–51

New School for Social Research, 138

Newton, Isaac, 190

New World of English Words, The (Phillips), 77

Nicomachean Ethics (Aristotle), 20–22, 24, 39–40, 57

Nietzsche, Friedrich, 13, 244n55

Nussbaum, Frederick, 124

Nye, David, 139

October Revolution, 188–89

Office of Scientific Research and Development (OSRD), 195, 198

Office of Technology Assessment (OTA), 227–28

Ogburn, William F., 153, 162, 201, 206, 208, 218–19, 276n76; cultural lag theory, 227; cultural theory of invention, 167; culture, concept of, 167; inventions, and social change, 170; inventions, and technological progress, 170–73; material culture, and inventions, 168–69; material culture, and technology, 170; sociology of technology, 167; theory of cultural change, 169; theory of invention, 168–69

Oldenziel, Ruth, 84, 125, 134

One-Dimensional Man (Marcuse), 224–25